Low Power Emerging Wireless Technologies

T0225392

Devices, Circuits, and Systems

Series Editor
Krzysztof Iniewski
CMOS Emerging Technologies Inc., Vancouver, British Columbia, Canada

PUBLISHED TITLES:

Atomic Nanoscale Technology in the Nuclear Industry
Taeho Woo

Biological and Medical Sensor Technologies
Krzysztof Iniewski

Electrical Solitons: Theory, Design, and Applications
David Ricketts and Donhee Ham

Electronics for Radiation Detection
Krzysztof Iniewski

**Graphene, Carbon Nanotubes, and Nanostuctures:
Techniques and Applications**
James E. Morris and Kris Iniewski

Integrated Microsystems: Electronics, Photonics, and Biotechnology
Krzysztof Iniewski

Internet Networks: Wired, Wireless, and Optical Technologies
Krzysztof Iniewski

Low Power Emerging Wireless Technologies
Reza Mahmoudi and Krzysztof Iniewski

Nano-Semiconductors: Devices and Technology
Krzysztof Iniewski

Optical, Acoustic, Magnetic, and Mechanical Sensor Technologies
Krzysztof Iniewski

Radiation Effects in Semiconductors
Krzysztof Iniewski

Semiconductor Radiation Detection Systems
Krzysztof Iniewski

Telecommunication Networks
Eugenio Iannone

Low Power Emerging Wireless Technologies

Edited by
Reza Mahmoudi
Krzysztof Iniewski

CRC Press
Taylor & Francis Group
Boca Raton   London   New York

CRC Press is an imprint of the
Taylor & Francis Group, an **informa** business

CRC Press
Taylor & Francis Group
6000 Broken Sound Parkway NW, Suite 300
Boca Raton, FL 33487-2742

First issued in paperback 2017

© 2013 by Taylor & Francis Group, LLC
CRC Press is an imprint of Taylor & Francis Group, an Informa business

No claim to original U.S. Government works
Version Date: 20121126

ISBN 13: 978-1-138-07634-1 (pbk)
ISBN 13: 978-1-4665-0701-2 (hbk)

Library of Congress Cataloging-in-Publication Data

Low power emerging wireless technologies / editors, Reza Mahmoudi and Krzysztof Iniewski.
  pages cm. -- (Devices, circuits, and systems)
  Includes bibliographical references and index.
  ISBN 978-1-4665-0701-2 (alk. paper)
  1. Wireless communication systems--Technological innovations. 2. Low voltage systems--Technological innovations. I. Mahmoudi, Reza. II. Iniewski, Krzysztof, 1960-

TK5103.2.L69 2013
621.382--dc23   2012034872

Visit the Taylor & Francis Web site at
http://www.taylorandfrancis.com

and the CRC Press Web site at
http://www.crcpress.com

Contents

Preface

Advanced concepts for wireless communications present a vision of technology that is embedded in our surroundings and practically invisible, but present whenever required. From established radio techniques like GSM, 802.11, or Bluetooth to those more recently emerging like ultra wideband (UWB) or smart dust motes, a common denominator for future progress is underlying integrated circuit (IC) technology. Although the use of deep submicron complementary metal oxide semiconductor (CMOS) processes allows for an unprecedented degree of scaling in digital circuitry, it complicates the implementation and integration of traditional radio frequency (RF) circuits. The explosive growth of standard cellular radios and radically different new wireless applications makes it imperative to find architectural and circuit solutions to these design problems.

Low power has always been important for wireless communications. With new developments in wireless sensor networks and wireless systems for medical applications, the power dissipation is becoming the number one issue. Wireless sensor network systems are being applied in critical applications in commerce, healthcare, and security. These systems have unique characteristics and face many implementation challenges. The requirement of long operating life for a wireless sensor node under limited energy supply imposes the most severe design constraints. This calls for innovative design methodologies at the circuit and system levels to address this rigorous requirement.

This book addresses the state of the art in wireless communication for 3G/4G cellular telephony, millimeter-wave applications, wireless sensor networks, and wireless medical technologies. New, exciting opportunities in body area networks, medical implants, satellite communications, automobile radar detection, and wearable electronics are discussed. It is a must for anyone serious about future wireless technologies. It is written by top international experts on wireless circuit design representing both an IC industry and academia. The intended audience is practicing engineers in the wireless communication field with some integrated circuit background. It will also be used as recommended reading and supplementary material in graduate course curricula.

The Editors

Dr. Krzysztof (Kris) Iniewski is manager of R&D at Redlen Technologies Inc., a start-up company in Vancouver, Canada. Redlen's revolutionary production process for advanced semiconductor materials enables a new generation of more accurate, all-digital, radiation-based imaging solutions. Kris is also a president of CMOS Emerging Technologies (www.cmoset.com), an organization of high-tech events covering communications, microsystems, optoelectronics, and sensors.

During his career, Dr. Iniewski has held numerous faculty and management positions at the University of Toronto, University of Alberta, Simon Fraser University, and PMC-Sierra Inc. He has published over 100 research papers in international journals and conferences. He holds 18 international patents granted in the United States, Canada, France, Germany, and Japan. Kris is a frequent invited speaker and has consulted for multiple organizations internationally. Dr. Iniewski has written and edited several books for IEEE Press, Wiley, CRC Press, McGraw–Hill, Artech House, and Springer. His personal goal is to contribute to healthy living and sustainability through innovative engineering solutions. In his leisure time, Kris can be found hiking, sailing, skiing, or biking in beautiful British Columbia. He can be reached at kris.iniewski@gmail.com.

Reza Mahmoudi studied electrical engineering at the Delft University of Technology, Delft, The Netherlands, where he joined the microwave component group and received the MSc degree in 1993 with his thesis entitled, "A Measurement System for Noise Parameters." He was employed as a full member of the same group from January 1, 1993 to December 7, 1999. He earned the Designers certificate from Delft in 1996 with his thesis entitled, "A Systematic Design Method for a Feed-Forward Error Control System." This work was the initial step leading to his PhD degree, concluded in November 2001, with the thesis entitled, "A Multi-Disciplinary Design Method for Second and Third Generation Mobile Communication System Microwave Components." Dr. Mahmoud has worked for Philips Discrete Semiconductors in Nijmegen, The Netherlands and Advanced Wave Research in El Segundo, California. From April 2003 to April 2011, Reza was an assistant professor in the Department of Electrical Engineering at Eindhoven University of Technology. Recently, he became an associate professor in the field of ultrahigh frequency front-end electronics in the same department. Reza's research efforts are concentrated in the areas of ultrahigh speed communication and high resolution imaging systems. He has contributed to 85 academic publications.

The Contributors

Mustafa Acar
NXP Semiconductors
Eindhoven, The Netherlands

Jos Bergervoet
NXP Semiconductors
Eindhoven, The Netherlands

William Biederman
Department of Electrical Engineering
    and Computer Science
University of California
Berkeley, California

Ji-Woong Choi
Department of Information and
    Communication Engineering
Daegu Gyeongbuk Institute of Science
    and Technology
Daegu, Korea

Davide Dardari
University of Bologna
Cesena, Italy

Harmke de Groot
Holst Centre/Imec–NL
Eindhoven, The Netherlands

Saverio De Vito
Ente Nuove Technologie
Energia e Ambiente
Centro Ricerche Portici
Naples, Italy

Pooyan Sakian Dezfuli
Department of Electrical Engineering
Eindhoven University of Technology
Eindhoven, The Netherlands

Guido Dolmans
Holst Centre/Imec–NL
Eindhoven, The Netherlands

Stefan Drude
NXP Semiconductors
Eindhoven, The Netherlands

Hassan Elwan
Newport Media Inc.
Lake Forest, California

Dmitriy Garmatyuk
Department of Electrical and Computer
    Engineering
Miami University
Oxford, Ohio

Benoit Gosselin
Department of Electrical and Computer
    Engineering
Université Laval
Quebec, Canada

Ali Hajimiri
Department of Electrical Engineering
California Institute of Technology
Pasadena, California

Payam Heydari
School of Engineering
University of California
Irvine, California

Farhad Sheikh Hosseini
Department of Electrical and Computer
    Engineering
Université Laval
Quebec, Canada

Li Huang
Holst Centre/Imec–NL
Eindhoven, The Netherlands

Inyup Kang
Mobile Solutions Lab
Samsung Electronics US R&D Center
San Diego, California

Kyle Kauffman
Air Force Institute of Technology
Wright-Patterson Air Force Base, Ohio

Hyukjoon Kwon
Mobile Solutions Lab
Samsung Electronics US R&D Center
San Diego, California

Jungwon Lee
Mobile Solutions Lab
Samsung Electronics US R&D Center
San Diego, California

Domine Leenaerts
NXP Semiconductors
Eindhoven, The Netherlands

Yu-Te Liao
Department of Electrical Engineering
National Chung-Cheng University
Chiayi, Taiwan

Pui-In Mak
The State-Key Laboratory of Analog
    and Mixed-Signal VLSI
University of Macau
Macau, China

Rui P. Martins
The State-Key Laboratory of Analog
    and Mixed-Signal VLSI
University of Macau
Macau, China

Y. T. Jade Morton
Department of Electrical and Computer
    Engineering
Miami University
Oxford, Ohio

Brian Otis
Department of Electrical Engineering
University of Washington
Seattle, Washington

Julien Penders
Holst Centre/Imec–NL
Eindhoven, The Netherlands

John Raquet
Advanced Navigation Technology Center
Oxford, Ohio

Sébastien Roy
Department of Electrical and Computer
    Engineering
Université Laval
Quebec, Canada

Ahmet Tekin
Waveworks Inc.
Irvine, California

Kai-Fai Un
The State-Key Laboratory of Analog
    and Mixed-Signal VLSI
University of Macau
Macau, China

Mark van der Heijden
NXP Semiconductors
Eindhoven, The Netherlands

Arthur van Roermund
Department of Electrical Engineering
Eindhoven University of Technology
Eindhoven, The Netherlands

Maja Vidojkovic
Holst Centre/Imec–NL
Eindhoven, The Netherlands

Hubregt J. Visser
Holst Centre/Imec–NL
Eindhoven, The Netherlands

Hua Wang
Department of Electrical Engineering
California Institute of Technology
Pasadena, California

1 Passive Imaging in Silicon Technologies

Payam Heydari

CONTENTS

1.1 INTRODUCTION

Passive millimeter-wave (PMMW) imaging is a method that forms images through the passive detection of natural millimeter-wave radiation (30–300 GHz) from objects. Although such systems have been developed and studied for decades, cost is still the major factor that limits the number of pixels in a real imager for commercial and medical use, resulting in lower resolution images. Benefiting from the aggressive feature size scaling, silicon-based technologies (e.g., CMOS, SiGe, BiCMOS) have become more and more popular in the realm of millimeter-wave (MMW) system design, which makes it possible to develop low cost, compact, high performance MMW imaging systems. This chapter provides a brief overview of the silicon implementation of the fully integrated W-band PMMW imaging receiver.

1.1.1 MOTIVATION

During the past decade, extensive research efforts have been put into developing silicon-based MMW systems for the target applications of short-range, high data rate wireless communication [1], short-range/long-range automotive radar [2], sensing, and imaging [3]. Figure 1.1 enumerates several frequency bands below 100 GHz, with their relevant applications, that have been approved by the Federal Communications Commission (FCC). The 24/77 GHz automotive radar sensors are mounted around the vehicle to detect surrounding objects in close range (<40 m) and long range (~150 m), which accommodates a wide variety of safety measures including collision avoidance, blind spot detection, airbag activation, and automatic cruise control. The 60 GHz band offers wide, unlicensed bandwidth from 57 to 66 GHz, stimulating many high data rate wireless applications such as wireless HDMI, wireless gigabit Ethernet, and wireless laptop docking stations. Note that although the 24 GHz band is not within the MMW range, it is included here, since it is very close to 30 GHz and it is designated for automotive radar applications.

Within the 30–300 GHz MMW frequency range, there are identifiable propagation windows located near 35, 94, 140, and 220 GHz, as shown in Figure 1.2, where the atmospheric absorption is relatively low—not only in clean air but also through smoke, dust, fog, and clothing. This notion makes a passive MMW imaging system an ideal candidate for various applications such as remote sensing, security surveillance (e.g., concealed weapon detection at the airport), and nondestructive inspection for biological tissues as well as industrial process control [4,5]. Compared to an active imaging system that employs a transceiver, a passive imaging system detects the thermal radiation from the objects and therefore exhibits less system complexity, lower cost, and also lower overall power consumption. Additionally, the noninvasive nature of passive imaging avoids any public health concern in medical and security applications. Figure 1.3 shows the visible (top row) and corresponding PMMW images of an individual with various weapons concealed by the sweatshirt shown in the visible picture at the start of the bottom row. The PMMW images were acquired indoors with a 94 GHz radiometer using a scanning 24 in. dish antenna.

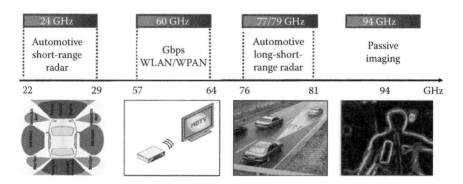

FIGURE 1.1 Miscellaneous MMW applications (frequency boundaries shown in the figure are not accurate).

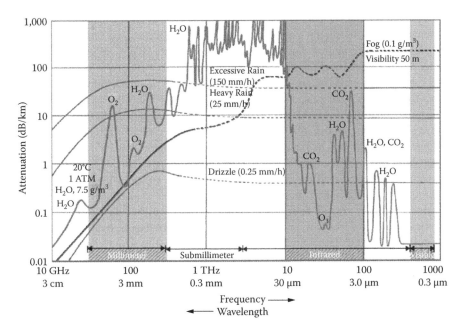

FIGURE 1.2 The attenuation of millimeter waves by atmospheric gases, rain, and fog [4].

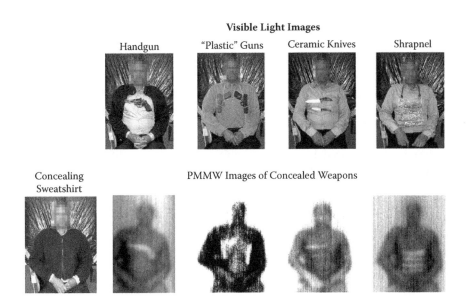

FIGURE 1.3 Passive MMW images [4].

PMMW imaging systems operating near the 94 GHz frequency window provide reasonable balance among capability of currently available silicon technologies, chip size, spatial resolution, and atmospheric attenuation. III-V compound semiconductor technologies have been commonly used as ideal platforms to realize MMW radiometers or passive imaging receivers that are based on multichip modules [6,7]. Recently, benefiting from the aggressive feature size scaling, silicon technology has shown the capability for implementation of W-band passive imaging receivers with fine image and temperature resolution [3,8–10]. However, these efforts are limited to a single receiver/pixel. Only recently, efforts have been made to design and implement multipixel imaging array [11–13]. The transition of W-band signals from chip to antenna remains a challenging task, particularly in the context of the multipixel imaging systems. To reduce the scanning time and enable video rate real-time imaging, focal-plane array (FPA) could be used with an array of detectors located at the focal plane of a focusing system [11–13]. Despite non-negligible loss at W-band frequency range, the use of an on-chip antenna can still be advantageous considering the nontrivial electrical interface and assembly cost to implement antenna-in-package or on-board in a multipixel FPA system [12,13]. This chapter provides an overview of the passive imaging design and implementation in silicon technologies.

1.1.2 Passive Imaging Fundamentals

This section provides an overview of passive millimeter-wave imaging systems. It covers the basic concept of passive imaging, the miscellaneous applications, the system architecture of a radiometer receiver, the commonly used figure of merit to evaluate passive imaging systems, and the state-of-the-art imaging receivers.

1.1.3 Total Power Radiometer

Figure 1.4 shows the block diagram of a total power radiometer, which consists of a low noise RF pre-amplification stage, a square-law power detector, a low frequency amplifier, and an integrator. The radiometer collects the radiated power ($P_E = k_B \Delta TB$) from the target object and produces an output voltage proportional to the incident power. Delta-T is the effective radiometric temperature [4], B is the receiver bandwidth, and k_B is the Boltzmann constant.

The sensitivity or the minimum detectable temperature change of an imaging receiver is characterized by a noise equivalent temperature difference (NETD), which is expressed by (1.1) for a total-power radiometer [13,14]:

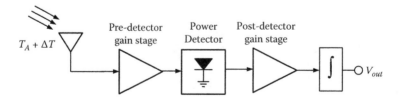

FIGURE 1.4 Block diagram of a total power radiometer.

$$\mathrm{NETD} = \mathrm{T}_s\sqrt{\frac{1}{\mathrm{B}\tau}} \tag{1.1}$$

where T_s is the system noise temperature, B is the bandwidth, and τ is the integration time, which is typically less than 30 ms (standard video imaging rate) in real-time imaging application.

1.1.4 Dicke Radiometer

To build a "useful" imaging system, the NETD needs to be below 1 K, while less than 0.5 K NETD is preferred for good imaging quality [6]. However, the NETD of a total-power radiometer suffers from gain fluctuation, as the RX cannot distinguish between the change in input signal power and the variation of front-end gain. The NETD in the presence of gain fluctuation is given as [14]

$$\mathrm{NETD} = \mathrm{T}_s\sqrt{\frac{1}{\mathrm{B}\tau}+(\frac{\Delta \mathrm{G}}{\mathrm{G}})^2} \tag{1.2}$$

where $\Delta G/G$ denotes the total gain fluctuation in percentage. For example, 0.1% gain fluctuation merely translates to an NETD of 3 K for an imaging RX with 3000 K noise temperature. This problem can be solved by periodically chopping above the gain fluctuation frequency using Dicke architecture [15].

Figure 1.5 shows the block diagram of a Dicke radiometer that employs two synchronized single-pole double throw (SPDT) switches: The one at the front end switches between the antenna and a reference load, while the one after the detector demodulates the signal by multiplying it with ±1 in the opposite phase with respect to the front-end switch. In addition to the gain fluctuation problem, the 1/f noise is another source of low frequency disturbance, affecting the system NETD in a similar way. Therefore, the chopping frequency also needs to be higher than the 1/f noise corner frequency.

A power or energy detector serves as the core of an imaging pixel. The common figure of merit to evaluate performance of a power detector is noise-equivalent power (NEP) $[W/\sqrt{Hz}]$, defined by (1.3) as a detector's output rms noise voltage v_n $[V/\sqrt{Hz}]$

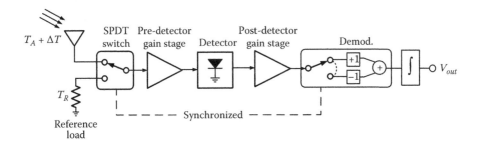

FIGURE 1.5 Block diagram of a Dicke radiometer.

divided by responsivity, R. The responsivity provides a measure of the detector gain and equals the output DC voltage divided by the input RF power—that is,

$$\text{NEP} = \frac{V_n}{\mathfrak{R}} = \frac{V_n}{V_{out,DC} / P_{in,RF}}$$

(1.3)

The NEP and responsivity definitions can be generalized to any power-detecting system—for instance, a power detector preceded by pre-amplification gain stage. In addition to (1.1), there are also other ways to calculate NETD, as reported in May and Rebeiz [3] and Tomkins, Garcia, and Voinigescu [8] and shown in (1.4) and (1.5). The gain fluctuation term is not included in (1.4) and (1.5) for simplicity.

$$\text{NETD} = \sqrt{\frac{T_s^2}{B\tau} + \left(\frac{\text{NEP}_{det}}{k_B GB}\right)^2 \frac{1}{2\tau}}$$

(1.4)

$$\text{NETD} = \frac{\text{NEP}_{sys}}{k_B B \sqrt{2\tau}}$$

(1.5)

NEP_{det} and NEP_{sys} are the noise-equivalent power of the detector (detector NEP) and imaging RX (system NEP), respectively; G is the total gain preceding the power detector; and k_B is the Boltzmann constant. Other parameters carry the same meaning as in (1.1). A brief discussion showing how these formulas are related to each other and how they are applied to different kinds of imaging receivers follows.

For a stand-alone detector acting as a simple imaging RX without any pre-amplification, the NEP rather than NF is the proper measure for noise performance, since the square-law detector is essentially a nonlinear circuit. Therefore, (1.5) should be used to calculate NETD. Note that in this case, NEP_{sys} equals NEP_{det}. For a direct detection imaging RX consisting of a pre-amplification gain stage (e.g., LNA) and a power detector [3,8,9], both the noise temperature (or NF) of the pre-amplification stage and noise from the detector (measured by NEP_{det}) contribute to overall system noise. Therefore, the system NETD is obtained either from system NEP, NEP_{sys}, using (1.5) or by superposing the noise contribution from the pre-amplification stage and the detector using (1.4). As clearly seen from (1.4), although the pre-amplification stage also contributes noise denoted by the first term under the square root, it is still required to suppress detector noise in order to achieve a less than 1 K NETD. For a frequency conversion type imaging RX architectures, such as the ones demonstrated in references 10, 12, and 13, where the detector noise is suppressed by high front-end gain, the first term under the square root of (1.4) (representing noise from the front end) dominates and thus (1.4) is simplified to (1.1). In this case, the system NETD calculated from (1.1) and (1.5) should reconcile. Note that a factor of two corresponding to Dicke radiometer needs to be added to all NETD calculations [15].

1.2 SINGLE-PIXEL SILICON-BASED PASSIVE IMAGING RECEIVER

Traditionally, radiometers operating at W-band frequency range have been implemented in III-V semiconductor technologies, using multichip systems with module-based level of integration [6]. Figure 1.6 shows a radiometer module that was designed using an imaging chip set consisting of an InP HEMT LNA and an Sb-based backward tunnel diode detector with a horn antenna input and E-plane probe transition. Thermal sensitivity and output noise measurements demonstrate excellent system performance with 0.45 K NETD for a 3.125 ms integration time.

SiGe technologies with $f_T > 200$ GHz have made possible the development of fully integrated, highly compact, passive imaging systems. May and Rebeiz [3] described the design of a W-band passive radiometer chip in a standard 0.12 μm SiGe BiCMOS technology, as shown in Figure 1.7. They presented a total-power radiometer that

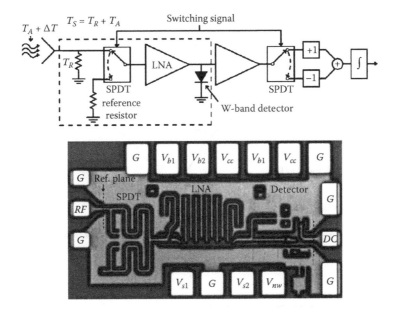

FIGURE 1.6 A W-band radiometer chipset in III-V technology [6].

FIGURE 1.7 A W-band single-chip SiGe BiCMOS Dicke radiometer [3].

achieved a temperature resolution of 0.69 K (30 ms integration time) with periodic calibration or chopping above 10 kHz. They also presented a switched Dicke radiometer chip, which addresses the 1/f noise issue of the total-power radiometer and can achieve a temperature resolution of 0.83 K with a 30 ms integration time.

Although the III-V diodes are not compatible with SiGe/CMOS technology, a high-responsivity, relatively low noise W-band SiGe detector circuit can still be constructed, as shown in Figure 1.8 [3]. The detector is a high-speed SiGe HBT biased in class B regime with a 94 GHz LC notch filter at the output port. The load resistance is chosen to be around 1 k in this design to obtain a high responsivity at video frequencies. The notch filter suppresses the generation of unintended nonlinearities by reducing RF variations in V_{CE} and reduces the effect of transistor collector capacitances.

For the total-power radiometer in May and Rebeiz [3], a five-stage common-emitter LNA was designed and fabricated. The bias currents are chosen for low noise in

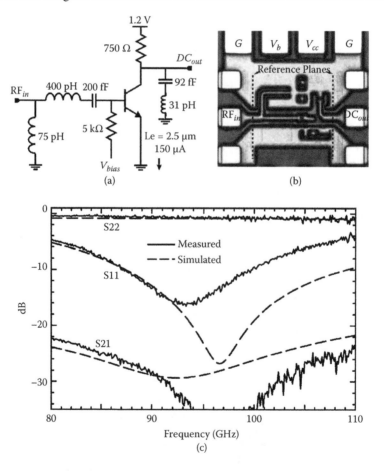

FIGURE 1.8 (a) Detector schematic, (b) chip micrograph (386 × 370 mm² including pads), and (c) measured S-parameters [3].

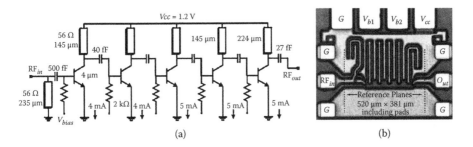

FIGURE 1.9 (a) The five-stage LNA schematic, and (b) the die photo.

stages 1 and 2 and for high gain in stages 3–5 (see Figure 1.9). The input transistor is sized to provide simple matching with a near-minimum noise figure. Shunt transmission line stubs and small metal–oxide–metal (MOM) capacitors provide interstage matching. Each stage uses five-layer MOM capacitors beneath the inductive transmission line loads to provide approximately 5 pF of total supply decoupling capacitance at W-band. The 56 Ω transmission line stubs have a signal width, $W = 11$ µm, and spacing, $S = 11$ µm, and have Q-factor of 9–1 9–15 from 80 to 100 GHz (including via resistances for connections to transistors) [3]. A deep trench is placed beneath the interstage matching capacitors to reduce substrate coupling. The use of a common-emitter stage, however, raises questions about stability of this LNA. The LNA exhibits higher peak gain of 27 and 24 dB gain across 82–100 GHz [3].

1.2.1 A PASSIVE IMAGING RX USING BALANCED LNA WITH EMBEDDED DICKE SWITCH

Typically, the Dicke switch is implemented as an SPDT switch (Figure 1.5) using PIN diodes [16]. This traditional architecture, with a switch right before the LNA, has the drawback of the front-end NF being degraded by the insertion loss of the SPDT switch. This directly results in an increase in NETD. This is not a major problem in III-V technologies, due to the availability of low loss PIN diode switches [17] and, more recently, zero-biased diode detectors [6]. However, silicon-based MMW switches exhibit unacceptably high insertion loss (~5 dB for an HBT switch in a standard SiGe technology). System-level analysis indicates that this 5 dB loss prior to the LNA will degrade the receiver NETD by a factor of approximately three, as shown in Figure 1.10. This degradation, coupled with the inherently high NF of silicon transistors, results in an NETD greater than 0.5 K (which is typically cited as the threshold for acceptable performance in indoor applications [2]). This drawback has been the major motivation behind the front-end architecture shown in Figure 1.11, which promises to eliminate this problem by embedding the Dicke switch functionality within a *balanced LNA* such that the switch insertion loss contributes minimal effect on the RX's NF [17]. In this design, the insertion loss of the input coupler directly adds to the front-end NF.

The system comprises a balanced LNA with an embedded Dicke switch, a power detector, and the baseband circuitry [17].

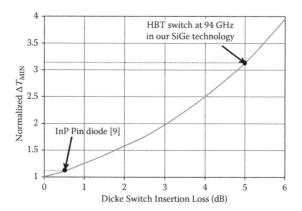

FIGURE 1.10   NETD versus Dicke switch insertion loss.

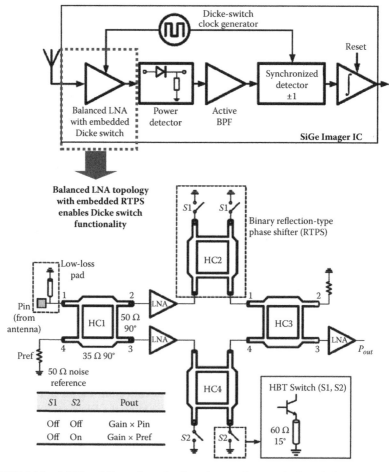

FIGURE 1.11   A W-band direct-detection imaging receiver employing a balanced LNA with embedded Dicke switch.

1.2.1.1 Design and Analysis of Balanced LNA with Embedded Dicke Switch

Figure 1.5 shows the schematic of the balanced LNA (BLNA) incorporating the embedded Dicke switch. Inspired by the GaAs topology first presented in Lo et al. [18], the circuit primarily comprises a balanced LNA with the addition of a reflection-type binary phase shifter in each branch. The operation of this BLNA can be understood using power-waves analysis [19]. Given the well-known S-parameter matrix of a branch-line coupler,

$$[S] = L_C \begin{bmatrix} 0 & \dfrac{-j}{\sqrt{2}} & \dfrac{-1}{\sqrt{2}} & 0 \\[2ex] \dfrac{-j}{\sqrt{2}} & 0 & 0 & \dfrac{-1}{\sqrt{2}} \\[2ex] \dfrac{-1}{\sqrt{2}} & 0 & 0 & \dfrac{-j}{\sqrt{2}} \\[2ex] 0 & \dfrac{-1}{\sqrt{2}} & \dfrac{-j}{\sqrt{2}} & 0 \end{bmatrix}$$

(L_C is the insertion loss of the branch coupler), using the superposition principle and assuming input, power waves on ports 1 (antenna port) and 4 (reference port) are expressed as

$$a_{pin1} = a_{IN1} \cdot e^{-j\theta_1}$$

$$a_{pin4} = a_{IN4} \cdot e^{-j\theta_4}$$

where $a_{IN,k}$ and θ_k denote the amplitude and phase of the power wave at the kth port, respectively; we compute the signal magnitude and phase at each point in the BLNA. Note that for this case, the power wave a_{IN4} represents the noise power of the 50 Ω noise reference. Since the insertion loss L_C appears in both gain paths of the BLNA, the power waves of intermediate and output ports are all normalized to L_C throughout power-wave analysis.

Port 2 of hybrid coupler HC1 will have the power wave

$$a_{pin2} = (a_{IN1} \cdot e^{-j\theta_1}) \cdot S_{21} + (a_{IN4} \cdot e^{-j\theta_4}) \cdot S_{24} \qquad (1.6)$$

$$= -j \frac{a_{IN1} \cdot e^{-j\theta_1}}{\sqrt{2}} - \frac{a_{IN4} \cdot e^{-j\theta_4}}{\sqrt{2}}$$

and port 3 of HC1 will have the power wave

$$a_{pin3} = (a_{IN1} \cdot e^{-j\theta_1}) \cdot S_{31} + (a_{IN4} \cdot e^{-j\theta_4}) \cdot S_{34} \qquad (1.7)$$

$$= -\frac{a_{IN1} \cdot e^{-j\theta_1}}{\sqrt{2}} - j \frac{a_{IN4} \cdot e^{-j\theta_4}}{\sqrt{2}}$$

For the case when both phase shifters are in the same state, both power waves go through identical paths consisting of LNA gain and phase shift as well as identical attenuation and phase shift due to the reflection-type phase shifters (RTPSs) in Figure 1.11. Therefore, the power wave at port 1 of HC3 will be

$$a_{pout1} = -j \frac{G_{LNA} \cdot L_{RTPS} \cdot a_{IN1} \cdot e^{-j(\theta_1 - \Phi)}}{\sqrt{2}} - \frac{G_{LNA} \cdot L_{RTPS} \cdot a_{IN4} \cdot e^{-j(\theta_4 - \Phi)}}{\sqrt{2}}$$

where G_{LNA} is the LNA's gain, L_{RTPS} denotes the RTPS's loss, and Φ represents the combined phase shift of both the LNA and RTPS. Similarly, the power wave at port 4 of HC3 will be

$$a_{pout4} = -\frac{G_{LNA} \cdot L_{RTPS} \cdot a_{IN1} \cdot e^{-j(\theta_1 - \Phi)}}{\sqrt{2}} - j \frac{G_{LNA} \cdot L_{RTPS} \cdot a_{IN4} \cdot e^{-j(\theta_4 - \Phi)}}{\sqrt{2}}$$

We then use superposition to compute the power delivered to port 3 of HC3 (i.e., the output port):

$$a_{pout3} = G_{LNA} L_{RTPS} \left(-j \frac{a_{IN1} \cdot e^{-j(\theta_1 - \Phi)}}{\sqrt{2}} - \frac{a_{IN4} \cdot e^{-j(\theta_4 - \Phi)}}{\sqrt{2}} \right) \cdot \frac{-1}{\sqrt{2}} \qquad (1.8)$$

$$+ G_{LNA} L_{RTPS} \left(-\frac{a_{IN1} \cdot e^{-j(\theta_1 - \Phi)}}{\sqrt{2}} - j \frac{a_{IN4} \cdot e^{-j(\theta_4 - \Phi)}}{\sqrt{2}} \right) \cdot \frac{-j}{\sqrt{2}}$$

$$= G_{TOTAL} \cdot a_{IN1} \cdot e^{-j(\theta_1 - \Phi - \pi/2)}$$

where G_{TOTAL} is the product of the LNA's gain and the RTPS's loss. As seen in (1.8), when the phase shifters are in the same state, only a power incident at port 1 of HC1 (a_{IN1}) is present at the BLNA's output port. However, when the phase shifters are in opposite states, one power wave will experience an additional 180° phase shift. In this case, the power delivered to the output port is expressed as

$$a_{pout3} = G_{LNA} L_{RTPS} \left(-j \frac{a_{IN1} \cdot e^{-j(\theta_1 - \Phi)}}{\sqrt{2}} - \frac{a_{IN4} \cdot e^{-j(\theta_4 - \Phi)}}{\sqrt{2}} \right) \cdot \frac{1}{\sqrt{2}} \qquad (1.9)$$

$$+ G_{LNA} L_{RTPS} \left(-\frac{a_{IN1} \cdot e^{-j(\theta_1 - \Phi)}}{\sqrt{2}} - j \frac{a_{IN4} \cdot e^{-j(\theta_4 - \Phi)}}{\sqrt{2}} \right) \cdot \frac{-j}{\sqrt{2}}$$

$$= G_{TOTAL} \cdot a_{IN4} \cdot e^{-j(\theta_4 - \Phi - \pi)}$$

This shows that when phase shifters are in opposite states, only power from the 50 Ω reference resistor (a_{IN4}) is delivered to the output port. By toggling between these two states, the desired chopping operation of the Dicke switch is achieved. It can also be shown that when the phase shifters are in opposite states, power from port 1 of HC1 is dissipated in the 50 Ω resistor connected to port 2 of HC3.

To compare NF performance of the BLNA in Figure 1.11 analytically with that of the traditional LNA + switch in Figure 1.7, suppose that each LNA stage in Figure 1.7 exhibits a gain of G_{LNA}. The noise factor of the LNA + switch $F_{SW\text{-}LNA}$ in Figure 1.7 is readily expressed as

$$F_{SW-LNA} = L_{SW} F_{LNA} \tag{1.10}$$

where L_{SW} represents the linear loss of the Dicke switch, and F_{LNA} represents the noise factor of the LNA in Figure 1.7. The noise factor of the BLNA with embedded Dicke switch is found to be

$$F_{BLNA} = L_C F_{LNA} (1 + \frac{L_C L_{RTPS} - 1}{G_{LNA} + 1}) \tag{1.11}$$

The loss of the hybrid couplers is two to three times (~4 dB) lower than that of SPDT switches. The improvement in noise factor IM_F is

$$IM_F = \frac{F_{SW-LNA}}{F_{BLNA}} = \frac{L_{SW}}{L_C \left(1 + \dfrac{L_C L_{RTPS} - 1}{G_{LNA} + 1}\right)} \tag{1.12}$$

Assuming 20 dB gain for each LNA; 6.5 and 6 dB losses for RTPS and Dicke switches, respectively; and 0.5 dB loss for hybrid coupler, approximately 2.5 dB improvement in NF will be achieved.

Figure 1.12 shows the measured gain and isolation from the antenna and reference ports for the two different phase-shifter states.

As expected, when both phase shifters are in the same state, the signal from the antenna is amplified while the reference signal is suppressed. Conversely, when the phase shifters are in opposite states, the reference input is amplified, while the antenna signal is suppressed. Note that the balanced structure ensures equal gains in both the antenna and reference modes. An additional LNA is used after the balanced structure in order to achieve a total predetection gain of 30 dB, as can be seen in Figure 1.12. The proposed BLNA with embedded Dicke switch achieves a minimum NF of around 9.7 dB.

The BLNA with embedded Dicke switch uses four couplers and three 5-stage LNAs. The LNA in a conventional SPDT structure uses two 5-stage LNAs to achieve almost the same predetection gain and therefore will consume a small chip area. Nevertheless, the BLNA, by virtue of its design, is able to achieve the required NETD for indoor imaging applications. On the other hand, the SPDT architecture (in currently available SiGe technology) essentially cannot. The use of a switch prior to the LNA in the conventional approach degrades the system noise temperature and, therefore, NETD (see Equation 1.4). It is noteworthy that the layout was done conservatively in order to avoid high-frequency EM coupling and to ensure first-pass

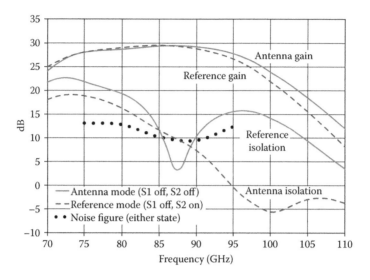

FIGURE 1.12   Measured gain, NF, and isolation for the BLNA.

success. A more aggressive layout approach (e.g., avoiding the use of quarter-wave length bias chokes) can be used to reduce the chip area.

1.2.1.2   Reflection-Type Phase-Shifter Design

The RTPS structure of Figure 1.13 was chosen because it provides broadband input and output matching. Additionally, the RTPS phase shift stays within ±10% of 180° for the majority of the W-band. Figure 1.13 shows the measured S-parameters as well as measured phase shift for the RTPS structure. The input and output return losses as well as insertion loss of this structure for both possible switch-state operations

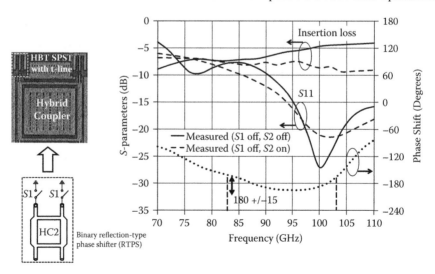

FIGURE 1.13   Measured RTPS S-parameters and phase shift.

have been measured. The output return loss was measured to be identical to the input return loss (due to the symmetric nature of the RTPS) and therefore was not included in Figure 1.13. The insertion loss for both operation states is better than −8 dB from 73–100 GHz. It should be noted, however, that this loss comes after 20 dB of LNA amplification and therefore does not contribute to front-end NF as a conventional Dicke switch architecture would.

1.2.1.3 Five-Stage LNA Design and Analysis

Figure 1.14 shows the maximum available gain (MAG) and minimum achievable NF (NF_{min}) of a common-emitter HBT in the technology used in this work. The device is optimally sized and biased such that it achieves the lowest NF_{min} for this technology. Simulations of the single HBT at 90 GHz show a MAG of 3.9 dB and an NF_{min} of 7.2 dB. A multistage amplifier designed in this technology, therefore, incorporates a first amplifier stage whose gain will not be high enough to reduce the NF contribution of the subsequent stages significantly. Using the well-known Friis equation for the cascaded NF of a multistage amplifier along with the transistor's MAG and NF_{min} values, the effect of the latter gain stages on the overall LNA NF can be estimated. It turns out that the second stage will add at least 1.2 dB to the overall LNA's NF. Adding a third and a fourth stage will contribute 0.4 and 0.1 dB to the overall NF, respectively, resulting in theoretical four-stage MAG and NF_{min} of 15.6 and 9.0 dB. The previous analysis assumes that each stage achieves maximum gain and minimum NF. However, by design, the first stage of an LNA will trade off a certain amount of available gain in order to achieve the best possible noise match at the input. This, along with loss in the matching networks, necessitates the use of a fifth gain stage in order to achieve the desired 15 dB LNA gain. The fifth stage has a negligible (<0.1 dB) contribution to the LNA NF.

The five-stage common emitter LNA schematic, used inside each LNA block of the BLNA circuit in Figure 1.11, is shown in Figure 1.15. The input matching networks of

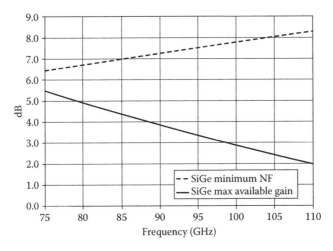

FIGURE 1.14   MAG and NFMIN for the device used in this work across the W-band.

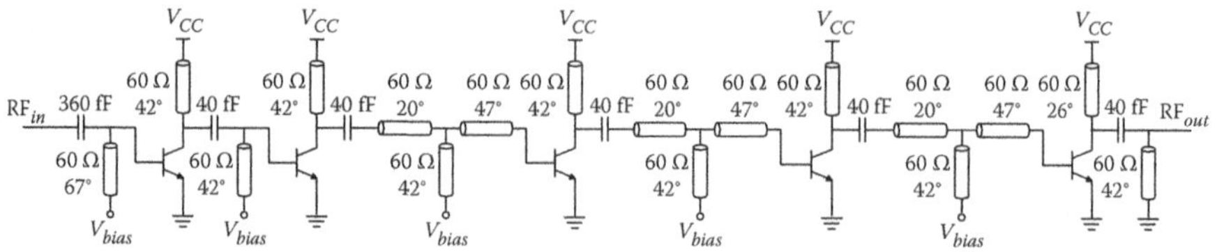

FIGURE 1.15 Five-stage LNA schematic.

the first two stages (i.e., the high-pass L- and the π-match networks at the input ports of the first and the second CE stages) are designed to achieve minimum NF.

These high-pass matching networks reduce the power gain at lower frequencies, where the HBT transistors exhibit naturally high gain, which helps the LNA's stability. Furthermore, it is desirable to keep the topologies of these matching networks as compact as possible in order to minimize any pregain losses, which will otherwise contribute to higher NF. To achieve the preceding goals, a different design methodology compared to standard silicon-based techniques for obtaining simultaneous power and noise match (presented in Nicolson and Voinigescu [20]) is used. Specifically, due to the low gain of the HBTs in the W-band, inductive emitter degeneration is not employed, thereby avoiding the associated reduction in gain. First, the current density that minimizes the HBT's NF_{min} is obtained. In HBT LNAs, as opposed to CMOS LNAs, maximizing f_T will not necessarily result in minimizing NF_{min} because, in HBT devices, an increase in bias current will lead to significantly higher shot noise. The minimum NF_{min} is thus achieved at a bias current lower than that for maximum f_T. The location of the optimum source reflection coefficient, Γ_{opt}, for noise match is plotted on the Smith chart, along with the device unmatched input reflection coefficient Γ_{in}. The device size is then swept such that Γ_{opt} moves sufficiently close to the 50 Ω point on the Smith chart, while at the same time, Γ_{in} moves to the same resistance (or conductance) contour as Γ_{opt}. This choice of the device size will require only a single stub at the input in order to move the device Γ_{in} toward Γ_{opt}, thereby achieving excellent noise and impedance match.

Following the methodology described before, an optimum HBT emitter area of 0.75 μm² is found. Figure 1.16 indicates simulated Γ_{in}, Γ_{opt}, and the 7.3 dB NF circle (which intersects the 50 Ω point on the Smith chart). As shown in Figure 1.16, Γ_{opt} is located on the circle with a constant VSWR of 1.4:1, which corresponds to −15 dB input return loss. A short-circuited stub at the input moves Γ_{in} along a constant conductance contour such that it achieves an input noise match within 0.1 dB of NF_{min} and an input return loss of −15 dB at 90 GHz.

Starting with the output of the second gain stage, all subsequent interstage matching networks employ a more complex T-match network topology realized using transmission lines (t-lines). This matching network offers more degrees of freedom than a single-stub matching network and therefore enables conjugate matching between the output of each stage and the input of the subsequent stage over a larger bandwidth than a single-stub matching network. As a result, maximum power transfer, and hence maximum gain, from the last three stages will be achieved. The insertion loss of the t-lines in the T-match should be minimized, since this loss will reduce the

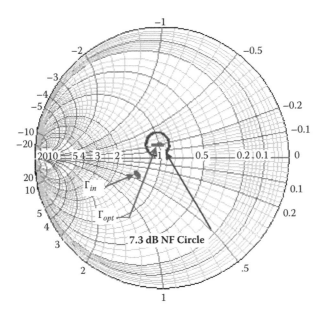

FIGURE 1.16 Γ_{IN}, Γ_{OPT}, and a 7.3 dB NF circle.

amplifier gain. To this end, t-lines were implemented as slow-wave coplanar wave-guide (CPW) structures. A slow-wave CPW (SW-CPW) t-line achieves roughly 60% higher phase shift compared to standard conductor-backed CPW t-lines, for a given length. This translates to reduced loss of the matching networks, as well as reduced chip area.

The LNA layout has been carefully designed to avoid parasitic feedback across each LNA stage. In particular, every input and output matching stub alternates its orientation in the layout in order to minimize EM coupling between the t-lines and further stabilize the LNA. The simulated k-factor was consistently greater than 10 across the W-band.

On-wafer LNA S-parameters were measured using a VNA with W-band frequency extenders. The VNA measurement results in Figure 1.17 are in good agreement with both theory and simulation, and they show a peak gain of 19 dB at 80 GHz, and better than −12 dB input return loss and −9 dB output return loss from 70 to 110 GHz. NF measurements, shown in Figure 1.17, were performed using a spectrum analyzer (Agilent E4448A) and an external down-converter. The NF was only measured up to 95 GHz due to limitations of the external down-converter. The input and output ports of the LNA were probed with a spectrum analyzer, and no oscillations were observed in the frequency range from 1 to 110 GHz, verifying the stability of the design. The core five-stage LNA draws 35 mA of current from a 1.8 V supply and consumes 1.0 mm² of chip area.

1.2.1.4   Detector Design and Analysis

The operation principle of a power detector in a direct-detection architecture is to convert the MMW input power to a constant output voltage. Therefore, a linear

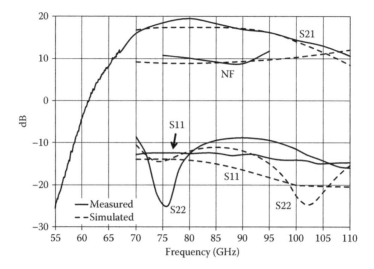

FIGURE 1.17   Measured and simulated five-stage LNA performance.

relationship between input power and output voltage needs to be established. This necessitates that the detector should operate in the square-law region. The detector in this work consists of a pair of common-emitter HBTs, as shown in Figure 1.18. The HBTs are biased via current mirrors and the MMW signal is applied to the base of one of the HBTs. The DC output voltage is taken differentially at the two collectors. Assuming that the input signal is a sinusoidal wave with frequency f and amplitude V_{im}, the output voltage is expressed as

$$V_{OUT} = R \cdot (I_1 - I_2) = R \cdot I_S e^{\frac{V_{BE,ON}}{V_T}} \left(e^{\frac{V_{im}\cos(2\pi ft)}{V_T}} - 1 \right) \tag{1.13}$$

where $V_{BE,ON}$ is the bias voltage at the base, $V_T = kT/q$ is the thermal voltage, and R is the load resistor. Expanding (1.13) using Taylor series, while truncating higher order terms and leaving the DC output, V_{OUT} becomes approximately equal to

$$V_{OUT} \approx I_{DC} R \frac{V_{im}^2}{4V_T^2} = I_{DC} R \frac{2 \cdot |Z_{in}| \cdot \mathrm{Re} \left(\dfrac{Z_{in}}{Z_{in} + R_S} \right) \cdot P_{IN}}{4V_T^2} \tag{1.14}$$

where I_{DC} is the DC current in each branch, P_{IN} is the input power, R_S is the source resistance, and Z_{in} is the input impedance of the HBT.

   An important detector figure of merit for imaging applications is the responsivity, which measures the change in detector output voltage per unit input power. From (1.11), the responsivity R can be calculated as

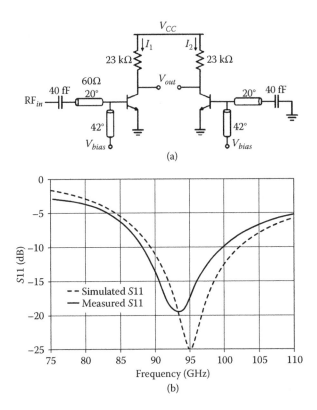

FIGURE 1.18 (a) W-band power detector schematic; (b) simulated and measured detector input return loss.

$$\Re = \frac{V_{OUT}}{P_{IN}} = \frac{I_{DC}R \cdot |Z_{in}| \cdot \mathrm{Re}\left(\dfrac{Z_{in}}{Z_{in} + R_S}\right)}{2V_T^2} = \alpha I_{DC}R \qquad (1.15)$$

where α (in W^{-1} units) accounts for MMW power transfer due to the input matching network. As can be inferred from (1.15), the responsivity is proportional to the DC current and the load resistor and inversely proportional to the square of the temperature.

The load resistor and HBT device generate three major types of noise at the detector output: shot noise, thermal noise, and flicker noise. Since the Dicke switch is necessarily designed to modulate the PMMW signal above the technology's 1/f corner frequency, flicker noise can be ignored. The output noise power density is then readily expressed as

$$\frac{\overline{V_n^2}}{\Delta f} = 2qI_{DC}R^2 + 4kTR \qquad (1.16)$$

The detector NEP is defined as the minimum input power required for a signal-to-noise ratio of unity at the detector output. From (1.15) and (1.16), the NEP (in W/Hz$^{1/2}$) is obtained as

$$NEP = \frac{\sqrt{V_n^2 / \Delta f}}{\mathfrak{R}} = \frac{\sqrt{2qI_{DC}R^2 + 4kTR}}{\alpha I_{DC}R} = \frac{1}{\alpha}\sqrt{\frac{2q}{I_{DC}} + \frac{4kT}{I_{DC}^2 R}} \qquad (1.17)$$

The foregoing analysis provides design insights for a differential HBT-based detector. Most notably, increasing the voltage drop across the load resistor will enhance both the responsivity and the NEP so long as the HBT stays in the forward-active region. Taking this design trend into account, the detector in Figure 1.18(a) is biased to have a collector current of 42 µA and a load resistance of 23 kΩ. As seen in Figure 1.18(a), an input matching network is also used in order to deliver maximum power to the detector input and provide a 50 Ω matched termination at the LNA output.

Figure 1.18(b) shows measured and simulated S_{11}. The detector's responsivity has been measured using a coherent test setup consisting of a signal generator as a variable input power source and an oscilloscope to measure the output voltage changes. The output spot noise power of the detector was measured using a spectrum analyzer at 1 MHz frequency (the frequency of our Dicke switch).

1.2.1.5   Measurement Results

Figure 1.19 shows the SiGe imaging RX chip micrograph. The power detector is followed by an active bandpass filter with an in-band gain of 20 dB and bandwidth of 0.1–10 MHz, which captures the first nine harmonics of the detector's output square wave. As mentioned before, all feedback capacitors are implemented using standard on-chip MIM capacitors in the SiGe process, which provides a capacitance density of 2.0 fF/µm^2.

In order to evaluate the passive imaging RX performance, relevant imaging parameters have been measured on wafer, with a system integration time of 30 ms. The responsivity of the RX chip is estimated by measuring the integrator's output voltage with the Dicke switch activated. A baseline calibration is performed before the responsivity measurement in order to estimate the input noise temperature when no signal is applied.

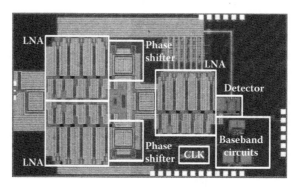

FIGURE 1.19   SiGe imaging RX chip micrograph.

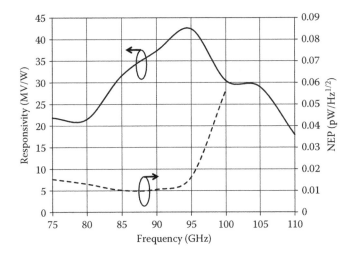

FIGURE 1.20 Measured system responsivity and NEP.

Figure 1.20 shows measured responsivity and NEP versus frequency of the passive imaging RX. The RX achieves a responsivity of 20–43 MV/W across the W-band. The minimum NEP of the imaging system is 10 fW/Hz$^{1/2}$, which is almost 10 times lower than that of the five-stage LNA + detector. The reason is because the entire integrated imager employs the BLNA of Figure 1.11, whose gain is 11 dB higher than that of the five-stage LNA. Although the base-band op-amp increases the signal level and the output noise, the improvement in NEP is primarily due to higher front-end gain.

Figure 1.21 shows the lab setup for the line-of-sight active measurement of the passive imaging RX. A 12–18 GHz signal generator drives a ×6 multiplier, providing a transmit power of –30 dBm at 90 GHz to a WR-10 horn antenna. The radiated

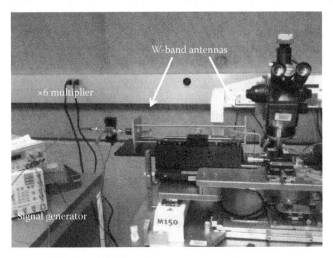

FIGURE 1.21 Imaging test setup.

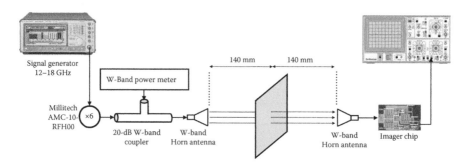

FIGURE 1.22   Imaging test setup diagram.

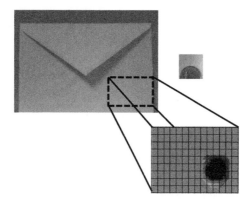

FIGURE 1.23   Image of an envelope containing a coin in visible light and the MMW image.

power from the antenna is used to illuminate the object of interest, which, in turn, increases the SNR at the RX input. The RX employs an off-chip narrow-beam horn antenna manufactured by Quinstar Technology. The pyramidal standard gain horn antenna has an aperture size of 26.2×20.3 mm^2, a midband gain of 24 dB, and a beam width of $11°$. The horn antenna is connected through a WR-10 waveguide to the wafer probe. A mechanical drawing of the imaging test setup is shown in Figure 1.22. A simple MMW image of an envelope containing a coin (US quarter) was created and demonstrated in Figure 1.23. The image was generated, one pixel at a time, by stepping the envelope position over a 60×40 mm^2 area in 5 mm increments. This coarse step was chosen due to the limited accuracy of manual movement of the envelope. The chip output voltage was read on an oscilloscope for each pixel.

1.3   FOCAL PLANE ARRAY IMAGING RECEIVERS

The previous section focused on design and implementation of a single receiver/ pixel. A real imaging system is, however, required to incorporate multipixel architecture. More precisely, to reduce the scanning time and enable video rate real-time imaging, focal-plane array (FPA) could be used with an array of detectors located at the focal plane of a focusing system [11–13].

In direct detection architecture, discussed in the previous section, the high gain requirement can be met by cascading several LNA stages [3]. Alternatively, one way to improve the sensitivity of a radiometer, without risking oscillation, is to spread gain across multiple frequencies. This can be done using a frequency-conversion type of receiver. Given the fact that the operation frequency is half of f_{max} for this design, in order to meet the stringent design requirement of the imaging receiver, the direct conversion architecture [10,12,13] is adopted with the LO frequency being placed in the middle of the RF band. Because the input signal is, in fact, broadband noise that contains no phase information, no I/Q path is needed to fulfill down-conversion. The design requirement for the zero-IF amplifier is also relaxed since the IF bandwidth is reduced to one-half of the RF bandwidth. Another advantage of employing direct conversion architecture is that the detector will operate at IF frequency instead of MMW frequency, which leads to lower detector NEP due to higher responsivity and lower output noise. And the lower detector NEP would in turn reduce the required predetection gain for the system to be limited by the front-end noise rather than the detector noise.

1.3.1 FPA Architecture

The proposed multipixel direct conversion imager architecture is shown in Figure 1.24. The multipixel imager architecture is presented in Figure 1.1. Each RX employs a local frequency tripler whose input is fed by a main 32 GHz PLL [21]. The PLL is shared among the four RXs with the PLL's output placed at the center of the chip to facilitate skew-minimized LO distribution. An advantage of the proposed scheme is that the LO signal can be generated/distributed at one-third the 96 GHz operation frequency. Each RX (Figure 1.1) consists of a folded slot antenna, an SPDT switch, a four-stage LNA, a single-balanced mixer, an injection-locked frequency tripler (ILFT), an IF VGA, a power detector, a bandpass VGA, and a synchronous demodulator. All signal paths are fully differential after the LNA. An off-chip 200 kHz clock is used to synchronize the SPDT switch with the baseband demodulator for Dicke operation.

1.3.2 Receiver Building Blocks

1.3.2.1 On-Chip Folded Slot Dipole Antenna

The small antenna form factor at MMW frequency makes it possible for on-chip antenna integration [22–25]. The antenna on-chip solution turns out to be an attractive option for a multipixel FPA imager, since it eliminates the complicated, lossy, and expensive (sub) MMW packaging [11,13]. Despite the benefits of integrating antenna on-chip, the combination of high dielectric constant ($\varepsilon_r = 11.9$) and low resistivity (8 $\Omega\cdot$cm) of silicon substrate poses a major challenge to on-chip antenna design. The bulk silicon substrate absorbs most of the radiation energy from the antenna into the undesired surface-wave mode due to its high permittivity, and the substrate's low resistivity leads to electric field loss, which is found to be the dominant loss factor. Several methods [23–25] have been proposed to overcome this problem and improve

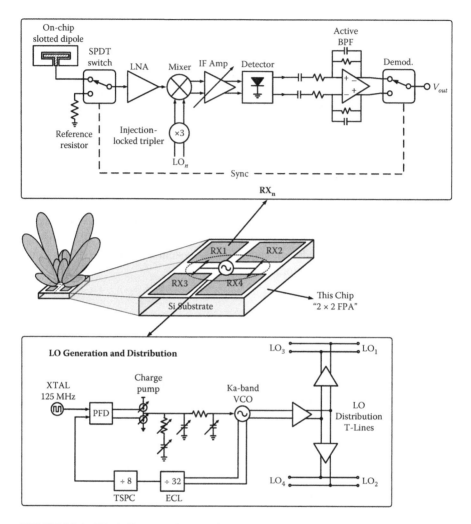

FIGURE 1.24 Block diagram of the 2×2 FPA.

antenna efficiency. Babakhani et al. [23] attach a hemispherical silicon lens to the bottom of the chip to realize backside radiation. In Wang et al. [24], an off-chip antenna director is placed half a wavelength away from the antenna to guide the electromagnetic wave. In Ahamdi and Naeini [25], a dielectric resonator is built on top of the chip, which enhances the radiation efficiency. However, these solutions either use additional off-chip components [24] or require complex postfabrication processes [23,25]; this makes the packaging even more challenging, reduces the yield, and increases the overall cost.

Figure 1.25 depicts the three-dimensional view of the SFDA with the CPW feed line used in this design. The folded-slot structure is favored since it provides wide bandwidth, CPW-friendly interface, low metallic loss [22], and high isolation. The advantages of using an SFDA can be better understood by comparing

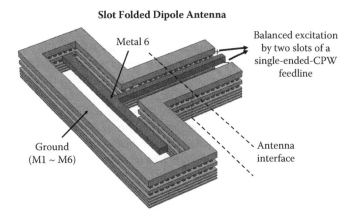

FIGURE 1.25   Three-dimensional view of the slot folded dipole antenna (SFDA).

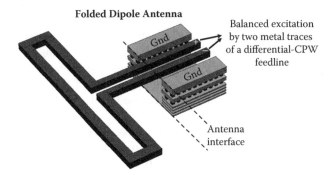

FIGURE 1.26   Three-dimensional view of the folded dipole antenna (FDA).

its complementary counterpart—that is, a folded dipole antenna (FDA), shown in Figure 1.26. The FDA is a differential type of antenna that requires balanced excitation from a differential feed line (cf. Figure 1.26). Although this differential type of antenna makes it possible to realize a fully differential RX, the front-end circuits of this chip (SPDT switch and LNA) use single-ended design for the purpose of power/area saving and easy characterization, since there is no four-port W-band VNA available in the lab. Therefore, if the FDA is employed, an additional balun is needed at the antenna interface, which degrades both gain and NF. By interchanging conductive and dielectric material, an FDA is converted to an SFDA, which is excited by a pair of balanced slots instead of balanced metal lines in an FDA. When a signal travels along a single-ended CPW line, a differential electromagnetic field is generated in the two slots between the signal line and the two ground sidewalls. These two slots can thus provide balanced excitation to the SFDA antenna (cf. Figure 1.25). In other words, the CPW feed line provides single-ended interface (through the metal trace) to the front-end circuits and, from the other side, provides differential interface (through the two slots) to the SFDA.

As shown in Figure 1.25, the ground walls surrounding the slot consist of all metal layers shorted by array of vias in parallel. This configuration leads to several advantages. First, it helps to meet metal density rules for all layers in the vicinity of the antenna. Second, the conductive loss from metals becomes significantly lower compared to FDA, since the current now flows through a much wider path. Third, the ground wall confines the electromagnetic field and prevents mutual coupling between antenna and front-end circuits. In contrast, the radiation pattern of FDA is susceptible to interference from surrounding metal lines or other circuit components. Moreover, the presence of the FDA will also negatively affect the performance of adjacent circuits, especially passive devices like inductors and transformers.

To improve the antenna efficiency further, a patterned deep trench mesh is embedded in the substrate underneath the antenna, as shown in Figure 1.27. In this way, the conductive bulk silicon substrate is decomposed into an array of isolated small squares near its surface. The deep trench available from the technology is 7–10 μm deep from the substrate's surface and is made of high resistivity material, and the substrate thickness of this process is 280 μm. The deep trench is commonly employed to surround the HBT transistor (as part of the p-cell) to reduce substrate coupling among active devices. The main advantage offered by the deep trench mesh is that it effectively increases the substrate resistance and thus reduces the substrate's electric field loss, which is the key factor for antenna efficiency. Note that although the depth of the deep trench is much smaller than that of the substrate, it is still effective in improving antenna efficiency because the electric field losses are more pronounced near the substrate's surface.

To examine the effect of deep trench lattice on antenna performance and validate the preceding analysis, two SFDA test structures have been fabricated and characterized. The two SFDAs have exactly the same design except that one employs deep trench lattice underneath, whereas the other one does not. The measurement results in Figure 1.28 demonstrate an increase in antenna gain by an average value of 2dB as a result of using deep trench lattice. This translates to an efficiency improvement of 1.6 times if antenna directivity is assumed to be the same. The deep trench lattice also reduces coupling between the antenna and other building blocks of the PMMW RX through silicon substrate. In simulation, the SFDA exhibits an efficiency of 16% with deep trench. Figure 1.29 shows the simulated radiation pattern of the SFDA.

FIGURE 1.27   Silicon substrate with embedded deep trench mesh.

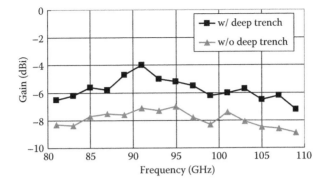

FIGURE 1.28 Measured peak gain of the SFDA with and without deep trench underneath.

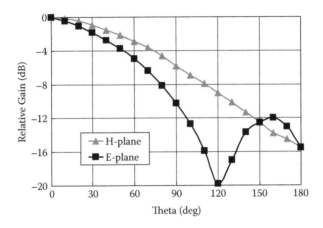

FIGURE 1.29 Simulated radiation pattern of the SFDA.

1.3.2.2 SPDT Switch

The SPDT switch, shown in Figure 1.30(a), is based on a $\lambda/4$ t-line approach similar to the one in Uzunkol and Rebeiz [26]. By turning on and off the shunt NMOS transistor connected to port 1 (or 2), the corresponding branch shifts between isolated and through state. The basic idea is as follows: When the NMOS transistor is turned on (gate voltage equals supply voltage), the drain terminal is shorted to ground. Since a $\lambda/4$ t-line converts "short" to "open," this branch presents an infinite impedance to port 3 and all the signals will go through the other branch. Because the insertion loss of the switch is directly added to the noise figure, it needs to be minimized for the sake of improving system sensitivity or NETD. In order to reduce insertion loss, we need to minimize the transistor's on-state resistance and at the same time maximize the off-state (gate voltage = 0 V) impedance.

The on-state resistance of a transistor can be reduced by choosing a large W/L ratio. The off-state impedance can be modeled using a parallel R-C network [26], where the capacitance is tuned out by a shorted t-line stub acting as an inductor and

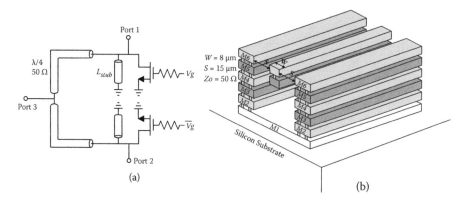

FIGURE 1.30 (a) Schematic of the SPDT switch; (b) geometry of the CPW line used in SPDT switch.

the off-state resistance is highly dependent on the substrate resistance. The substrate resistance can be increased by two layout techniques: (1) surrounding the NMOS device by deep trench, and (2) reducing the number of substrate contacts and placing the contacts several micrometers away from the transistor. Shown in Figure 1.30(b) is the CPW line used in the SPDT switch design with its geometry labeled on the figure. The characteristic impedance is 50 Ω, and the signal line combines M5 and M6 in parallel to reduce loss. The simulated loss at 95 GHz is 0.8 dB/mm. The CPW line uses a ground plane (M1) underneath and two ground side walls to isolate the signal line from the environment.

1.3.2.3   LNA

The LNA is a four-stage, single-ended cascode topology (Figure 1.31a). In order to reduce the footprint, lumped inductors (rather than t-lines) are used for output as well as interstage matching. Input matching is designed for minimum NF and is realized using the t-line to avoid layout discontinuity at the interface between the switch and the LNA. The first LNA stage is biased at minimum NF current density, whereas the last three stages are biased for highest gain. Occupying only 380×300 μm^2 (excluding pad), the LNA breakout circuit exhibits a measured peak gain of 25 dB, 3 dB bandwidth of 20 GHz (86–106 GHz), and a minimum NF of 8.3 dB, as shown in Figure 1.31(b). The LNA draws 21 mA from a 2.5 V supply.

1.3.2.4   LO Generation/Distribution and Tripler

The 32 GHz LO from the PLL is distributed to a local tripler within each RX through a symmetric differential CPW T-line H-tree network with two buffer stages (Figure 1.24). It is found from simulation that to ensure a locking range of >5 GHz for the injection-locked frequency tripler (ILFT), the LO power level at the input of each local tripler should be 1 dBm. The schematic of the ILFT is shown in Figure 1.32. The injection-locking operation is realized by feeding the third harmonic of the input signal generated by Q3 and Q4 into the emitters of a differential Colpitts oscillator. To minimize mutual as well as substrate noise coupling, ground shielded CPW lines

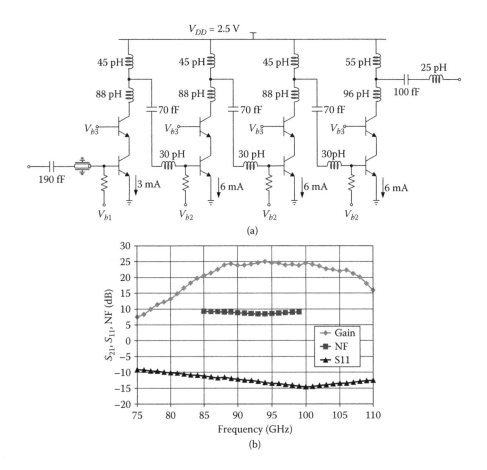

(a)

(b)

FIGURE 1.31   (a) Schematic and measured performance of LNA; (b) measurement results.

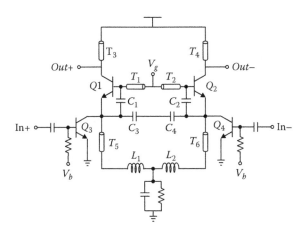

FIGURE 1.32   Schematic of ILFT.

(T1 ~ T6) are used at base, collector, and emitter to provide the tank, the load, and part of the degeneration inductances. Additional emitter degeneration inductance is realized by spiral inductor (L1 and L2) to save area. The 96 GHz output signals are taken out from the collector terminals of Q1 and Q2 through a differential cascode buffer. The tripler circuit consumes 70 mW from a 2.5 V supply.

1.3.2.5   Mixer and IF/BB Circuitry

A single-balanced mixer is adopted for better noise performance. Further LO rejection is provided by the IF VGA, which is based on Cherry-Hopper topology with 10–20 dB of variable gain. Finally, power detection is realized using a common-emitter-based square law detector.

1.3.3   SYSTEM PERFORMANCE

The front-end RX achieves 34–44 dB variable gain and 11.3 dB minimum NF, as shown in Figure 1.33. The PLL and tripler together exhibit a measured tuning range of 92.67–98.2 GHz, which is limited by the locking range of the ILFT. Using a 125 MHz crystal oscillator (CVHD-950) as reference, the PLL + tripler's phase noise at 96 GHz is –93 dBc/Hz at 1 MHz offset (Figure 1.34).

A chip-on-board solution was used to evaluate the system performance. Figure 1.35 shows the measured responsivity (antenna de-embedded) and the NEP referred to the input of the switch for RX1. The average responsivity and NEP over 3 dB BW are 285 MV/W and 8.1 fW/√Hz, respectively. Imaging functionality of the 2 × 2 FPA was verified using an incoherent source in a transmission mode measurement setup [11,13], shown in Figure 1.36. Image of a metallic object inside an envelope is constructed by mechanically scanning the chip and mapping the output voltage of each pixel to corresponding grayscale intensity.

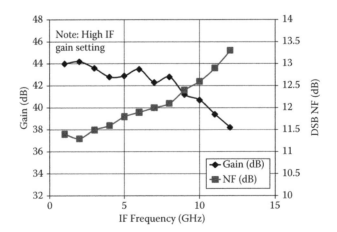

FIGURE 1.33   Measured receiver front-end gain and NF.

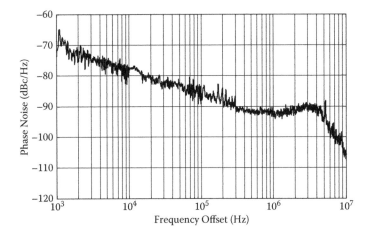

FIGURE 1.34 Measured 96 GHz phase noise.

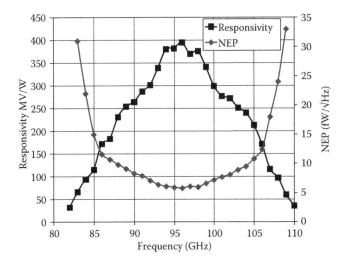

FIGURE 1.35 Measured single RX responsivity and NEP.

The complete measured performance of the 2×2 FPA is summarized in Table 1.1. The imaging chip achieves a calculated NETD of 0.48 K with 30 ms integration time. The chip is fabricated in a 0.18 μm SiGe BiCMOS process and occupies an area of 3.5×3 mm^2 (Figure 1.37).

ACKNOWLEDGMENTS

The author thanks all PhD members of the NCIC Labs, including Zhiming Chen, Chun-Cheng Wang, Vipul Jain, and Leland Gilreath. The author also thanks TowerJazz Semiconductor for chip fabrication. This work has been supported in part by an NSF grant under contract ECCS-1002294 and an SRC grant under contract 2009-VJ-1962.

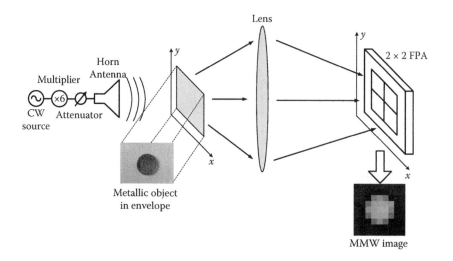

FIGURE 1.36 Imaging system measurement setup and MMW image.

TABLE 1.1
Performance Summary

Antenna		PLL + tripler	
Antenna gain	−4 dBi	Phase noise	−93 dBc/Hz at 1 MHz
Bandwidth ($S_{11} \leq 10$ dB)	77–110 GHz	Tuning range	92.67–98.2 GHz
		Spur	−55 dBc
Front end (switch + LNA + mixer + IF)		**Radiometer system**	
Bandwidth	86–106 GHz	RX noise temp.[b]	4000 K
Switch loss/isolation	2.8 dB/−21 dB	Responsivity[b]	48 MV/W including the on-chip antenna loss
Power gain[a]	34–44 dB	NEP[b]	8.1 fW/Hz$^{1/2}$
Noise figure[a]	11.3 dB	NETD[c]	3 K including the on-chip antenna loss
LO-to-RF isolation	≤65 dB	**Power consumption**	
Output P 1 dB	0 dBm	4 × RX	4 × 152.5 mW (1.8/2.5 V)
Technology	0.18 μm BiCMOS	PLL and buffers	85 mW (1.8/2.5 V)
Die size	3.5 mm × 3 mm	Total	695 mW

[a] Includes switch loss.

[b] Average over 3 dB bandwidth.

[c] A factor of 2 was included for Dicke switching; assumes 30 ms integration time; calculated with measured NEP.

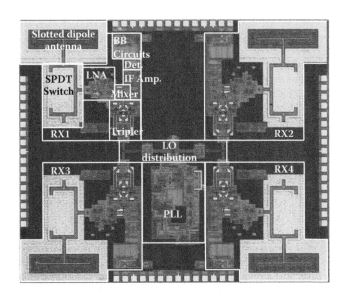

FIGURE 1.37 Die micrograph of the 2 × 2 FPA.

REFERENCES

1. C. Marcu, D. Chowdhury, C. Thakkar, J.-D. Park, L.-K. Kong, M. Tabesh, Y. Wang, et al., A 90 nm CMOS low-power 60 GHz transceiver with integrated baseband circuitry. *IEEE Journal of Solid-State Circuits,* vol. 44, no. 12, pp. 3434–3447, Dec. 2009.
2. V. Jain, F. Tzeng, L. Zhou, P. Heydari, A single-chip dual-band 22–29 GHz/77–81 GHz BiCMOS transceiver for automotive radars. *IEEE Journal of Solid-State Circuits,* vol. 44, no. 12, pp. 3469–3485, Dec. 2009.
3. J. W. May, G. M. Rebeiz, Design and characterization of W-Band SiGe RFICs for passive millimeter-wave imaging. *IEEE Transactions Microwave Theory & Technology,* vol. 58, no. 5, pp. 1420–1430, May 2010.
4. L. Yujiri, M. Shoucri, P. Moffa, Passive millimeter wave imaging. *IEEE Microwave Magazine,* vol. 4, no. 3, pp. 39–50, Sept. 2003.
5. R. Appleby, R. N. Anderton, Millimeter-wave and submillimeter-wave imaging for security and surveillance. *IEEE Proceedings*, vol. 95, no. 8, pp. 1683–1690, Aug. 2007.
6. J. J. Lynch, H. P. Moyer, J. H. Schaffner, Y. Royter, M. Sokolich, B. Hughes, Y. J. Yoon, J. N. Schulman, Passive millimeter-wave imaging module with preamplified zero-bias detection. *IEEE Transactions Microwave Theory & Technology,* vol. 56, no. 7, pp. 1592–1600, July 2008.
7. W. R. Deal, L. Yujiri, M. Siddiqui, R. Lai, Advanced MMIC for passive millimeter and submillimeter wave imaging. *IEEE ISSCC Digest Technical Papers,* pp. 572–573, Feb. 2007.
8. A. Tomkins, P. Garcia, S. P. Voinigescu, A passive W-band imaging receiver in 65 nm bulk CMOS. *IEEE Journal of Solid-State Circuits,* vol. 45, no. 10, pp. 1981–1991, Oct. 2010.
9. L. Gilreath, V. Jain, P. Heydari, Design and analysis of a W-band SiGe direct-detection-based passive imaging receiver. *IEEE Journal of Solid-States Circuits,* vol. 46, no. 10, Oct. 2011.

10. L. Zhou, C.-C. Wang, Z. Chen, P. Heydari, A W-band CMOS receiver chipset for millimeter-wave radiometer systems. *IEEE Journal of Solid-States Circuits,* vol. 46, no. 2, pp. 378–391, Feb. 2011.

11. E. Ojefors, U. R. Pfeiffer, A. Lisauskas, H. G. Roskos, A 0.65 THz focal-plane array in a quarter-micron CMOS process technology. *IEEE Journal of Solid-State Circuits,* vol. 44, no. 7, pp. 1968–1976, July 2009.

12. C.-C. Wang, Z. Chen, H.-C. Yao, P. Heydari, A fully integrated 96 GHz 2 × 2 focal-plane array with on-chip antenna. *IEEE RFIC Symposium Digest,* pp. 357–360, June 2011.

13. Z. Chen, C.-C. Wang, P. Heydari, A BiCMOS W-band 2 × 2 focal-plane array with on-chip antenna. *IEEE Journal of Solid-State Circuits,* vol. 47, 2012.

14. M. Tiuri, Radio astronomy receivers. *IEEE Transactions on Antennas and Propagation,* vol. 12, no. 7, Dec. 1964.

15. R. H. Dicke, The measurement of thermal radiation at microwave frequencies. *Reviews Scientific Instruments,* vol. 17, pp. 268–275, July 1946.

16. C. Martin et al., Rapid passive MMW security screening portal. *Proceedings of SPIE Defense and Security 2008,* vol. 6948.

17. V. Ziegler et al., Low-power consumption InGaAs PIN diode switches for V-band applications. *Japan Journal of Applied Physics,* vol. 38, pp. 1208–1210, 1999.

18. D. C. W. Lo et al., Novel monolithic millimeter wave multi-functional balanced switching low noise amplifiers. *IEEE Transactions Microwave Theory and Techniques,* vol. 42, no. 12, pp. 2629–2634, Dec. 1994.

19. G. Gonzalez, *Microwave transistor amplifiers: Analysis and design,* 2nd ed. Englewood Cliffs, NJ: Prentice Hall, Aug. 1996.

20. S. T. Nicolson, S. P. Voinigescu, Methodology for simultaneous noise and impedance matching in W-band LNAs. *Proceedings IEEE Compound Semiconductor Integrated Circuits Symposium,* pp. 279–282, Nov. 2006.

21. Z. Chen, C.-C. Wang, P. Heydari, W-band frequency synthesis using a Ka-band PLL and two different frequency triplers. *IEEE RFIC Symposium Digest,* pp. 83–86, June 2011.

22. A. Arbabian, S. Callender, S. Kang, B. Afshar, J.-C. Chien, A. M. Niknejad, A 90 GHz hybrid switching pulsed-transmitter for medical imaging. *IEEE Journal of Solid-State Circuits,* vol. 45, no. 12, pp. 2667–2681, Dec. 2010.

23. A. Babakhani, X. Guan, A. Komijani, A. Natarajan, A. Hajimiri, A 77 GHz phased-array transceiver with on-chip antennas in silicon: Receiver and antennas. *IEEE Journal of Solid-State Circuits,* vol. 41, no. 12, pp. 2795–2806, Dec. 2006.

24. C.-H. Wang, Y.-H. Cho, C.-S. Lin, H. Wang, C.-H. Chen, D.-C. Niu, J. Yeh, et al., A 60 GHz transmitter with integrated antenna in 0.18 μm SiGe BiCMOS technology. *IEEE ISSCC Digest Technical Papers,* pp. 659–660, Feb. 2006.

25. M. R. N. Ahamdi, S. S. Naeini, On-chip antennas for 24, 60, and 77 GHz single package transceivers on low resistivity silicon substrate. *IEEE Symposium Antennas and Propagation,* pp. 5059–5062, 2007.

26. M. Uzunkol, G. M. Rebeiz, A low-loss 50–70 GHz SPDT switch in 90 nm CMOS. *IEEE Journal of Solid-State Circuits,* vol. 45, no. 10, pp. 2003–2007, Oct. 2010.

2 Challenges in Wireless Chemical Sensor Networks

Saverio De Vito

CONTENTS

2.1 INTRODUCTION

The capability to detect and quantify chemicals in the atmosphere, together with their concentration, is a source of valuable information for many safety and security applications ranging from pollution monitoring to detection of explosives and drug factories. Actually, several gases are considered responsible for respiratory illness in citizens; some of them (e.g., benzene) are known to induce cancers in cases of prolonged exposure even at low concentrations [1]. Volatile organic compounds (e.g., formaldehyde) released as off-gas by furniture adhesives or cleaning agents or by smoking in indoor environments reach concentration levels that are orders of magnitude higher than in outdoor settings. Hazardous gas, like explosive or flammable ones, are also sources of increasing concerns for security reasons due to their possible use in terrorist attacks of military or civil installations. Moreover, some of them are currently in use or are foreseen to be used as energy carriers for automotive transport, so their diffusion is expected to grow significantly; for example, hydrogen-powered car refilling stations could become very common in the near future [2]. The estimation of chemical distribution is hence significantly relevant for citizens' safety and security; it is to be considered a potential life saving assets. It is also recognized as a technological enabler for the arising "smart cities" paradigm;

35

just as an example, the knowledge of pollutant distribution is of paramount importance for the definition of integrated urban and mobility plans designed to face pollution generated by fossil fuel cars.

However, chemical monitoring in outdoor and indoor environments is heavily affected by the peculiarity of the chemical propagation process [3]. To be concise, fluid dynamic effects like diffusion and turbulence, as main propagation drivers, make it very difficult to predict gas concentration in space and time domains. A single point of measure is usually totally ineffective, calling for distributed approaches to chemical detection and concentration estimation. In many applications, this, in turn, requires the monitoring task to be fulfilled by a network of wireless (sometimes mobile) modules, with the wireless term being related to connectivity/communication and/or power supply. These architectures have been recently termed wireless chemical sensor networks (WCSNs). Just as an example, the plume generated by an H_2 spill in a hydrogen-based car refilling station could move in rather unpredictable paths and the probability of a fixed, single solid-state chemical sensor being hit by it with a significant concentration in a timely way could be negligible in many circumstances. At the same time, provided the plume could be detected, it would be impossible to assess the position of the spill. A distributed WCSN could dramatically improve the detection chance, while an autonomous moving sensor exploring the station would have the best chance to locate the spill.

When it comes to pollution monitoring, the commercial state of the art proposes the use of conventional spectrometer-based stations characterized by significant costs and relevant size. High costs make it very difficult to achieve the appropriate density in the measurement mesh of a cityscape and thus to obtain statistically significant estimations despite problem complexities (e.g., the influence of canyon effects). As a consequence, public policy makers are not allowed to take appropriate mobility management decisions to avoid or mitigate air pollution. The use of low cost, compact, and wireless multisensory devices can offer a viable solution; furthermore, because of their limited footprint, they can be deployed almost everywhere, despite limitations imposed in cities' historical centers and cultural heritages. In this framework, researchers recently began to tackle the three-dimensional (3D) unconstrained chemical sensing scenario with two approaches. The first one is actually based on the use of a moving detector. Together with appropriate modeling information, these detectors can follow random or goal-oriented paths, exploring a particular environment before being hit by a chemical plume [4,5]. After that, by using the search algorithms for chemical spills, often exploiting biomimetic approaches, they try to detect the source of contamination in the so-called source declaration problem. The other approach basically relies on the use of multiple, low cost, and autonomous distributed fixed detectors that try to cooperate in reconstructing a chemical image of the sensed environment [6,7].

Advantages of the networked approach are identifiable in flexibility, scalability, enhanced signal-to-noise ratio, robustness, and self-healing. Several sensor nodes can be placed in different locations—each one with its own characteristics in terms of environmental conditions (air flow, temperature, humidity, different gas concentration, etc.) contributing to describe more thoroughly the environment in which they are embedded. Each smart chemical sensor comprising the distributed architecture has its own communication capabilities and its information is available for more than one client. The network can adapt itself to a variable number of chemical sensors, improving reliability. If a sensor fails, the network can estimate its response on the basis of the previous behavior and of the response of the closest sensors while being able to self-heal the network structure by reconstructing routing trees [8]. On the other hand, an autonomous moving robot guarantees enhanced coverage and could be more effective in the source declaration problem. Of course, the two approaches may be combined in the use of an autonomous fleet of chemical sensing robots.

In this process, a number of challenges repeatedly occurs when engineers try to design real-world operating WCSN systems, limiting their practical application. Actually, designing appropriate strategies for single module calibration and sensor stability, calibration transfer, efficient power usage, and 3D gas concentration mapping seems to be the most common challenge to face in indoor and outdoor settings.

This chapter reviews these challenges, together with the solutions proposed by our group during its commitment to the wireless chemical sensing topic.

2.2 MODULE CALIBRATION AND SENSOR STABILITY

The characteristics of an optimal sensor for distributed chemical sensing should include low-power operational capability, low cost, long-term reliability, and stability; additionally, it should also be easy to integrate with simple signal conditioning schemes. Depending on the application, the sensor should possibly express good specificity properties, high sensitivities, and very low detection limits. By far, this depiction applies more to an ideal device than to a real one; in fact, no current chemical sensor technology seems close to obtaining such results simultaneously.

Most chemical sensors suffer from nonspecificity (i.e., their response to the target gas is heavily affected by so-called interferents). In fact, interferents produce response changes that are indistinguishable from the one induced by target gas, hampering detection and quantification capabilities. Chemical multisensor devices, also called electronic noses, practically exploit the partial overlapping responses of an array of nonspecific sensors to detect and estimate concentrations of several gases simultaneously. Often this capability extends to gases toward which none of the enrolled sensors have been targeted. In this case, the response of a sensor subset can provide information by exploiting partial specificities or, more rarely, peculiar ratios among chemical species in fixed ratio mixture scenarios (proxy sensing). Due to the intrinsic complexities of model-based approaches, calibration is mainly achieved by the use of multivariate techniques based on statistical regressors like artificial neural networks (ANNs). In this case, a data set containing the response of the sensor array to a representative subset of gas concentrations can be used to train a statistical regressor to detect and estimate gas concentrations in mixtures. Unfortunately, when it comes to complex mixtures and dynamic environments like the ones involved in city air pollution monitoring, it is nearly impossible to generate such a representative data set with synthetic mixtures and in-lab measurements, thus strongly limiting the application of solid-state multisensory devices in this framework.

Starting in 2007 we began investigating the use of "on-field calibration methodologies" that exploit on-field recorded samples captured from multisensor devices and colocated conventional stations to build adequate data sets for the training of ANN-based models [9]. In fact, we have investigated the use of a number of samples belonging to intervals ranging from 1 to less than 100 days to build representative data sets. Results show that a relatively compact data set recorded during 10 consecutive days allows the trained regressor to achieve rather interesting performances when estimating the concentration of several pollutants for a significant number of months to come. In practice, using such methodology near real-time estimation of benzene concentration could be performed with a mean relative error (MRE) of less than 4% (see Figure 2.1). However, it was evident that combined effects of sensor drifts and concept drifts induced by seasonal cycles and human activities caused the calibration to lose accuracy over time.

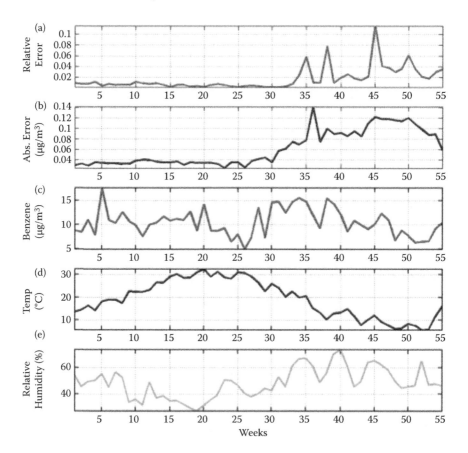

FIGURE 2.1 Benzene concentration estimation results in the 10-day training length run (weekly averages): (a) MRE; (b) MAE; (c) benzene concentrations as measured by a conventional station; (d) temperature measured by multisensor device; (e) relative humidity measured by the multisensor device. Results are affected by what we expect to be seasonal meteorological effects, evident after the 30th week (starting in November) superimposed to a slow degradation due to sensor aging effects (drifts). After the 50th week, the absolute error shows a definite recovery trend

Sensor recalibration is needed and has been found to be very efficient but, since the number of multisensory devices to deploy is very high, conventional recalibration procedures employing reference gases or on-field approaches cannot be easily implemented. State-of-the-art drift correction algorithms, acting to subtract drift contribution to sensor responses, can obtain very interesting results but need a significant number of samples themselves in order to ground the value of their free parameters. Recently, we have started to investigate the use of semisupervised learning techniques as a way to solve this problem adaptively that could be classified as a dynamic pattern recognition problem (see Figure 2.2) [10]. In fact, distributed architectures often are built with sensing nodes that sport relevant but unused computing capabilities; this can provide diagnostic services and allow the possibility to self-improve the stability of sensor metrological characteristics. In particular, this can be significant for implementing drift adaptation strategies, accommodating this paramount problem in solid-state chemical sensing.

In networked configurations, sensor recalibration could also be performed cooperatively by temporarily adding other moving reference sensors or, thanks to data fusion techniques, using mutual recalibration strategies. Practically, the development of application-specific algorithms could allow mutual calibration by exploiting networked cooperation in a totally unmanned fashion. In a work from Tsujita and co-workers [11], a network of NO_2 sensors recalibrate their baseline response by exploiting the identification of baseline conditions (very low NO_2 concentrations) by

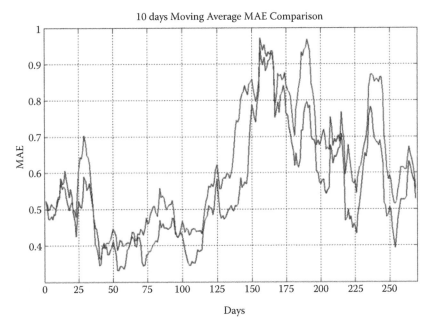

FIGURE 2.2 Drift counteraction with semisupervised learning. CO estimation comparison with SSL (secure sockets layer) algorithm (dark line) and standard neural network algorithm (light line) based on only 24 training samples (1 day). The SSL approach achieved an 11.5% performance gain with respect to the 1-year long averaged MAE score.

cooperating sensors' responses and particular conditions (dawn at low humidity and reference temperature).

Anyway, sensor drift still remains a significant issue that prevents the spread of the use of multisensing devices in urban pollution scenarios. Future directions of research include a combination of advanced signal processing techniques and sensing capability improvement reducing both nonspecificities and sensor drift. Calibration transfer strategies are also to be developed to cope with sensor production issues like inhomogeneity.

2.3 ENERGY EFFICIENCY IN WCSN

The process of the development of battery-operated WCSNs cannot avoid addressing the power consumption issue. State-of-the-art metal oxide (MOX) chemical sensors require high working temperatures for best sensitivity and specificity [12]. This issue limits the possibility of using them in wireless chemical sensing motes because their average power consumption is in the range of hundreds of milliwatts with continuous operation allowing only for a limited life span in battery-operated nodes. Instead, polymer-based chemiresistors, resonators, and mass sensors (QMBs, SAWs) are usually operated at room temperature [13]. Although they are not as common as their MOX counterparts, their low power operation capability can be recognized as a huge advantage with respect to the other technologies, especially when VOC (volatile organic compound) detection is concerned [14,15]. Unfortunately, they may not be as efficient at very low concentrations and most of them are significantly affected by humidity and heavy drift [16].

For a decade, polymer/nanocomposite reactivity to chemicals has also been applied to the development of passive resonant sensors that are capable of wireless remote operations in a very simple way. LED/polymer-based optical sensing can be an extremely interesting solution for all applications where high limits of detection are not an issue because of very low cost, reliability, and very low power demand [17]. However, in order to obtain suitable sensitivity in most applications, laser sources should be used, and this may increase both costs and power needs significantly.

From the architectural point of view, current commercial e-noses are not designed to tackle distributed chemical sensing problem especially with regards to power management, however during last years a number of novel approaches have been proposed and experimented. This field is hence rapidly evolving exploiting the plethora of results obtained by researchers in wireless technology field. As an example, Bicelli et al. [18] have investigated the use of commercial low power MOX sensors in a WSN network for indoor gas detection applications. They first suggested employing a specially designed pulsed heating procedure in an attempt to achieve a significant increase in the wireless sensor battery life (about 1 year) with a sample period in the range of 2 minutes. Their results showed a serious trade-off between total power consumption and actual response time [18]. Pan et al. realized a single w-nose for online monitoring of livestock farm odors integrating meteorological information and wind vector [19]; detection performances and power consumption have not been reported, so it was not possible to estimate autonomous life expectation. Becher et al. recently presented a four MOX sensor-based WCSN network for flammable detection in military docks, but again, power needs restricted the application to the availability of power mains [20].

During the last years, our WCSN commitment focused on the development of a wireless electronic nose platform called TinyNose [21], based on commercially available WSN motes and controlled by software components relying on the TinyOS operating system [22]. The platform makes use of four room-temperature operating low power polymeric chemical sensors that extend the entire sensory network life span at the cost of a low sensitivity. When low power sensors are used, radio power becomes significant in the overall power budget of the platform. In a 2011 work, we focused on the possibility of using computational intelligence algorithms in order to further extend the network life span by implementing a sensor censoring strategy (see references 23–25) focused on allowing the transmission of informative data packets neglecting uninformative data [26]. The target scenario for the application of the proposed methodology was the distributed monitoring of indoor air quality and, specifically, the detection of toxic/dangerous VOC spills. The same architecture has been used to develop and test an ad hoc algorithm for real-time 3D gas mapping with a WCSN in indoor simulated environments so as to extract the needed semantic content from the deployed network [27].

Specifically, our platform relies on the use of nonconductive polymer/carbon black sensors whose sensing mechanism of response is, at its simplest level, based on swelling. When the isolating polymer film within which conductive particles have been dispersed is exposed to a particular vapor, it swells while absorbing a varying amount of organic vapors depending on polymer type. The swelling disrupts conductive filler pathways in the film by pushing particles apart and the electric resistance of the composite increases [15].

In order to obtain a suitable voltage signal (i.e., showing proportionality to sensor resistance variation), a simple amplified resistance to voltage converter signal conditioning system was implemented. Suitable choice of circuit parameters allows the proper operation of the board within a wide range of base resistance. In Figure 2.2, a simplified functional scheme of the board is shown.

The overall sensing subsystem was then directly connected to four A/D input ports provided by the core mote of the proposed platform, the commercial Crossbow TelosB mote—a research-oriented mote platform that has shown its operative potential over time [28]. Powered by 2 AA batteries, the chosen platform is based on a low power T I MSP430F1611 RISC microcontroller w featuring 10 kB of RAM, 48 kB of flash, and 128 B of information storage. The embedded low power radio, Chipcom CC2420, represents the kernel for 802.15.4 protocols' stack support.

The heterogeneity of the practical applications and the limited energy availability required a careful design of the software (sw) structure and protocols for the management of the wireless e-nose platform. The TinyOS operating system guaranteed the possibility to keep the focus on domain-specific optimization and, in particular, the implementation of the sensor censoring strategy.

TinyOS is an open-source operating system for WSN applications developed by the University of California at Berkeley [22]. Essentially, its component library includes network protocols, distributed services, basic sensor drivers, and data acquisition tools—all of which can be further refined for custom applications. Specific software modules allow for the basic management of the sensor node, like local I/O and radio transmissions. Furthermore, the power management model of TinyOS allows for the automatic management of module subsystems switching among active, idle, and sleep phases, achieving a generalized power management strategy

A C-derived language, NesC, is the reference programming language for TinyOS-based programming. NesC relies on a component-based programming module.

Figure 2.3 shows a UML deployment diagram of the overall platform software architecture from which its three-layer design is clearly apparent. The first layer encompasses all the embedded sw components developed in NesC and providing local control of the pneumatic section of the electronic nose (if available), sensor interface control, data acquisition and processing, and, eventually, data transmission capabilities. At the second layer, a PC-based component coded in Java captures data packets from a sensorless node that act as a gateway toward the IP-based network and storage facilities. At the third level, multiple graphical user interfaces (GUIs), also coded in Java, can provide visualization and recording features while remotely controlling relevant parameters for embedded sw operations (e.g., duty cycle parameters).

Actually, the overall architecture has been designed to host two pattern recognition and sensor fusion layers. The first layer defines methods to connect a sensor's raw data processing component that will allow for local situation awareness. It was connected to the local sensor fusion component that allows for the local estimation of pollutant concentrations by using a trained neural network (NN) algorithm. The second

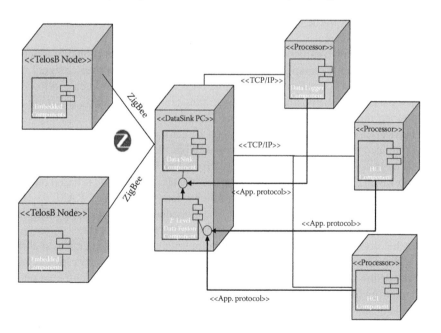

FIGURE 2.3. UML deployment diagram of the software architecture for the proposed platform.

layer provides second-level sensor fusion services, allowing for integrating estimation coming from all the deployed nose. This level is responsible for the cooperative reconstruction of an olfactive image of the environment in which they are deployed.

2.4 DUTY CYCLE AND POWER CONSUMPTION

The TinyNose node has been designed for continuous real-time monitoring of volatiles with a programmable duty cycle, including sensor data acquisition, processing, and transmission toward a data sink. Duty cycle parameters and, in particular, the length of each phase and the sampling frequency are fully programmable by the application designer; sample frequency can be set dynamically at run time by GUI. In order to characterize the power consumption fully, we can easily separate the duty cycle into four separate phases, each one having different power needs:

- Sleep phase. In this phase, each node is put to sleep. MCU and radio are turned into stand-by mode.
- Sensing phase. Sensors driving, data acquisition, and ADC conversion are carried out.
- Computing phase. Data processing is performed in order to prepare data to be transmitted toward a data sink.
- Transmission/reception phase. The actual data transfer takes place.

Battery life can be optimized by controlling the duration of each phase and the activation of sensors driving electronics.

At any instant, a single module power consumption can be computed as a function of its microcontroller power state; whether the radio, pump, and sensor driving electronics are on; and what operations the active MCU subunits are performing (analog to digital conversion). By using appropriate programming models (e.g., relying on split-phase operations), it is possible to best utilize the features provided by TinyOS to keep the power consumption to a minimum. In particular, with regard to radio stack management, we have chosen to rely on the LPL (low power listening) algorithm [22]. As such, node radio can be programmed to switch on periodically just long enough to detect a carrier on the channel. If a carrier has been detected, then the radio remains on long enough to detect a packet to be routed to the data sink for mesh shaped networking. After a timeout, the radio can be switched off.

In order to assess the base consumption of the TinyNose platform, a measuring setup based on a Tektronix TDS 3032 digital oscilloscope was set to measure Vshunt on 10 Ω shunt resistance and to let us derive current. Actually, in its simplest configuration with sample frequency set at 1 Hz, the sensor node, as mentioned before, remains in sleeping mode for most of the time (T_{RS} time interval). In this phase, a current, I_{RS}, measured as 12 μA, is drawn. During the T_{WR} wake-up period, the sensor node turns on the chip radio awakening from the sleep mode. In this stage, we measured a 5 mA mean current draw. After that, the system sets up itself for data capture and conversion with a mean current draw of 19 mA. Even in this stage, provided that data are sent every acquisition cycle, radio activity is the main source of power consumption. In fact, power consumption, in the active phase, is dominated by radio activity until switch-off timeout inset; after ADC converter timeout, it can be measured as 18 mA. Eventually, the radio frequency circuit is turned off again and the state changes from transmission mode to sleeping mode.

Figure 2.4 shows the detailed evolution of the TelosB drawn current during acquisition, processing and data transmission of sensor data, while Table 2.1 shows the consumption in terms of measured currents during each operating phase. Because of the high consumption of the prototype signal condition board (38 mA), a digital signal drives a switch that lets the electronics board power supply to be switched on only during the data capture time slice. This strategy is made possible by a polymer sensor sensing mechanism; in fact, polymer swelling is only negligibly affected by actual sensor power up so that sensors do not need warming up before their resistance can be sampled (see Table 2.2).

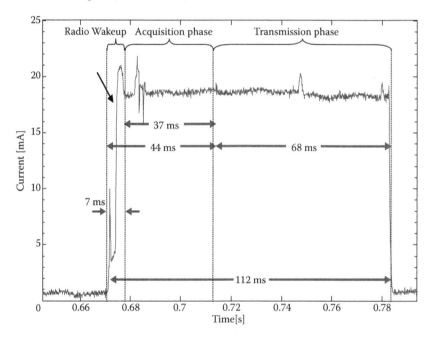

FIGURE 2.4   Current absorbed by TelosB in its active operating phase—in particular during the radio wake-up, sensing, and transmission phases in a complete cycle.

TABLE 2.1
Core Mote Consumption in Different Phases of the Duty Cycle

Operation phase	Time (m)	Current request (mA)
Radio sleep (RS)	$T_{RS} = 888$	$I_{RS} = 0.012$
Radio wake-up (RW)	$T_{RW} = 7$	$I_{RW} = 5$
Acquisition (A)	$T_A = 37$	$I_A = 19$
Computing (C)	$T_C = 25$	$I_C = 2.5$
Transmission (T)	$T_T = 68$	$I_T = 18$

TABLE 2.2
Platform Signal Conditioning Board Consumption

Operation phase	Time (ms)	Current request (mA)
Operating condition	$T_{ON} = 30$	38

Usually, module life span can be estimated using the mean current consumption—namely, Icc,mean—of the wireless sensor, considering battery capacity, conversion efficiency, and power supply output voltage gain.

The mean current value (Icc,mean) obtained under the basic operating conditions without applying sensor censoring can be computed according to Equation (2.1) by using the current measured in the proposed setup:

$$I_{cc,mean} = \frac{T_{RS}}{T} I_{RS} + \frac{T_{RW}}{T} I_{RW} + \frac{T_A}{T} I_A + \frac{T_T}{T} I_T \qquad (2.1)$$

Neglecting conversion efficiency, battery life (BL) can be computed as a function of the battery capacity, C, and total current draw. For our w-nose, C is equal to 3500 mAh, while total current is equal to the sum of mean current absorbed by conditioning board Ib, mean (30/1000 * 38 = 1.14 mA) and the Icc, mean (1.97 mA computed using Equation 2.1). Expressing BL as the ratio C/(Icc,mean + Ib,mean), we can estimate that the overall e-nose battery life, with sampling and transmission frequency of 1 Hz, is roughly 47 days—a rather interesting value for a four-sensor four-nose. Extending the sample period to 10 s, hence losing real-time characteristics, this basic setting will account for a battery life in excess of 1 year.

2.5 POWER SAVING USING SENSOR CENSORING

The additional power required for the execution of the local data processing components should be seriously taken into account in order to evaluate the benefits of censoring strategies. However, the outcome depends basically on the rate of significant events' occurrence; in most chemical sensing scenarios, the probability, p, of a significant event to occur (e.g., chemical spills) is expected to be very low, while the timely transmission of relevant data is needed in security applications.

In order to perform an experimental check of the sensor censoring concepts in WCSN scenarios, we designed and implemented an ad hoc lab-scale experiment. A sensor fusion component that was trained for a distributed pollutant detection application was developed and the power savings obtained by using sensor censoring were evaluated. Actually, we assumed a general chemical sensing problem characterized by the presence of two pollutants whose toxic/dangerous concentration limits were different. To save battery energy, the single mote should be able to decide whether to transmit or not the sampled data on the basis of the concentration of the two gases estimated by sensor responses. The e-nose sensor arrays were exposed to different concentration levels of acetic acid in a controlled environment setup, and their responses, sampled by the motes, were recorded to build a suitable data set for the training of an ANN component to be run onboard. ANN architecture was chosen on the basis of considerations about its flexibility, high capacity, compact knowledge representation (low space footprint), and low computational demands. The onboard component was then loaded with network weights obtained by training an identical model in MATLAB with the recorded data set so as to reach a reasonable point-to-point real-time estimation of concentrations of different analytes.

The proposed architecture performance was evaluated by dividing validation set mean absolute error (MAE) by analyte concentration range span, obtaining a 6% (standard deviation: 10%) value for the acetic acid concentration estimation problem and 11% (standard deviation:11%) for the ethanol problem (see Figure 2.5).

Following a measurement approach, the execution of the NN sensor fusion component accounts for additional 2.5 mA consumption over a total time span of 25 ms (function call overhead included). See Tables 2.3 and 2.4.

In order to reassess total power consumption, we write Equation (2.1) again by taking into account the computing phase:

$$I_{cc,mean}^{NN} = I_{RS}\left(1 - \frac{T_{RW} + T_A + T_T + T_C}{T}\right) + I_{RW}\frac{T_{RW}}{T} + I_A\frac{T_A}{T} + I_T\frac{T_T}{T} + I_C\frac{T_C}{T} \quad (2.2)$$

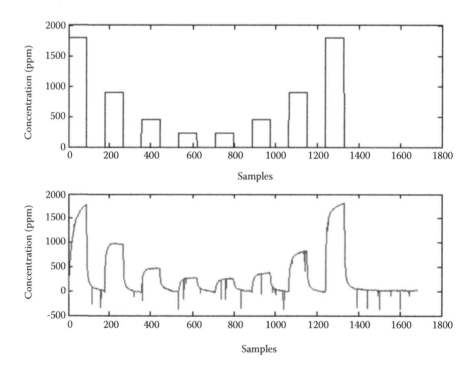

FIGURE 2.5(A) Acetic acid concentration estimation (bottom) performed by the FFNN (feed forward neural network) component plotted against true concentration (top). The x-axis depicts time (samples) while the y-axis depicts real and estimated concentration values.

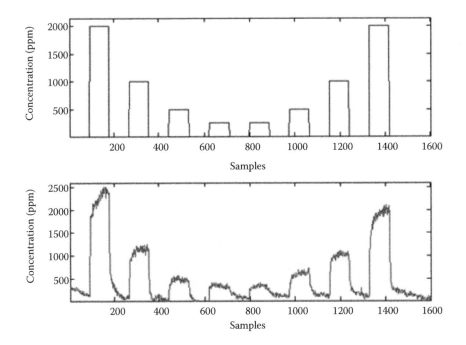

FIGURE 2.5(B) Ethanol concentration estimation (bottom) performed by the FFNN component plotted against true concentration (top). The x-axis depicts time (samples) while the y-axis depicts real and estimated concentration values.

This equation provides the mean current supplied by the batteries in case of significant event occurrence: The mote performs signal conditioning and data sampling, data fusion by means of NN, and data transmission every T seconds.

Using a Bernoulli random variable X ~ B (1, p), which would model the result of the computation performed by the NN to verify whether the acquired data have been transmitted on the radio channel, Equation (2.2) can be rewritten as

$$I_{cc,mean}^{NN} = I_{RS}\left(\frac{T_{RS}'p + T_{RS}''(1-p)}{T}\right) + I_{RW}\frac{T_{RW}}{T}p + \qquad (2.3)$$

$$+ I_A''\frac{T_A}{T}(1-p) + I_A'\frac{T_A}{T}p + I_T\frac{T_T}{T}p + I_C\frac{T_C}{T}$$

where T_{RS}' and T_{RS}'' are, respectively:

$$\begin{cases} T_{RS}' = T - T_{RW} - T_A - T_T - T_C \\ T_{RS}'' = T - T_A - T_C \end{cases} \qquad (2.4)$$

By exploiting Equation (2.3), we can finally discuss the advantages of NN-based sensor censoring for the implementation of power saving strategies in the proposed architecture.

TABLE 2.3

Memory Footprint Increase in Embedded Component Resulting from Linking of Sensor Fusion Neural Component

Algorithm	Bytes in ROM	Bytes in RAM
Basic	20,380	574
Basic + NN comp.	27,340	910

TABLE 2.4

Computational Complexity in terms of functional calls

Functional	Calls per each NN estimation
Tanh	10
Multiplication	76
Sum	82

The worst case is obtained when $p = 1$; that is, when all samples refers to significant events, then the mean current computed using Equation (2.3) is obviously greater than the one calculated by Equation (2.1) (i.e., $I_{cc,mean}^{NN} = 2.013$ mA). However, by equating Equations (2.1) and (2.3) and selecting p as independent variable, we obtain:

$$p = 0.97 \tag{2.5}$$

The obtained value represents the percentage threshold of a significant event under which the NN-based sensor censoring becomes more efficient with regard to power management. This computed threshold level makes the use of the proposed approach feasible for most of the analyzed distributed chemical sensing scenarios.

In monitoring industrial chemical spills, even considering false-positive generation, it is reasonable to expect values of p that reflect only a few significant samples a day. Of course, censoring criteria (i.e., the "spiking" threshold) should be chosen exploring the trade-off among sensitivity and the false-positive rate caused by estimation errors, also considering the danger/toxicity level of the target gas. For an experimental evaluation, using the previously mentioned setup and allowing a small slack to the network estimation of ethanol (detection threshold on NN response = 10 ppm with respect to a 2000 ppm max exposure level), we obtain a false positive rate of 5% for ethanol and less than 1% for acetic acid. In a case of no positive events during node lifetime—a reasonable expectation in safety oriented leakage detection scenario—a proposed node can be expected to experiment a maximum operative life span that is very near to the intrinsic limit now dominated by the power needed by a signal conditioning board. In particular, considering $p = 0.01$, the expected lifetime computed by taking into account Equation (2.3) reaches 113 days (110 days for $p = 0.05$)—a rather interesting value for a four-sensor wireless electronic nose with real-time operating characteristics.

2.6   3D RECONSTRUCTION

As stated in the introduction, the capability to reconstruct 3D gas mappings is becoming more and more interesting for the possible applications in city air pollution mapping, hazardous gas spills' localization, and energy efficiency (e.g., efficient control of heating, ventilation, and air conditioning [HVAC] automation in buildings). Recently, we investigated the possibility of obtaining 3D quantitative indoor air quality assessment with the TinyNose architecture [27]. The outcome revealed the capability of building gas concentration mapping of acetic acid–ethanol mixtures (as VOC pollutant simulants) in ambient air with the mesh of four w-noses coupled with a two-stage sensor fusion system. In order to focus on the 3D reconstruction problem, a set of four TinyNose w-noses equipped with a four MOX sensor array has been considered. A basic performance estimation procedure has been conducted on a single TinyNose equipped with the novel MOX-based sensor array. The node was exposed using a controlled climatic chamber to different concentrations of acetic acid and ethanol at different relative humidity percentages (see Table 2.5). By using sensor responses, a two-slot tapped delay neural network (TDNN) with 10 hidden layer neurons for instantaneous concentration estimation obtained a MAE/range value of 2.34% (0.75 ppm) for acetic acid and 6.5% (9.8 ppm) for ethanol (see Figure 2.6).

However, following an on-field approach to calibration, we decided to design a second experimental setup involving the deployment of a network of four instances of TinyNose in an ad hoc glass box (volume = 0.36 m^3) and evaluate their capability to reconstruct a real-time 3D chemical concentration image of the two previous pollutants. Different amounts of the two chemicals (see Table 2.6) were introduced and diluted until complete evaporation with the use of a standard PC fan. Based on the reasonable hypothesis of uniform concentration distribution over the box, at steady state, the response of the node was sampled (1 Hz) to build a suitable data set for the training of the TinyNoses. This was based on a two-level classifier/regressor scheme. In this way, each of the w-noses was made capable of estimating the local pollutant concentration at its deployment location. Further, 10 runs of steady-state sample acquisition with the same procedure were then used to build a suitable test set. The test set MAE, averaged for all four w-noses, reached 3.15 and 4.36 ppm for ethanol and acetic acid, respectively. These figures allow us to locate the expected absolute error on the real-time local concentration estimation under a 5 ppm threshold. This

TABLE 2.5
Controlled Chamber Experimental Setup

RH (%)	Gas concentration ranges	
	Acetic acid (ppm)	Ethanol (ppm)
[20, 30, 50]	[0, 5, 7, 10, 15, 20, 25, 30, 32]	[0, 15, 30, 70, 90, 115, 130, 150]

Notes: The different gas mixtures used during the experimental setup were synthesized by using all combinations of the reported concentrations of acetic acid and ethanol at different relative humidity concentrations for a total of 216 cycles. The baseline mixture was set at RH = 50%.

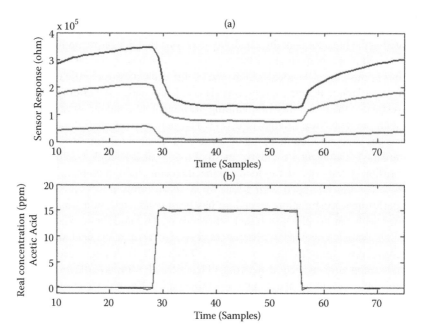

FIGURE 2.6 (a) Sensor array responses in controlled chamber setup; (b) trained TDNN responses and ground truth comparison during the correspondent complete exposure cycle.

TABLE 2.6
Glass Box Experimental Setup

Gas concentration ranges	
Acetic acid (ppm)	**Ethanol (ppm)**
[0, 5, 10, 15, 20]	[0, 6, 12, 17, 23]

Notes: The different gas mixtures used during the glass box experimental setup were synthesized by using all combinations of the reported concentrations of acetic acid and ethanol. For each combination, two different exposure cycles were executed. The steady-state sensor array response to each exposure cycle was recorded to build the training data set. Ambient RH values and temperatures (not controlled) were also recorded to be part of the on-board feature set.

---

value is valid for concentration estimation in a mixture and thus in the presence of interferents, provided that concentration levels have a slow variation rate.

In order to reconstruct a 3D chemical image of the glass box, the kernel-DV algorithm (see Nakamoto and Ishida [3]), originally developed by the Lilienthal group for use with mobile robot acquisitions, was adapted for real-time cooperative sensor fusion. The algorithm basically used a 3D Gaussian kernel to propagate localized measurement to a 3D environment based on confidence values depending on the point distance from the actual measurement points. Estimation based on the kernel propagation was balanced with a default averaged value (homogeneous gas distribution) by the use of the confidence value that is normalized by a scale factor. Eventually, should confidence value fall to 0 (points located far from all actual measurement points), the algorithm reverts to the homogeneous gas distribution hypothesis. In the last setup, 17 μg of ethanol was allowed to evaporate within the glass box in one of the left-down box corner. By using the neural calibration obtained before and encoded in the on board computational intelligence component, the single nodes were able to estimate local concentration of both gases. Their estimations were transmitted and collected at a data sink, where a sensor fusion component was coded to reconstruct an instantaneous 3D chemical image of the box.

Overall performance of the 3D reconstruction algorithm depends on the value of three base parameters of the kernel-DV algorithm (i.e., cell mesh width, the kernel width σ, the confidence scale parameter, and, of course, by the w-nose deployment positions. Cell mesh width only trades off 3D reconstruction resolution with computational costs; for this reason, a fixed value that allows for real-time reconstruction has been selected. The confidence scale parameter depends on kernel width, so, for a fixed deployment configuration of the w-noses, an automated procedure has been designed to choose the appropriate kernel width parameter value on the basis of a leave-one-mote-out approach. Actually, in scanning a parameter value array ([0.05, 0.10, 0.15, 0.2]) for each parameter setting, all but one sensing node have been used to estimate the concentration value of the two analytes, with the adapted kernel-DV algorithm at the remaining node position. Figure 2.7(a) shows the comparison between local instantaneous concentration estimation at node position 3 and estimation carried out by the 3D reconstruction algorithm at the same position. Without affecting generalization, the mean absolute difference between the estimated value and the

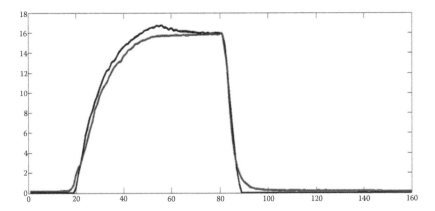

FIGURE 2.7(A)   Instantaneous ethanol concentration estimation as computed at node 3 (top) compared with the estimation obtained by 3D reconstruction, using the remaining three nodes, at the same location. The experimental setup foresees the deployment of four w-noses in the 0.36 m³ glass box and the release of 17 mg (20.9 ppm) of ethanol near one corner of the box.

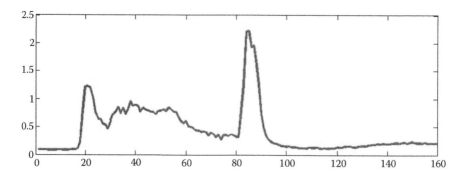

FIGURE 2.7(B) Instantaneous absolute difference among local ethanol concentration and 3D reconstruction algorithm with kernel width s set at 0.1 value. The instantaneous difference reported here was averaged throughout the leave-one-mote-out procedure executed for the exposure to 17 mg (20.9 ppm) of ethanol. The computed MAE value was used for the optimization of the 3D reconstruction algorithm.

actual value as estimated by the remaining node, during all the exposure time, has been defined as the performance value to be optimized by the brute force approach.

Figure 2.7(b) shows the averaged instantaneous absolute difference among local and 3D reconstruction-based estimations for the four motes in the ethanol case. Peaks can be spotted in the rise and fall of the concentration levels during transients. The peaks can be explained by the different concentrations experimented with by the motes during transients due to their different positions. In this case, the peaks' magnitude could be effectively reduced by tuning mote positioning and density of the measurement mesh in the sensed environment. Figures 2.8 and 2.9 depict, respectively, w-nose positioning and an instantaneous reconstruction of the concentration of the two pollutants. This preliminary result shows the possibility of effectively using a w-nose deployment for real-time 3D quantitative air quality analysis in the presence of a pollutant mixture. The use of mote cross validation has

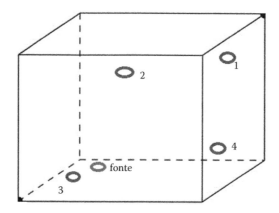

FIGURE 2.8   Positioning of the four w-noses and gas source within the glass box.

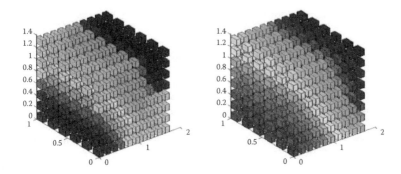

FIGURE 2.9 Instantaneous 3D ethanol (right) and acetic acid (left) concentration reconstruction (computed at data sink) using a four w-nose deployment in the glass box experimental setup.

also been shown for the sensor fusion algorithm parameter tuning and performance evaluation.

2.7 CONCLUSIONS

WCSN architectures can provide a new insight on important and potentially life-saving phenomena like the detection of pollutants and/or hazardous gas. Their practical diffusion is still limited by several issues, like power management and sensors' nonspecificity and instability. Ad hoc 3D diffusion models can be devised to obtain high level semantic information about gas concentrations with applications to spill localization or energy efficiency (HVAC automation).

In both cases computational intelligence, together with distributed sensing techniques, may allow for viable solutions—even in the case of using low cost equipment exploiting redundancy and the power of machine learning. Since 2006, our research group, together with several researchers all around the globe, has been committed to the development of solutions for real-world applications that could be engineered to the market. Despite the growing number of contributions in this field, the main issues are still there and appear to be solved only for specific segments of a complete application. Our efforts have produced a complete platform, called TinyNose, that allows us to address all the applicative segments of WCSN architecture. Currently the development of WCSNs is rapidly progressing toward real-world applications; however, the massive on-field testbed deployments needed to test their reliability in indoor and outdoor scenarios are still to be realized.

ACKNOWLEDGMENTS

Core work for this chapter has been conducted by researchers at UTTP-MDB lab of the ENEA Centro Ricerche Portici. For cooperation and support, I am indebted to my co-workers and in particular to E. Massera, G. Fattoruso, G. Di Francia, and M. Miglietta.

REFERENCES

1. D. Dockery, C. A. Pope, X. Xu, F. Speizer, J. Schwartz. An association between air pollution and mortality in six US cities. *New England Journal of Medicine* 329:1753–1759 (1993).
2. C. Grimes, K. G. On, O. K. Varghese, X. Yang, G. Mor, M. Paulose, E. C. Dickey, C. Ruan. A sentinel sensor network for hydrogen sensing. *Sensors* 3:69–82 (2003).
3. T. Nakamoto, H. Ishida. Chemical sensing in spatial/temporal domains. *Chemical Review* 108 (2): 680–704 (2008).
4. J. Achim, A. L. Lilienthal, T. Duckett. Airborne chemical sensing with mobile robots. *Sensors* 6:1616–1678 (2006).
5. J. Achim, A. L. Lilienthal, T. Duckett. Building gas concentration grid maps with a mobile robot. *Robotics and Autonomous Systems* 48 (1): 3–16 (2004).
6. S. De Vito et al. TinyNose: Developing a wireless e-nose platform for distributed air quality monitoring applications. *IEEE Conference on Sensors,* Oct. 26–29, Lecce, Italy, pp. 701–704 (2008).
7. R. Shepherd, S. Beirne, K. T. Lau, B. Corcoran, D. Diamond. Monitoring chemical plumes in an environmental sensing chamber with a wireless chemical sensor network. *Sensors and Actuators B: Chemical* 121 (1): 142–149 (2007).
8. I. F. Akyildiz, W. Su, Y. Sankarasubramaniam, E. Cayirci. Wireless sensor networks: A survey. *Computer Networks* 38:393–422 (2002).
9. S. De Vito, E. Massera, M. Piga, L. Martinotto, G. Di Francia. On-field calibration of an electronic nose for benzene estimation in an urban pollution monitoring scenario. *Sensors and Actuators B: Chemical* 129 (2): 750–757 (2007).
10. S. De Vito, G. Fattoruso, M. Pardo, F. Tortorella, G. Di Francia. Semi-supervised learning techniques in artificial olfaction: A novel approach to classification problems and drift counteraction. *IEEE Sensors Journal* 12:3215–3224 (2012).
11. W. Tsujita, A. Yoshino, H. Ishida, T. Moriizumi. Gas sensor network for air-pollution monitoring. *Sensors and Actuators B: Chemical* 110:304–311 (2005).
12. N. Barsan, U. Weimar. Conduction model of metal oxide gas sensors. *Journal of Electroceramics* 7 (3): 143–167 (2001).
13. K. Arshak, E. Moore, G. M. Lyons, J. Harris, S. Clifford. A review of gas sensors employed in electronic nose applications. *Sensor Review* 24 (2): 181–198 (2004).
14. L. Quercia, F. Loffredo, B. Alfano, V. La Ferrara, G. Di Francia. Fabrication and characterization of carbon nanoparticles for polymer based vapor sensors. *Sensors and Actuators B* 100:22–27 (2004).
15. E. J. Severin, B. J. Doleman, N. S. Lewis. An investigation of the concentration dependence and response to analyte mixtures of carbon black/insulating organic polymer composite vapor detectors. *Analytical Chemistry* 72:658–668 (2000).
16. S. C. Ha, Y. Yang, Y. S. Kim, S. H. Kim, Y. J. Kim, S. M. Cho. Environmental temperature independent gas sensor array based on polymer composite. *Sensors and Actuators B* 108:258–264 (2005).
17. D. Diamond, S. Coyle, S. Scampagnani, J. Hayes. Wireless sensor networks and chemo-/biosensing. *Chemical Reviews* 108 (2): 652–679 (2008).
18. S. Bicelli et al. Model and experimental characterization of the dynamic behavior of low-power carbon monoxide MOX sensors operated with pulsed temperature profiles. *IEEE Transactions Instrumentation and Measures* 58 (5): 1324–1332 (2009).
19. L. Pan et al. A wireless electronic nose network for odors around livestock farms. *Proceedings of 14th M2VIP* 2007, Xiamen, pp. 211–216 (2007).
20. C. Becher, P. Kaul, J. Mitrovics, J. Warmer. The detection of evaporating hazardous material released from moving sources using a gas sensor network. *Sensors and Actuators B: Chemical* 146 (2): 513–520 (2010).

21. S. De Vito et al. Power savvy wireless e-nose network using in-network intelligence. *Proceedings of 13th ISOEN, AIP Conference Proceedings* 1137:211–214. ISBN: 978-0-7354-0674-2.
22. http://www.tinyos.net
23. C. Rago, P. Willett, Y. Bar-Shalom. Censoring sensors: A low-communication rate scheme for distributed detection. *IEEE Transactions Aerospace Electronics Systems* 32 (2): 554–568 (1996).
24. S. Appadwedula, V. V. Veeravalli, D. L. Jones. Decentralized detection with censoring sensors. *IEEE Transactions on Signal Processing* 56 (4): 1362–1373 (2008).
25. E. Y. Chang, A. Jain. Adaptive sampling for sensor networks. *Proceedings of First International Workshop on Data Management for Sensor Networks, DMSN,* August 30, Toronto, Canada, 72:10–16 (2004).
26. S. De Vito, P. Di Palma, C. Ambrosino, E. Massera, G. Burrasca, M. L. Miglietta, G. Di Francia. Wireless sensor networks for distributed chemical sensing: Addressing power consumption limits with on-board intelligence. *IEEE Sensors Journal* 11 (4): 947–955 (2011).
27. S. De Vito et al. Cooperative 3D air quality assessment with wireless chemical sensing networks. *Procedia Engineering* 25:84–87
28. J. Polastre, R. Szewczyk. Telos: Enabling ultra-low power wireless research. *Proceedings International Symposium on Information Processing in Sensor Networks,* pp. 364–369 (2005).

3 Radio Frequency (RF) Radiation Exposure and Health

Hubregt J. Visser

CONTENTS

Through the history of wireless communication I show the explosive growth in time of mobile telephony and explain how this affects the perception of fear of the general public for electromagnetic radiation. Then, after explaining the physics of electromagnetic radiation, I discuss the interaction with the human body. I explain how RF and microwave radiation cannot be directly carcinogenic. Then I discuss the heating effect of RF and microwave radiation and the exposure limits to prevent adverse health effects for humans. After explaining how a correlation between electromagnetic fields and electromagnetic hypersensitivity has not been established, I will discuss the probability and possibility of a causal relationship between long-term RF radiation exposure and cancer. I will discuss the viewpoint of the World Health Organization that considers RF radiation as possibly carcinogenic for humans and those of leading expert organizations and conclude with current and future epidemiologic research.

3.1 THE HISTORY OF WIRELESS COMMUNICATION

In 1873, James Clerk Maxwell (1831–1879) published his famous work, "A Treatise on Electricity and Magnetism" [1], unifying electricity and magnetism. In experimentally proving Maxwell's theory of electromagnetics, in 1866 the German scientist Heinrich Rudolf Hertz (1857–1894) constructed—without being aware of it—the first ever radio system, consisting of a transmitter and a receiver. The Italian Guglielmo

Marconi (1874–1937), upon reading about Hertz's experiments, did see the wireless data transport potential. After a lot of experimentation, he succeeded, in 1895, in transmitting Morse code signals over distances up to and beyond 1.5 km. In 1900 he was granted a patent for tuning the transmitter and receiver to a certain frequency. In that same year Reginald Fessenden (1866–1932) in the United States was able, for the first time, to transmit and receive intelligible speech wirelessly. Up to then, radio had been used only for wireless telegraphy. In 1905, John Ambrose Fleming (1849–1945) invented the thermionic valve or diode, a reliable, compact, rectifying detector. This thermionic valve paved the way for the invention of the triode or audion by the American Lee de Forest (1873–1961) in 1906. In the triode, a third electrode is added to the diode, thus making amplification of signals possible [2].

The radio was taken to the Boer War (1899–1902) and the First World War (1914–1918) and improved. After the First World War, in the 1920s, radio broadcasting started on a regular basis and in 1922 the British Broadcasting Corporation (BBC) was formed. Television was introduced in the 1930s but transmission was stopped in the United Kingdom (UK) at the outbreak of the Second World War. After the World War, radio and television broadcasting expanded. In 1947, 54,000 television receiver sets were registered in the UK and 44,000 in the United States. At that same time 40 million radio receiver sets were present in the United States. The number of radio sets in the world now (2012) is estimated to be about 2.5 billion [3] and the number of television sets in the world today is estimated to be about 1.4 billion [4].

In 1990, phase I of the global system of mobile (GSM) communications was published and on the first of July 1991 the world's first GSM call was made. The number of GSM base stations worldwide was estimated in 2006 to be over 1.4 million [5] and 4.6 billion mobile phone subscriptions globally were estimated in 2011 [6]. We see that the growth of mobile telephony in 20 years has been much larger than that of radio and television broadcasting in a century. Therefore, it is not surprising that the general public has become concerned that exposure to radio frequency (RF) radiation (radio waves) could lead to negative health effects and consequently has developed a strong opposition against the placement of new base stations.

This concern and opposition are, however, not just the consequence of the lack of knowledge of electromagnetic fields, RF radiation, and the effects on the human body. Risk perception is a more complex process, involving age, sex, and cultural and educational backgrounds [7].

3.2   THE PUBLIC PERCEPTION OF RF RADIATION EXPOSURE RISKS

Risk is the probability that a person will be harmed by a particular hazard. A hazard is an object or a set of circumstances that can potentially harm a person's health [7]. Here, the hazard is exposure to RF radiation. The risk is that this RF radiation harms a person's health. The perception of risk comprises a number of factors that influence a person's decision to take a risk or reject it. If the benefits severely outweigh the risks of a certain action (here: being exposed to RF radiation), the risk will be perceived as acceptable.

This explains why, for a lot of people, carrying a mobile phone is acceptable while the erection of a new GSM base station in the neighborhood is not. The benefits of carrying a mobile phone are clear. The benefits of having another base station nearby are not. Other characteristics of this situation also play a role in the risk perception [7]:

- **Voluntary versus involuntary exposure.** Mobile phone users generally perceive the risks from exposure from the RF radiation of the handset, which is voluntary, as low. The risks from exposure from the RF radiation of the base station, which is involuntary, are generally perceived as high. For people who do not use mobile phones, the risk perception may be even higher.
- **Fair versus unfair.** Closely related to the preceding point is the situation wherein people that are exposed to RF radiation—from, for example, a GSM base station while not having or using a mobile phone—consider this as unfair and are likely to perceive the risks as high.
- **Control versus lack of control.** When people have no influence in the placing of new base stations, the risks of RF radiation exposure tend to be perceived as being high.
- **Familiarity versus unfamiliarity.** The perceived risks decrease with the level of understanding the potential health effects of exposure to RF radiation.
- **Dread versus confidence.** Some potential health effects are more feared than others. Thus, even a small possibility of a relation between RF radiation exposure and cancer receives significant public attention.

The first three characteristics of public risk perception can be brought into proportion by involving the public in the decision-making process of (local) governments. Therefore, it is necessary that the latter two characteristics are dealt with first: An effective system of public information and communications needs to be established among scientists, government, the industry, and the public [7]. Within this perspective I will, in the remainder of this chapter, explain the physics of RF radiation and the known interactions with human tissue and discuss past and present epidemiologic studies.

3.3   RF RADIATION

An accelerating or decelerating charged particle will induce a changing magnetic field around the direction of acceleration/deceleration. On a macroscopic level, a net change in electric current will induce a changing magnetic field around this changing electric current. This changing magnetic field will create a changing electric field—the electric field lines encircling the magnetic field lines. The changing electric field in its turn will induce a changing magnetic field—the field lines encircling the electric field lines—and so on. This process—known as electromagnetic field propagation or electromagnetic radiation—is shown in Figure 3.1.

The representation in Figure 3.1 is schematic. The process is, in reality, three dimensional and continuous. The keyword in the explanation of electromagnetic radiation has been *changing*. A static or nonchanging electric current in a wire will produce a surrounding magnetic field, but this field will be static too and will not induce an electric field. Thus, a direct current will not produce a radiating electromagnetic field.

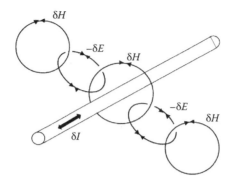

FIGURE 3.1   Schematic view of electromagnetic field propagation (radiation) stemming from a changing electric current. **E** denotes the electric field, **H** denotes the magnetic field. **δI, δE,** and **δH** denote a changing current, electric field, and magnetic field, respectively.

Man-made electromagnetic fields are mostly created by sinusoidal currents [2]. Consequently, the electromagnetic fields are also sinusoidal (see Figure 3.2a).

As a function of time, the electric and magnetic fields vary sinusoidally. The period T is the inverse of the frequency f. Then, as a function of distance, the propagating electric and magnetic fields vary sinusoidally (see Figure 3.2b). The wavelength λ is obtained by dividing the velocity of light c by the frequency f. Far away from the source (antenna), the electric and magnetic fields are in phase and perpendicular to one another and perpendicular to the direction of propagation [8] (see Figure 3.2b).

We may now order electromagnetic radiation to an increasing frequency or (through the $\lambda = c/f$ relation) to a decreasing wavelength. Then, going up in frequency, we encounter successively: RF or radio waves, microwaves, infrared radiation, visible light, ultraviolet radiation, x-rays, and gamma rays.

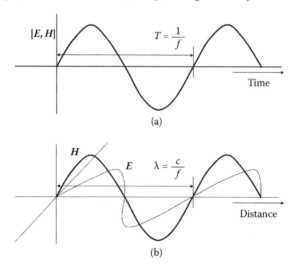

FIGURE 3.2   Sinusoidal electromagnetic fields. (a) The electric and magnetic fields vary sinusoidally in time. (b) The electric and magnetic fields vary sinusoidally over distance.

So, we see that RF radiation, (visible) light, and radioactive decay are all forms of the same physical phenomenon.

In Table 3.1, the electromagnetic spectrum has been roughly ordered in terms of frequency and wavelength.

Another way of ordering the electromagnetic spectrum is based on the photon energy E_p (the subscript $_p$ has been added to avoid confusion with the electric field amplitude). In doing so, we make use of the particle characteristics of electromagnetic radiation. The photon energy E_p is related to the frequency f through $E_p = hf$, where h is Planck's constant. Planck's constant may be expressed in joule seconds or in electronvolt seconds: One electronvolt is the amount of energy gained by the charge of a single electron moved across an electric potential difference of 1 V. Although it is very practical, the electronvolt is not a unit belonging to the International System of Units (SI). The corresponding SI unit is joule. 1 eV equals 1.602176565e19 J. Planck's constant is then h = 6.62606896e-34 Js = 4.13566733e-15 eVs. The ordering by photon energy is given in Table 3.2.

With the physics of RF radiation explained, we can now move forward to the interaction of RF radiation with human tissue and possible (adverse) effects on health.

TABLE 3.1

Electromagnetic Spectrum, Frequencies, and Wavelengths

Type of radiation	Frequency (Hz)	Wavelength (m)
RF and microwave	0.0–0.3e12	>1.0e-3
Infrared light	0.3e12–4.0e14	7.5e-7–1.0e-3
Visible light	4.0e14–7.5e14	4.0e-7–7.5e-7
Ultraviolet light	7.5e14–3.0e16	1.0e-8–4.0e-7
X-rays	>3.0e16	<1.0e-8
Gamma rays	>1.0e20	<3.0e-12

Note: e stands for exponent, a power of 10.

TABLE 3.2

Electromagnetic Spectrum, Frequencies, and Energies

Type of radiation	Frequency (Hz)	E (J)	E (eV)
RF and microwave	0.0–0.3e12	0.0–2.0e-22	0.0–1.2e-3
Infrared light	0.3e12–4.0e14	2.0e-22–2.7e-19	1.2e-3–1.7
Visible light	4.0e14–7.5e14	2.7e-19–5.0e-19	1.7–3.1
Ultraviolet light	7,5e14–3.0e16	5.0e-19–2.0e-17	3.1–124
X-rays	>3.0e16	>2.0e-17	>124
Gamma rays	>1.0e20	>6.6e-14	>4.1e5

Note: e stands for exponent, a power of 10.

3.4 HEALTH EFFECTS OF RF RADIATION

In discussing the possible (negative) effects of RF radiation on human health, we have to distinguish between short-term and long-term effects. Beneficial effects of intentional RF exposure of biological tissue (e.g., among others, wound healing effects and bone growth stimulation) are left out of the discussion. The interested reader is referred to Pirogova, Vojisavljevic, and Cosic [9] for more information on these aspects.

3.4.1 SHORT-TERM EFFECTS

The short-term effects to be discussed are the carcinogenicity of RF radiation, heating effects, and electromagnetic hypersensitivity.

3.4.1.1 Direct Carcinogenicity Effects

When the photon or quantum energy exceeds the level of a few electronvolts, this energy is high enough to eject an electron from an atom. When this happens, a positively charged particle (i.e., an ion) is left behind (see Figure 3.3).

Electromagnetic radiation of a high enough energy level is therefore called *ionizing*. The boundary between non-ionizing and ionizing radiation lies somewhere in the ultraviolet region of the electromagnetic spectrum. Ionizing radiation is capable of mutating deoxyribonucleaic acid (DNA) and thus may cause cancer. As we can conclude from Table 3.2, RF and microwave radiation cannot be directly carcinogenic, no matter what the intensity of the radiation is. A higher intensity of non-ionizing radiation means that more photons are emitted, but the quantum energy of every photon remains the same (i.e., too low to eject an electron from an atom). Thus, with respect to direct carcinogenic risks, walking outside on a summer day is more dangerous than being subjected to RF radiation. With respect to long-term and indirect effects, current scientific evidence has not conclusively linked RF and microwave radiation to the development of cancer. I will discuss past and present epidemiologic studies after finishing this section on short-term effects and will continue with heating effects.

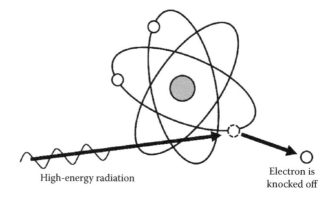

High-energy radiation   Electron is knocked off

FIGURE 3.3 Schematic view of how high-energy electromagnetic radiation may eject an orbiting electron from an atom.

3.4.1.2   Heating through RF Exposure

Microwave heating is caused by dipole rotation. Molecules having an electric dipole moment (like H_2O) will align themselves in an electric field. When this field alternates, as in an electromagnetic field, the molecules will reverse direction and in this process they will distribute energy to adjacent molecules in the form of heat. This principle is applied in microwave ovens where high-power microwaves are being employed [10].

The effect of heating through RF and microwaves has been known since the mid-1930s, but since the Second World War, which saw the birth of high-power radio detection and ranging (radar), tales of radar personnel suffering from RF and microwave radiation heat effects (lenticular opacities, sterilizing effects) forced the US Navy and Air Force to identify safe levels of microwave exposure for servicemen [11–13]. The early guidelines were developed based on animal experiments in which cataracts in rabbit eyes, dog testicular degenerations, and lesions in rabbit brains were developed using different frequencies and power levels. Based on these experiments and especially on the threshold level for forming cataracts in eyes, the first damage risk criterion was agreed upon at 100 mW/cm² in 1953 [11]. In the first general RF exposure standard, issued by the American National Standards Institute (ANSI) in 1966, a safety factor of 10 was introduced and the exposure limit was set to 10 mW/cm² from 10 MHz to 100 GHz [14].

This first risk criterion developed by the US military in the 1950s has evolved into today's frequency-dependent exposure limits [15]. These limits are based on the fact that the human body can handle heat up to a level of 4 W/kg without risking permanent damage. Human cells die at a temperature of 107°F (41.7°C). The human circulatory system functions as a cooling device, much like the radiator of a car. Based on these limits and applying a safety factor of 10, the International Commission on Non-Ionizing Radiation Protection (ICNIRP) has issued guidelines for limiting exposure to time-varying electric, magnetic, and electromagnetic fields up to 300 GHz [15]. A summary of the ICNIRP exposure limits is shown in Table 3.3.

The ICNIRP health-effect guidelines have been accepted by more than 30 countries and, for the mobile telephony frequency spectrum, are consistent with, among others, the IEEE standard [13,16].

Since the 4 W/kg is based on a full-body exposure, for appendages, the exposure limits are increased in most standards up to 20 times this value due to the blood

TABLE 3.3
ICNIRP Exposure Limits

Frequency range	Plane wave power density (W/m²)	
10–400 MHz	Occupational exposure	10
	General public exposure	2
400–2000 MHz	Occupational exposure	f/40 f: frequency in MHz
	General public exposure	f/200 f: frequency in MHz
	Occupational exposure	50
	General public exposure	10

cooling properties. Due to the limited blood flow through the eyes and the male's testes, the full-body exposure limits apply for these organs.

3.4.1.3   Electromagnetic Hypersensitivity

With the explosive growth of mobile and short-range communication over the last two decades, a group of individuals has evolved that is reporting several health problems that are presumed to be caused by exposure to electromagnetic field radiation. The reputed sensitivity to electromagnetic fields is known as electromagnetic hypersensitivity [13,17]. Electromagnetic hypersensitivity is also known as idiopathic environmental intolerance attributed to electromagnetic fields, electrohypersensitivity, electrosensitivity, and electrical sensitivity.

The reported nonspecific symptoms differ from individual to individual. The symptoms most commonly experienced are skin redness, tingling and burning sensations, sleeplessness, fatigue, tiredness, concentration difficulties, dizziness, nausea, heart palpitation, and digestive disturbances. The symptoms are not part of any recognized syndrome [17]. Sufferers of electromagnetic hypersensitivity are self-described and respond to non-ionizing electromagnetic fields of a power density well below the recommended (thermal) limits. The symptoms are real and can be a disabling problem for an individual. Electromagnetic hypersensitivity is therefore taken very seriously by the World Health Organization [18].

In a number of studies, individuals have been subjected to electromagnetic radiation under controlled laboratory conditions. The frequencies and power levels were similar to what they claimed to be the cause of their symptoms. Rubin, Munshi, and Wessely [19] conducted a systematic review of 31 known experiments in 2005. In all the experiments, the individuals were exposed to genuine and sham electromagnetic fields under single- or double-blind conditions. Seven studies were found that did report an association, while 24 could not find any association with electromagnetic fields. However, of the seven positive studies, two could not be replicated even by the original authors, three had serious methodological shortcomings, and the final two presented contradictory results. Follow-up studies in 2008 [20] and 2010 [21] confirmed the conclusion of Rubin et al. [19]:

> The symptoms described by "electromagnetic hypersensitivity" sufferers can be severe and are sometimes disabling. However, it has proved difficult to show under blind conditions that exposure to EMF can trigger these symptoms. This suggests that "electromagnetic hypersensitivity" is unrelated to the presence of electromagnetic fields, although more research into this phenomenon is required.

The symptoms experienced by electromagnetic hypersensitive individuals might arise from other environmental factors, like flicker from fluorescent lights, glare and other visual problems with video display units, poor ergonomic design of computer workstations, poor indoor air quality, or stress in the workplace or living environment. There are also some indications that the symptoms may be due to pre-existing psychiatric conditions as well as stress reactions as a result of worrying about electromagnetic health effects, rather than the electromagnetic exposure itself [17]. In Rubin, Nieto-Hernandez, and Wessely [21], the *nocebo* effect has been identified as a possible trigger for developing electromagnetic hypersensitivity symptoms: The expectation of symptoms upon being subject to electromagnetic radiation will result in the forming of these symptoms when thought to be subject to electromagnetic radiation.

Since a correlation with electromagnetic fields has not been established, in the light of the preceding discussion, the proposed treatment for individuals suffering from electromagnetic hypersensitivity is directed toward reducing the symptoms and functional handicaps. This should be done in close cooperation with a qualified medical specialist (to address the medical and psychological aspects of the symptoms) and a hygienist (to identify and, if necessary, control factors in the environment that are known to have adverse health effects of relevance to the patient) [17].

3.4.2 Long-Term Effects

With long-term effects we mean an increased risk of developing cancer other than by ionizing electromagnetic radiation. With the widespread use of cell phones and the introduction of GSM base stations, a growing public concern about such possible long-term adverse health effects has grown. Some media have a tendency to promote poorly conducted studies with "positive" results that, upon a closer examination, turn out to be false positives. The latter are seldom mentioned. With the original "news" covering, a tone of fear is set, obscuring an open discussion on an important but complex issue. Media being more concerned about providing sensation rather than news will leave the lay audience with a perception that complexity is used by scientists to cover up some "hidden" facts.

3.4.2.1 Past Epidemiologic Studies

An Internet search on the key words "cancer clusters" and "base stations" will result in a multitude of articles reporting the existence of cancer clusters around RF broadcast and GSM transmitters. Since the distribution of cancer in the population is a random process, it is, from a statistical viewpoint, not very surprising that such clusters appear. Also, since base stations are erected everywhere, they must also appear near existing cancer clusters [13]. It takes a carefully planned and executed epidemiologic study to gather scientific evidence on how the distribution of cancer in the population might be related to electromagnetic field exposure.

The largest, most prominent, and most scientifically sound epidemiologic study on mobile phones and cancer completed today is the Interphone Study [22,23]. The study was conducted in 13 countries (Australia, Canada, Denmark, Finland, France, Germany, Israel, Italy, Japan, New Zealand, Norway, Sweden, and the UK) using a common core protocol. Interphone was the largest case-control study to date investigating risks related to mobile phone use and to other potential risk factors for the tumors of interest. It included 2,765 glioma (a tumor that starts in the brain or spine), 2,425 meningioma (a tumor growing from the coverings of the brain and spine, or meninges), 1,121 acoustic neuroma (a tumor that develops on the nerve that connects the ear to the brain), and 109 malignant parotid gland (a salivary gland) tumor cases and 7,658 controls. The conclusions of the Interphone Study are the following [23]:

Overall, no increase in risk of glioma or meningioma was observed with use of mobile phones. There were suggestions of an increased risk of glioma at the highest exposure levels, but biases and error prevent a causal interpretation. The possible effects of long-term heavy use of mobile phones require further investigation.

There was no increase in risk of acoustic neuroma with ever regular use of a mobile phone or for users who began regular use 10 years or more before the reference date. Elevated odds ratios observed at the highest level of cumulative call time could be due to chance, reporting bias or a causal effect. As acoustic neuroma is usually a slowly growing tumor, the interval between introduction of mobile phones and occurrence of the tumor might have been too short to observe an effect, if there is one.

Although it is the largest, most prominent, and most scientifically sound epidemiologic study on mobile phones and cancer completed today, the study still suffered from inconsistencies, as outlined by the researchers. The causes for such inconsistencies for data obtained through interviews and questionnaires are the following [24]:

- **Recall bias.** Individuals who have developed cancer may recall their telephone use more in accordance with the anticipated causal relation.
- **Inaccurate reporting.**
- **Morbidity and mortality.** Individuals surviving first treatments are often impaired, which may affect their responses to questions. Next of kin of individuals that have died may not have accurate data on the cell phone use of the diseased.
- **Participation bias.** Individuals with cancer may be more likely to enroll in a research study than individuals without.
- **Changing technologies and methods of use.** Modulation forms and frequencies may change as well as the use of the phone, such as against the head or hands free.

3.4.2.2   Standpoints of Expert Organizations

Based largely on the data provided by the Interphone Study, the International Agency for Research on Cancer (IARC), which is part of the World Health Organization (WHO), recently [25] classified radio frequency electromagnetic fields as possibly carcinogenic to humans (group 2B). This category is used when a causal association is considered credible, but when chance, bias, or confounding cannot be ruled out with reasonable confidence [26]. The different WHO classifications are stated in Table 3.4 [25].

TABLE 3.4
WHO Carcinogenicity Group Classifications

Group	Classification
Group 1	The agent is carcinogenic to humans
Group 2A	The agent is probably carcinogenic to humans
Group 2B	The agent is possibly carcinogenic to humans
Group 3	The agent is not classifiable as to its carcinogenicity to humans
Group 4	The agent is probably not carcinogenic to humans

The reason for reclassifying RF electromagnetic fields (from group 2A to group 2B) is based on the Interphone Study indications of an increased risk of glioma for those who reported the highest 10% of cumulative hours of cell phone use, although there was no consistent trend of increasing risk with greater duration of use. The researchers concluded that biases and errors limit the strength of these conclusions and prevent a causal interpretation [26].

More research is required since current epidemiologic studies can only deal with tumors that have evolved within a limited time frame. Mobile phones have only been widely in use since the early 1990s. Many cancers are not detectable until many years after the interactions that led to the tumor [26]. However, results of animal studies consistently show no increased cancer risk for long-term exposure to radio frequency fields.

The standpoints of other expert organizations are the following [24]:

- **American Cancer Society (ACS).** The ACS states that the IARC classification means that there could be some risk associated with cancer, but the evidence is not strong enough to be considered causal and needs to be investigated further.
- **National Institute of Environmental Health Sciences (NIEHS).** The NIEHS states that the weight of the current scientific evidence has not conclusively linked cell phone use with any adverse health problems, but more research is needed.
- **US Food and Drug Administration (FDA).** The FDA states that the majority of human epidemiologic studies have failed to show a relationship between exposure to radio frequency energy from cell phones and health problems.
- **US Centers for Disease Control and Prevention (CDC).** The CDC states that, although some studies have raised concerns about the possible risks of cell phone use, scientific research as a whole does not support a statistically significant association between cell phone use and health effects.
- **Federal Communications Commission (FCC).** The FCC states that there is no scientific evidence that proves that wireless phone use can lead to cancer or to other health problems, including headaches, dizziness, or memory loss.

The general consensus seems to be that more research is needed. With respect to epidemiologic studies, the earlier mentioned causes for inconsistencies should be avoided.

3.4.2.3 Current and Future Epidemiologic Studies

In March 2010, the COSMOS study was launched in Europe [27,28]. COSMOS is an international cohort study investigating possible health effects from long-term use of mobile phones and other wireless technologies. Evidence to date suggests that short-term (less than 10 years) exposure to mobile phone emissions is not associated with an increase in brain and nervous system cancers. The COSMOS study will enroll about 250,000 cell phone users aged 18 years or older and will follow them—among other means, through their mobile phone records—for 20 to 30 years. COSMOS is an international cohort study taking place in the United Kingdom, Finland, Denmark, the Netherlands, Sweden, and France.

3.5   CONCLUDING REMARKS

We may conclude that, based on scientific evidence to date, adverse health effects have not been established to have been caused by electromagnetic radiation from mobile phones or GSM base stations.

At the frequencies used by mobile phones, most of the energy is absorbed by the skin and other superficial tissues, resulting in negligible temperature rise in the brain or any other organs of the body [24]. To an even greater extent, the same applies for GSM base stations, for which the power densities at the position of a human are much lower.

To date, research does not suggest any consistent evidence of adverse health effects (such as effects on brain electrical activity, cognitive functions, sleep, heart rate, and blood pressure) from exposure to radio frequency fields at levels below those that cause tissue heating [24].

Research has not been able to provide support for a causal relationship between exposure to electromagnetic fields and self-reported symptoms, or electromagnetic hypersensitivity [24].

The accumulated evidence from research does not establish the existence of adverse long-term health effects from the signals produced by mobile phones and base stations. The lack of data for mobile phone use over time periods longer than 15 years warrants further research of mobile phone use and brain cancer risk [24].

However, we do know that the human body absorbs about five times more of the signals from FM radio and television broadcasting stations (approximately 100–800 MHz) than it does from a GSM base station. Those broadcasting stations have been in use for over 50 years now. To date no adverse health effects originating from these broadcasting stations have been established [13].

So, it seems that the ICNIRP or related limits for electromagnetic power densities suffice to avoid adverse health effects. However, countries and regions have adopted and are about to adopt more stringent guidelines. The reasons for these more stringent guidelines are not dictated by science, but appear to be political decisions based merely on fear of the unknown. A better understanding of RF radiation and the interactions with the human body will help to put everything back in perspective.

REFERENCES

1. J. C. Maxwell, *A treatise on electricity and magnetism.* Cambridge, UK: Cambridge University Press, 2010.
2. H. J. Visser, *Array and phased array antenna basics.* Chichester, UK: John Wiley & Sons, 2005.
3. http://www.nationmaster.com/graph/med_rad-media-radios
4. http://www.nationmaster.com/graph/med_tel-media-televisions
5. World Health Organization, Electromagnetic fields and public health. Fact sheet no. 304, May 2006.
6. World Health Organization, Electromagnetic fields and public health: Mobile phones. Fact sheet no. 193, June 2011.
7. World Health Organization, Electromagnetic fields and public health: Public perception of EMF risks. Fact sheet no. 184, May 1998.

8. H. J. Visser, *Antenna theory and applications.* Chichester, UK: John Wiley & Sons, 2012.
9. E. Pirogova, V. Vojisavljevic, and I. Cosic, Biological effects of electromagnetic radiation. *Biomedical Engineering,* In-Tech, Vienna, Austria, 2009.
10. J. M. Osepchuk, A history of microwave heating applications. *IEEE Transactions on Microwave Theory and Techniques,* vol. 32, no. 9, pp. 1200–1224, Sept. 1984.
11. W. W. Mumford, Some technical aspects of microwave radiation hazards. *Proceedings of the IRE,* pp. 427–446, Feb. 1961.
12. J. M. Osepchuk and R. C. Petersen, Historical review of RF exposure standards and the International Committee on Electromagnetic Safety (ICES). *Bioelectromagnetics,* Supplement 6, pp. S7–S16, 2003.
13. P. A. Valberg, T. E. van Deventer, and M. H. Repacholi, Workgroup report: Base stations and wireless networks—Radiofrequency (RF) exposures and health consequences. *Environmental Health Perspectives,* vol. 115, no. 3, pp. 416–424, March 2007.
14. American Standards Association, Safety levels of electromagnetic radiation with respect to personnel. USASI Standard C95, 1966.
15. International Commission on Non-Ionizing Radiation Protection (ICNIRP), Guidelines for limiting exposure to time-varying electric, magnetic and electromagnetic fields (up to 300 GHz). *Health Physics,* vol. 74, no. 4, pp. 494–522, April 1998.
16. The Institute of Electrical and Electronics Engineers, IEEE standard for safety levels with respect to human exposure to radiofrequency electromagnetic fields, 3 kHz to 300 GHz. Standard C95.1-2005.
17. World Health Organization, Electromagnetic fields and public health: Electromagnetic hypersensitivity. Fact sheet no. 296, Dec. 2005.
18. K. H. Mild, M. Repacholi, E. van Deventer, and P. Ravazzani (eds.), *Proceedings of International Workshop on EMF Hypersensitivity,* World Health Organization, 2006.
19. G. J. Rubin, J. D. Munshi, and S. Wessely, Electromagnetic hypersensitivity: A systematic review of provocation studies, *Psychosomatic Medicine,* vol. 67, pp. 224–232, 2005.
20. C. Cinel, R. Russel, A. Boldini, and E. Fox, Exposure to mobile phone electromagnetic fields and subjective symptoms: A double-blind study. *Psychosomatic Medicine,* vol. 70, pp. 345–348, 2008.
21. G. J. Rubin, R. Nieto-Hernandez, and S. Wessely, Idiopathic environmental intolerance attributed to electromagnetic fields (formerly electromagnetic hypersensitivity): An updated systematic review of provocation studies. *Bioelectromagnetics,* vol. 31, pp. 1–11, 2010.
22. International Agency for Research on Cancer, Interphone Study report on mobile phone use and brain cancer risk. World Health Organization, press release no. 200, May 17, 2010.
23. C. Wild, IARC report to the Union for International Cancer Control (UICC) on the Interphone Study. IARC, World Health Organization, Oct. 3, 2011.
24. National Cancer Institute, Fact sheet: Cell phones and cancer risk. National Institutes of Health, available at http://www.cancer.gov/cancertopics/factsheet/Risk/cellphones
25. International Agency for Research on Cancer, IARC classifies radiofrequency electromagnetic fields as possibly carcinogenic to humans. World Health Organization, press release no. 208, May 31, 2011.
26. World Health Organization, Electromagnetic fields and public health: Mobile phones. Fact sheet no. 193, June 2011.
27. COSMOS, available at: www.ukcosmos.org
28. J. Schüz, P. Elliot, A. Auvinen, H. Kromhout, A. H. Poulsen, C. Johansen, J. H. Olsen, et al., An international prospective cohort study of mobile phone users and health (COSMOS): Design considerations and enrolment. *Cancer Epidemiology,* vol. 35, no. 1, pp. 37–43, Feb. 2011.

4 Receiver Front Ends with Robustness to Process Variations

Pooyan Sakian Dezfuli, Reza Mahmoudi, and Arthur van Roermund

CONTENTS

The increasing demand for compactness and speed of digital circuits and the necessity of integration of the digital back end electronics with radio frequency front ends calls for exploiting deep submicron technologies in RF circuit design. However, scaling into the deep submicron regime, mainly in CMOS (complementary metal oxide semiconductor) technologies, accentuates the effect of process spread and mismatch on the fabrication yield [1]. Furthermore, design for manufacturability requires all manufacturing and process variations to be considered in the design procedure. Statistical circuit-level methods based on modeling data provided by fabrication foundries (e.g., Monte Carlo) are extensively used to evaluate the effect of process spread and are utilized by simulation tools to design circuits with the desired performance over the specified range of process variation [2]. However, most of these statistical methods are based on random variation of design variables, which need long simulation times for large-scale circuits, like a full receiver. Furthermore, as the size and complexity of designs are increased, less insight is obtained from these random statistical methods.

Recently, system-level design techniques have been developed to determine the specifications of individual blocks of a receiver for minimum overall power consumption and for a given noise and nonlinearity performance [3–5]. However, system-level design guidelines in the literature do not sufficiently address the sensitivity of the receiver to process variations. In this chapter, for a given required overall performance, the block-level budgeting is performed in such a way that the effect of process variation on the overall performance is minimized. A system-level sensitivity analysis is performed on a generic receiver that can pinpoint the sensitive building blocks and show how to reduce the overall sensitivity to the performance of individual building blocks. Based on the presented analysis, the optimum plan for block-level specifications in terms of sensitivity to variability of components can be determined.

In this chapter, each building block of the receiver is described by three process-sensitive parameters: noise, nonlinearity, and voltage gain. In our analysis these parameters are defined in a way that they include the loading effect of the preceding and following blocks; in other words, the parameters are determined when the block is inside the system. The sensitivity of the overall performance of the receiver to variations of noise, nonlinearity, and gain of building blocks is calculated. In fact, the variations of noise, nonlinearity, and gain of building blocks represent variations in any circuit-level or process-technology level parameter that can affect the goal function of the receiver (BER as defined in the next section). Therefore, the presented methods include all relevant sources of variability. Since the number of parameters is limited to three for each block, faster computations are possible. On the other hand, any part of the receiver that can be characterized by these three parameters can be identified as a building block in the analysis. Therefore, the analysis is flexible in defining the building blocks of the system. To guarantee passing the yield test for all interferers, this analysis is carried out for the specified worst-case interferer.

The presented methods can include RF circuits, analog baseband circuitry, and ADC in the analysis. They are generic and can be used for any circuit topology or process technology, but there are some limitations:

1. The presented methods are applicable to narrowband systems or to wide-band systems, the noise, nonlinearity and gain of which can be represented by an equivalent value across the whole band of operation. Therefore, some types of filters or frequency-dependent components cannot be covered by this analysis.

2. The interferers are assumed either to be completely filtered or to be amplified by almost the same gain as the desired signal is amplified with.

3. The operation bandwidth of the components preceding the mixer is assumed to be narrow enough so that the harmonics of the local oscillator (LO) do not down-convert the noise of these components. In other words, only the fundamental harmonic of the LO down-converts the noise of the preceding stages of the mixer. This assumption allows using the Friis formula to calculate the noise figure of a cascade of stages including a zero-IF mixer. For nonzero-IF mixers, additional calculations are done to include the effect of image noise.

Performing the analysis on systems violating any of these limitations may invalidate the obtained results.

In Section 4.1, bit error rate (BER) is defined as the goal function of the receiver and is described as a function of total noise, total nonlinearity, and input impedance of the receiver (for the given type of modulation). In Section 4.2, the performance requirements of a typical 60 GHz receiver for indoor applications are described. Total noise and total nonlinearity are described as a function of block-level noise, nonlinearity, and gain in Section 4.3—providing a connection between BER and block-level performance parameters. In Section 4.4, the relationships derived in Sections 4.1 and 4.3 are used to determine the sensitivity of the overall performance of the receiver to the performance of the building blocks. It is shown that, for all blocks, the first-order sensitivity to the gain of the block can be nullified, whereas the sensitivity to noise and nonlinearity of all the blocks cannot be minimized simultaneously. In Section 4.5, different approaches to minimizing the sensitivities derived in Section 4.4 are investigated. In Section 4.6 the design of two 60 GHz receivers including and excluding the ADC is explored using different system-level design guidelines presented in this chapter.

4.1 BIT ERROR RATE, NOISE, GAIN, AND NONLINEARITY

Defined as the ratio of erroneous received bits to the total number of received bits, bit error rate (BER) is an essential performance measure for every receiver involved with digital data. The BER in a receiver is, for a given type of modulation, directly determined by the signal-to-noise-plus-distortion ratio (SNDR). Normally, one can determine the required SNDR for the desired BER by system-level simulation of the baseband demodulator. Having the required SNDR and the specified minimum detectable signal (MDS), the maximum total noise-plus-distortion (NPD) can be calculated by

$$10\log\left(\frac{NPD}{10^{-3}}\right) = MDS - SNDR \qquad (4.1)$$

where NPD is in watts, MDS is in dBm, and SNDR is measured in decibels at the output of the receiver. NPD is selected as the goal function of our analysis and since it is defined as the sum of the noise and nonlinearity distortion, it can be described as a function of noise and nonlinearity by

$$NPD = N_{Antenna} + N_{i,tot} + \sum_{q=2}^{\infty} P_{IMDqi,tot} \qquad (4.2)$$

where $N_{Antenna}$ is the noise coming from the antenna, $N_{i,tot}$ is the equivalent input-referred available noise power of the receiver, and $P_{IMDqi,tot}$ is the equivalent input-referred available power of in-band qth order intermodulation distortion due to an out-of-band interferer.

We have assumed that there is no correlation between noise and distortion and between distortion components of different orders. We have also assumed that the noise and distortion have equal influence on the BER. However, in general that is not the case and the influence of the distortion on the BER depends on many factors, such as the modulation type of the interferer. Therefore, if the modulation type is known, (4.2) should be modified in a way that reflects the weight of the distortion. As illustrated in Figure 4.1, a noisy and nonlinear circuit can be described by an ideal noiseless and linear circuit with equivalent noise and nonlinearity distortions referred to the input. In this case, as shown in Figure 4.1(b), input-referred noise and distortions are represented by voltage sources. This representation can also be applied to a zero-IF mixer by considering its RF port as the input and its IF port as the output. In order to describe NPD in terms of equivalent voltages, we need to convert the available power of noise and distortion in (4.2) to their equivalent voltages. This results in

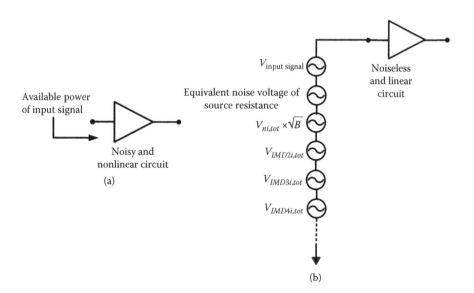

FIGURE 4.1 (a) A noisy and nonlinear circuit that can serve here both as the total receiver or one of its building blocks; (b) corresponding noiseless and linear circuit with equivalent noise and distortion voltage at input.

$$NPD = N_{Antenna} + N_{i,tot} + \sum_{q=2}^{\infty} P_{IMDqi,tot} \qquad (4.3)$$

$$NPD = kTB + \frac{(R_s + R_{in})^2}{4R_s R_{in}^2} \times B\bar{V}_{ni,tot}^2 + \frac{(R_s + R_{in})^2}{4R_s R_{in}^2} \times \sum_{q=2}^{\infty} V_{IMDqi,tot}^2$$

where

k = 1.38 × 10⁻²³ J/K is the Boltzmann constant

T is the absolute temperature

B is the effective noise bandwidth

R_{in} is the input impedance

R_s is the source impedance

$V_{ni,tot}$ is the equivalent input-referred noise voltage in V/√Hz

$V_{IMDqi,tot}$ is the equivalent input-referred voltage of qth order intermodulation distortion in volts, as shown in Figure 4.1(b)

Without loss of generality and for simplicity, the input impedance of the receiver is assumed to have only a real part.

Once the required $V_{ni,tot}$ and $V_{IMDqi,tot}$ of the whole receiver are determined, overall system specifications, including noise factor (NF_{tot}) and the voltage of qth order input intercept point ($V_{IPqi,tot}$), can be calculated:

$$NF_{tot} = \frac{4kT \times R_s + \bar{V}_{ni,tot}^2 \left(\dfrac{R_{in} + R_s}{R_{in}}\right)^2}{4kT \times R_s} \qquad (4.4)$$

$$\frac{1}{V_{IPqi,tot}^{q-1}} = \frac{V_{IMDqi,tot}}{V_{interferer}^q} \qquad (4.5)$$

where $V_{interferer}$ is the worst-case out-of-band interferer signal voltage at the input. Substituting $V_{IMDqi,tot}$ from (4.5) in (4.3) yields the NPD as a function of $V_{IPqi,tot}$ and $V_{ni,tot}$:

$$NPD = kTB + \frac{(R_s + R_{in})^2}{4R_s R_{in}^2} \times \left(B\bar{V}_{ni,tot}^2 + \sum_{q=2}^{\infty} \frac{V_{interferer}^{2q}}{V_{IPqi,tot}^{2q-2}} \right). \qquad (4.6)$$

The NPD can also be affected by IQ mismatch (only in receivers with I and Q paths) and phase noise. IQ phase or amplitude imbalance results in some cross talk between I and Q channels. This cross talk appears as a distortion and affects the NPD. Block-level budgeting of the noise, gain, and nonlinearity does not have a significant (if any) impact on the IQ mismatch. Therefore, if the amount of the IQ cross talk is known, its impact can be treated as a constant distortion added to the NPD and in this way it can be incorporated in the analysis. Extensive research has been done on IQ mismatch cancelation methods over the past years. Digital IQ imbalance compensation methods are widely used to suppress the impact of IQ mismatch. Details of these methods are beyond the scope of this chapter.

The phase noise originating from the frequency synthesizer can cause interchannel and in-band interference. The interchannel interference can be modeled as a distortion added to the NPD, assuming a worst-case adjacent-channel interference. However, the impact of in-band interference is signal dependent and cannot be modeled with just a constant additive distortion because the effect of phase noise is multiplicative and not additive, like thermal noise. Nevertheless, the impact of phase noise is hardly dependent on the noise, gain, and nonlinearity of the building blocks of the receiver. Therefore, the way the block-level budgeting is performed in the receiver can hardly influence the impact of the phase noise on the receiver performance. Minimization of the phase noise impact can be best accomplished in the frequency synthesizer section (e.g., phase-locked loop). In the rest of the chapter, the focus will be on the impact of the noise, gain, and nonlinearity distortion of individual blocks of the receiver on the NPD. Therefore, IQ mismatch and phase noise, which are rather independent of the block-level budgeting of the receiver, will not be addressed any further in this chapter.

4.2 PERFORMANCE REQUIREMENTS

The frequency band of 57–66 GHz, as allocated by the regulatory agencies in Europe, Japan, Canada, and the United States, can be used for high rate wireless personal area network (WPAN) applications [6]. According to the IEEE 802.15.3c standard, three different modes are possible for the physical layer of such a network: single-carrier mode, high speed interface mode, and audio/visual mode. The single-carrier mode supports various types of modulation schemes including $\pi/2$ QPSK, $\pi/2$ 8-PSK, $\pi/2$ 16-QAM, precode MSK, precoded GMSK, on/off keying (OOK), and dual alternate mark inversion (DAMI). The high speed interface mode is designed for non-line-of-sight operation and uses OFDM. The audio/visual mode is also designed for non-line-of-sight operation, uses OFDM, and is considered for uncompressed, high definition video and audio transport.

The whole band is divided into four channels with center frequencies located at 58.32, 60.480, 62.640, and 64.8 GHz, each with a bandwidth of 2.16 GHz [6]. A transceiver complying with the single-carrier mode should support at least one of these channels. A transceiver complying with the high speed interface mode should support at least the channel centered at 60.480 GHz or the one centered at 62.640 GHz. The audio/visual mode is in turn divided into two modes of low data rate (LRP: in the order of 5 Mbps) and high data rate (HRP: in the order of 1 Gbps). A transceiver complying with HRP mode should support at least the channel centered at 60.480 GHz. On the other hand, in the LRP mode, the bandwidth of the channels is about 98 MHz (i.e., in each of the mentioned channels, three LRP channels, with 98 MHz bandwidth, are defined around the center).

Different physical layer definitions are a result of different possible application demands. For example, a kiosk application would require 1.5 Gbps data rate at a 1 m range. This data rate at such a short range can be easily provided by the single-carrier mode with less complexity and thus lower cost compared to physical layer definitions, which use OFDM. On the other hand, the audio/visual mode is best fitted to uncompressed video streaming applications. An ad hoc system for connecting computers and devices around a conference table can be best implemented by the high speed interface mode because, in this case, all the devices in the WPAN are expected to have bidirectional, non-line-of-sight, high speed, low latency communication [6].

According to the IEEE 802.15.3c standard, the limit for effective isotropic radiated power (EIRP) at this frequency band is 27 dBi for indoors and 40 dBi for outdoors in the United States. The EIRP limit in Japan and Australia is 57 dBi and 51.8 dBi, respectively. It is worth remembering that EIRP is the sum of the transmitter output power and its antenna gain.

In the single-carrier mode, the frame error rate must be less than 8%, with a frame payload length of 2,048 octets. The minimum detectable signal varies between −70 and −46 dBm, depending on the data rate. The maximum tolerable power level of the incoming signal, which meets the required error rate, is −10 dBm.

In the high speed interface mode, for a BER of 10^{-6}, the minimum detectable signal varies between −50 and −70 dBm, depending on the required data rate. The maximum tolerable power level of the incoming signal, which meets the required error rate, is −25 dBm.

In the audio/visual mode, a BER of less than 10^{-7} must be met with a bit stream generated by a special pseudorandom sequence defined in the standard. The audio/visual mode has two different data rate options: high data rate in the order of gigabits per second and low data rate in the order of megabits per second. For low data-rate and high data-rate applications, the minimum detectable signal of the receiver should be −70 and −50 dBm, respectively. The maximum tolerable power level of the incoming signal is −30 and −24 dBm for low data rate and high data rate modes, respectively.

4.3 BLOCK-LEVEL IMPACT

To address the impact of block-level performance on the overall performance, we first start with a rather general case including the image noise and a precise calculation of the total nonlinearity and then we derive a more useful special case.

4.3.1 GENERAL CASE

In this section the building blocks of the receiver are described with three parameters including noise, nonlinearity, and gain, as shown in Figure 4.2(a). The system consists of N stages with at least one mixer as the Lth stage. The gain of each stage is defined as the ratio between the output voltage and the input voltage of each stage when the stage is inside the system, as shown in Figure 4.2(b). The output noise voltage of each stage is observed by assuming that all the other stages and also the antenna are noiseless. Then the input noise voltage of the stage is calculated via dividing the output noise voltage by the gain, as shown in Figure 4.2(c). A similar procedure is used to define the nonlinearities of stages, as described later in more detail. The stages prior to the mixer are characterized by two additional parameters: $A_{vn,k,image}$ is the gain by which the noise voltage of the previous stage (the antenna

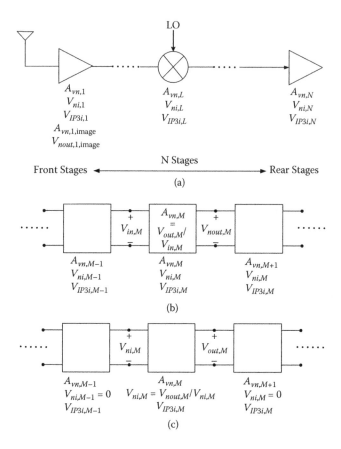

FIGURE 4.2 (a) N nonlinear and noisy stages cascaded in the receiver described by their noise, nonlinearity and gain. (b) Gain is defined as the ratio between the output voltage of the stage to its input voltage when the stage is in the system. (c) Noise of a stage is defined by assuming that all the other stages and antenna are noiseless and observing the noise voltage at the output of the stage.

in the case of k = 1) at the image band is amplified and $V_{nout,k,image}$ is the noise at the output of the kth stage, generated by the stage itself, residing at the image band. Defining $V_{nout,k}$ as the output noise of the kth stage, generated by the stage itself, $V_{ni,k}$ is defined as $V_{nout,k}$ divided by voltage gain ($A_{vn,k}$). In order to calculate the overall equivalent input-referred noise voltage ($V_{ni,tot}$) as a function of the noise and gain of individual stages, the total noise at the output of the receiver, generated by the receiver itself (and not by the antenna), is expressed in (4.7).

The meanings of all parameters are described in Table 4.1. Two dummy variables—$A_{vn,0}$ and $A_{vn,N+1}$—are defined as unity to facilitate the representation of calculations:

$$\bar{V}^2_{nout,tot} = \begin{aligned}
&\bar{V}^2_{nout,1} \times \prod_{j=2}^{N} A^2_{vn,j} && +\bar{V}^2_{nout,1,image} \times \prod_{j=2}^{L} A^2_{vn,j,image} \times \prod_{j=L+1}^{N} A^2_{vn,j} \\
&\bar{V}^2_{nout,2} \times \prod_{j=3}^{N} A^2_{vn,j} && +\bar{V}^2_{nout,2,image} \times \prod_{j=3}^{L} A^2_{vn,j,image} \times \prod_{j=L+1}^{N} A^2_{vn,j} \\
&\qquad\vdots && \qquad\vdots \\
&+\bar{V}^2_{nout,L-1} \times \prod_{j=L}^{N} A^2_{vn,j} && +\bar{V}^2_{nout,L-1,image} \times A^2_{vn,L,image} \times \prod_{j=L+1}^{N} A^2_{vn,j} \\
&+\sum_{k=L}^{N} \bar{V}^2_{nout,k} \times \prod_{j=k+1}^{N+1} A^2_{vn,j}
\end{aligned} \tag{4.7}$$

Defining two auxiliary variables, as in (4.8) and (4.9); replacing $V_{nout,k}$ with $V_{ni,k}A_{vn,k}$; and making some simplifications, the total input-referred noise of the receiver is obtained in (4.10).

$$c_{\text{Im}\,N,k} = \frac{\bar{V}^2_{nout,k,image}}{\bar{V}^2_{nout,k}} \tag{4.8}$$

$$c_{\text{Im}\,A,k} = \frac{A^2_{vn,k,image}}{A^2_{vn,k}} \tag{4.9}$$

$$\bar{V}^2_{ni,tot} = \frac{1}{1+\prod_{j=1}^{L} c_{\text{Im}\,A,j}} \left[\sum_{k=1}^{L-1} \frac{\left(1+c_{\text{Im}\,N,k} \prod_{j=k+1}^{L} c_{\text{Im}\,A,j}\right)\bar{V}^2_{ni,k}}{\prod_{j=0}^{k-1} A^2_{vn,j}} + \sum_{k=L}^{N} \frac{\bar{V}^2_{ni,k}}{\prod_{j=0}^{k-1} A^2_{vn,j}} \right] \tag{4.10}$$

TABLE 4.1
Some of the Notations Used in the Formulae

Parameter	Notation for	
	kth Stage	**Whole receiver**
Noise voltage at the output	$V_{nout,k}$	$V_{nout,tot}$
Image noise voltage at the output	$V_{nout,k,image}$	$V_{nout,tot,image}$
Gain: the ratio between output voltage and input voltage at the signal band	$A_{vn,k}$	$A_{vn,tot}$
Image gain: the ratio between output voltage and input voltage at the image band	$A_{vn,k,image}$	$A_{vn,tot,image}$
Equivalent input-referred noise voltage	$V_{ni,k} = \dfrac{V_{nout,k}}{A_{vn,k}}$	$V_{ni,tot} = \dfrac{V_{nout,tot}}{A_{vn,tot}}$

For obtaining (4.10), the relationship between the effective noise bandwidth (B) in (4.6) and the signal bandwidth (B_{sig}) is defined as follows:

$$B = B_{sig} \times \left(1 + \prod_{j=1}^{L} c_{\mathrm{Im}A, j}\right) \tag{4.11}$$

The contribution of the kth stage to the total noise is calculated from

$$C_{Noise}(k) = \frac{\overline{V}_{ni,k}^2}{\left(1 + \prod\limits_{j=1}^{L} c_{\mathrm{Im}A, j}\right)\prod\limits_{j=0}^{k-1} A_{vn,j}^2} \times \begin{cases} \left(1 + c_{\mathrm{Im}N,k} \prod\limits_{j=k+1}^{L} c_{\mathrm{Im}A, j}\right), & k < L \\ 1, & k \geq L \end{cases} \tag{4.12}$$

This means that the noise contribution of each stage is inversely proportional to the square of the voltage gain of its preceding stages.

In order to calculate the overall voltage of the qth order input intercept point of the receiver ($V_{IPqi,tot}$) in terms of that of the individual stages, the phase relationships between the nonlinearities and the gains of the stages must be introduced into the calculations.

To start the analysis, the nonlinearity and gain parameters of the receiver and its stages are expressed in phasor form, as shown in Table 4.2. The total input-referred voltage of the qth order intermodulation distortion can be expressed in terms of the distortion of the individual stages by

$$V_{IMDqi,tot}\angle\varphi_{IMDqi,tot} = \sum_{k=1}^{N} \frac{V_{IMDqi,k}\angle\varphi_{IMDqi,k}}{\prod\limits_{j=0}^{k-1} A_{vn,j}\angle q\varphi_{vn,j}} \tag{4.13}$$

Then the distortions can be expressed in terms of intercept points:

$$V_{IMDqi,k}\angle\varphi_{IMDqi,k} = \frac{V_{interferer}^q \times \prod\limits_{j=0}^{k-1} A_{vn,j}^{q-1}\angle q\varphi_{vn,j}}{V_{IPqi,k}^{q-1}\angle(q-1)\varphi_{IPqi,k}}. \tag{4.14}$$

TABLE 4.2
Some of the Notations in Phasor Form

	Phasor notation for	
Parameter	kth Stage	Whole receiver
Input-referred voltage of the qth order intermodulation distortion	$V_{IMDqi,k} < \phi_{IMDqi,k}$	$V_{IMDqi,tot} < \phi_{IMDqi,tot}$
Voltage of the qth order input intercept point	$V_{IPqi,k} < \phi_{IPqi,k}$	$V_{IPqi,tot} < \phi_{IPqi,tot}$
Voltage gain including the loading effect	$A_{vn,k} < \phi_{vn,k}$	$A_{vn,tot} < \phi_{vn,tot}$

Therefore, the total intercept point can be expressed in terms of the intercept points of the individual stages by substituting the distortions from (4.14) into (4.13):

$$\frac{1}{V_{IPqi,tot}^{q-1} \angle (q-1)\varphi_{IPqi,tot}} = \sum_{k=1}^{N} \left(\frac{\prod_{j=0}^{k-1} A_{vn,j}^{q-1}}{V_{IPqi,k}^{q-1}} \angle (q-1)\left(\sum_{j=0}^{k-1} \varphi_{vn,j} - \varphi_{IPqi,k} \right) \right) \qquad (4.15)$$

or

$$\frac{1}{\left| V_{IPqi,tot} \right|^{q-1}} = \sqrt{\left(\sum_{k=1}^{N} \left(\frac{\prod_{j=0}^{k-1} A_{vn,j}^{q-1}}{V_{IPqi,k}^{q-1}} \cos\left((q-1)\left(\sum_{j=0}^{k-1} \varphi_{vn,j} - \varphi_{IPqi,k} \right) \right) \right) \right)^2 + \left(\sum_{k=1}^{N} \left(\frac{\prod_{j=0}^{k-1} A_{vn,j}^{q-1}}{V_{IPqi,k}^{q-1}} \sin\left((q-1)\left(\sum_{j=0}^{k-1} \varphi_{vn,j} - \varphi_{IPqi,k} \right) \right) \right) \right)^2} \qquad (4.16)$$

4.3.2 Special Case: Zero-IF Receiver and Worst-Case Nonlinearity

In the special case of a zero-IF receiver with worst-case nonlinearity superposition, the preceding relationships can be simplified substantially. This special case is the focus of most of the analysis in the rest of this work.

In a zero-IF mixer with a complex mixer and I/Q signal paths, (4.10)–(4.12) simplify to (4.17)–(4.19):

$$\bar{V}_{ni,tot}^2 = \sum_{k=1}^{N} \frac{\bar{V}_{ni,k}^2}{\prod_{j=0}^{k-1} A_{vn,j}^2} \qquad (4.17)$$

$$B = B_{sig} \qquad (4.18)$$

$$C_{Noise}(k) = \frac{\bar{V}_{ni,k}^2}{\prod_{j=0}^{k-1} A_{vn,j}^2} \qquad (4.19)$$

If the intermodulation distortions of the consecutive stages are added in-phase, resulting in a worst-case scenario, the expression for the intercept point in (4.15) simplifies to [3]

$$\frac{1}{V_{IPqi,tot}^{q-1}} = \sum_{k=1}^{N} \frac{\prod_{j=0}^{k-1} A_{vn,j}^{q-1}}{V_{IPqi,k}^{q-1}} \tag{4.20}$$

The preceding relationships are based on the assumption that the interferer and the in-band signal are amplified with almost the same gain. However, if the interferer is far from the in-band signal in the frequency domain, it can be filtered at the input of the receiver such that it generates no distortion.

Based on (4.20), the contribution of the kth stage to the total qth order nonlinearity distortion is equal to

$$C_{Distortion,q}(k) = \frac{\prod_{j=0}^{k-1} A_{vn,j}^{q-1}}{V_{IPqi,k}^{q-1}} \tag{4.21}$$

This means that the nonlinearity distortion contribution of each stage is directly proportional to the combined voltage gain of its preceding stages. This in fact creates a trade-off in defining the gain of stages. The parameters C_{Noise} and $C_{Distortion}$ will play a central role in the rest of our analysis.

4.4 SENSITIVITY TO BLOCK-LEVEL PERFORMANCE

In analogy with the previous section, the sensitivities of the overall performance to the block-level performance are calculated in both general and specific cases.

4.4.1 GENERAL CASE

To find a block-level budgeting for optimum robustness of the receiver to variability of block-level performance, one has to minimize the sensitivity of the total performance to the performance of individual blocks. In fact, one way to make the receiver robust to process variations is to make it robust to performance degradations of its building blocks. In this section, the sensitivity of total NPD to the noise, nonlinearity, and gain of individual blocks is calculated. The normalized single-point sensitivity of $F(x)$ to the variable x is defined by the following operator [2]:

$$S_x^{F(x)} \triangleq \frac{\partial F(x)}{\partial x} \times \frac{x}{F(x)} \tag{4.22}$$

This should be calculated at the selected nominal values of x and $F(x)$. As mentioned earlier, to achieve a certain BER, a specific NPD requirement has to be met. Therefore, variations of NPD can cause variations in BER. The variations of NPD can be described as a function of the variation of the performance parameters of the individual stages:

$$\Delta NPD = \sum_{k=1}^{N} \frac{\partial NPD}{\partial A_{vn,k}} \Delta A_{vn,k} + \sum_{k=1}^{N} \frac{\partial NPD}{\partial V_{ni,k}} \Delta V_{ni,k} + \sum_{q=2}^{\infty} \sum_{k=1}^{N} \frac{\partial NPD}{\partial V_{IPqi,k}} \Delta V_{IPqi,k} \qquad (4.23)$$

$$+ \frac{\partial NPD}{\partial Z_{in}} \Delta Z_{in}$$

The random variables $\Delta A_{vn,k}$, $\Delta V_{ni,k}$, $\Delta V_{IPqi,k}$, and ΔZ_{in}, which represent the performance variations of the individual stages and the input impedance, are usually correlated. However, regardless of the amount of correlation between these random variables, one can minimize ΔNPD by minimizing (ideally, nullifying) the derivatives in (4.23), which are proportional to the sensitivity functions. Furthermore, the random variables $\Delta A_{vn,k}$, $\Delta V_{ni,k}$, and $\Delta V_{IPqi,k}$ are in general different for different stages (i.e., different values of k). If they are significantly larger for a specific stage, it is advisable to focus on reducing the sensitivity of the total performance to the performance of that stage. However, such knowledge requires circuit-level information about each stage, which may be achieved during the circuit design. Therefore, in this analysis, we attempt to reduce all the sensitivities, assuming that the random variables $\Delta A_{vn,k}$, $\Delta V_{ni,k}$, and $\Delta V_{IPqi,k}$ are in the same order for different stages. The sensitivity functions of NPD to performance parameters of each stage ($V_{ni,k}$, $V_{IPqi,k}$, and $A_{vn,k}$) are listed in (4.25), (4.26), and (4.27), respectively. The derivatives are taken using the chain rule applied to (4.6), (4.10), and (4.16). To simplify the equations, an auxiliary variable is defined in the following:

$$R_{eq} \triangleq \frac{4 R_s R_{in}^2}{\left(R_s + R_{in} \right)^2} \qquad (4.24)$$

$$S_{V_{ni,k}^2}^{NPD} = \frac{B}{\left(1 + \prod\limits_{j=1}^{L} c_{Im A,j} \right) R_{eq} \times NPD} \times \frac{\bar{V}_{ni,k}^2}{\prod\limits_{j=0}^{k-1} A_{vn,j}^2} \times \begin{cases} \left(1 + c_{Im N,k} \prod\limits_{j=k+1}^{L} c_{Im A,j} \right), & k < L \\ 1, & k \geq L \end{cases} \qquad (4.25)$$

$$S_{\left(\frac{1}{V_{IPqi,k}^{q-1}} \right)}^{NPD} = \frac{(q-1) V_{interferer}^{2q}}{R_{eq} \times NPD} \times \frac{\prod\limits_{j=0}^{k-1} A_{vn,j}^{q-1}}{V_{IPqi,k}^{q-1}} \times \qquad (4.26)$$

$$\left[\cos\left((q-1) \left(\sum_{j=0}^{k-1} \varphi_{vn,j} - \varphi_{IPqi,k} \right) \right) \sum_{m=1}^{N} \left(\frac{\prod\limits_{j=0}^{m-1} A_{vn,j}^{q-1}}{V_{IPqi,m}^{q-1}} \cos\left((q-1) \left(\sum_{j=0}^{m-1} \varphi_{vn,j} - \varphi_{IPqi,m} \right) \right) \right) \right.$$

$$\left. + \sin\left((q-1) \left(\sum_{j=0}^{k-1} \varphi_{vn,j} - \varphi_{IPqi,k} \right) \right) \sum_{m=1}^{N} \left(\frac{\prod\limits_{j=0}^{m-1} A_{vn,j}^{q-1}}{V_{IPqi,m}^{q-1}} \sin\left((q-1) \left(\sum_{j=0}^{m-1} \varphi_{vn,j} - \varphi_{IPqi,m} \right) \right) \right) \right]$$

It is clear from (4.25)–(4.27) that the sensitivities to the noise and nonlinearity of the stages cannot be made zero, whereas the sensitivity to the gain of individual stages can be made zero.

$$
S_{A_{m,k}}^{NPD} = \begin{aligned}
& \frac{2}{R_{eq} \times NPD} \times \sum_{q=2}^{N}\left[\left[\left(\sum_{m=1}^{N}\frac{(q-1)V_{interferer}^{2q}\prod_{j=0}^{m-1}A_{vn,j}^{q-1}}{V_{IPqi,m}^{q-1}}\cos\left((q-1)\left(\sum_{j=0}^{m-1}\varphi_{vn,j}-\varphi_{IPqi,m}\right)\right)\right)\left(\sum_{m=k+1}^{N}\frac{\prod_{j=0}^{m-1}A_{vn,j}^{q-1}}{V_{IPqi,m}^{q-1}}\cos\left((q-1)\left(\sum_{j=0}^{m-1}\varphi_{vn,j}-\varphi_{IPqi,m}\right)\right)\right)\right]\right. \\
& \left.+\left(\sum_{m=1}^{N}\frac{(q-1)V_{interferer}^{2q}\prod_{j=0}^{m-1}A_{vn,j}^{q-1}}{V_{IPqi,m}^{q-1}}\sin\left((q-1)\left(\sum_{j=0}^{m-1}\varphi_{vn,j}-\varphi_{IPqi,m}\right)\right)\right)\left(\sum_{m=k+1}^{N}\frac{\prod_{j=0}^{m-1}A_{vn,j}^{q-1}}{V_{IPqi,m}^{q-1}}\sin\left((q-1)\left(\sum_{j=0}^{m-1}\varphi_{vn,j}-\varphi_{IPqi,m}\right)\right)\right)\right], \quad k\ L-1 \\
& -\frac{2}{R_{eq}\times NPD}\sum_{j=k+1}^{L-1}\frac{B\left(1+c_{\ln N,j}\prod_{m=j+1}^{L}c_{\ln A,m}\right)\overline{V}_{ni,j}^{2}}{\left(1+\prod_{j=1}^{L}c_{\ln A,j}\right)\prod_{m=0}^{j-1}A_{vn,m}^{2}} \quad \frac{2}{R_{eq}\times NPD}\sum_{j=k}^{N}\frac{B\overline{V}_{ni,j}^{2}}{\left(1+\prod_{j=1}^{L}c_{\ln A,j}\right)\prod_{m=0}^{j-1}A_{vn,m}^{2}}
\end{aligned}
$$

(4.27)

$$
\begin{aligned}
& \frac{2}{R_{eq}\times NPD}\times\sum_{q=2}^{N}\left[\left[\left(\sum_{m=1}^{N}\frac{(q-1)V_{interferer}^{2q}\prod_{j=0}^{m-1}A_{vn,j}^{q-1}}{V_{IPqi,m}^{q-1}}\cos\left((q-1)\left(\sum_{j=0}^{m-1}\varphi_{vn,j}-\varphi_{IPqi,m}\right)\right)\right)\left(\sum_{m=k+1}^{N}\frac{\prod_{j=0}^{m-1}A_{vn,j}^{q-1}}{V_{IPqi,m}^{q-1}}\cos\left((q-1)\left(\sum_{j=0}^{m-1}\varphi_{vn,j}-\varphi_{IPqi,m}\right)\right)\right)\right]\right. \\
& \left.+\left(\sum_{m=1}^{N}\frac{(q-1)V_{interferer}^{2q}\prod_{j=0}^{m-1}A_{vn,j}^{q-1}}{V_{IPqi,m}^{q-1}}\sin\left((q-1)\left(\sum_{j=0}^{m-1}\varphi_{vn,j}-\varphi_{IPqi,m}\right)\right)\right)\left(\sum_{m=k+1}^{N}\frac{\prod_{j=0}^{m-1}A_{vn,j}^{q-1}}{V_{IPqi,m}^{q-1}}\sin\left((q-1)\left(\sum_{j=0}^{m-1}\varphi_{vn,j}-\varphi_{IPqi,m}\right)\right)\right)\right], \quad k\ L-1 \\
& -\frac{2}{R_{eq}\times NPD}\sum_{j=k+1}^{N}\frac{B\overline{V}_{ni,j}^{2}}{\left(1+\prod_{j=1}^{L}c_{\ln A,j}\right)\prod_{m=0}^{j-1}A_{vn,m}^{2}}
\end{aligned}
$$

4.4.2 Special Case: Zero-IF Receiver and Worst-Case Nonlinearity

If the receiver is zero-IF with a complex mixer and I/Q signal paths in IF and the design is done for worst-case scenario, in which the intermodulation distortions of the individual stages are assumed to add up in phase, (4.25)–(4.27) are simplified to (4.28)–(4.30):

$$
S_{V_{ni,k}^{2}}^{NPD} = \frac{B}{R_{eq}\times NPD}\times\frac{\overline{V}_{ni,k}^{2}}{\prod_{j=0}^{k-1}A_{vn,j}^{2}}
$$

(4.28)

$$
S_{\left(\frac{1}{V_{IPqi,k}^{q-1}}\right)}^{NPD} = \frac{2V_{interferer}^{2q}}{R_{eq}\times V_{IPqi,tot}^{q-1}\times NPD}\times\frac{\prod_{j=0}^{k-1}A_{vn,j}^{q-1}}{V_{IPqi,k}^{q-1}}
$$

(4.29)

$$
S_{A_{vn,k}}^{NPD} = \frac{2}{R_{eq}\times NPD}\times\left(\sum_{q=2}^{\infty}\left(\frac{(q-1)V_{interferer}^{2q}}{V_{IPqi,tot}^{q-1}}\times\sum_{j=k+1}^{N}\frac{\prod_{m=1}^{j-1}A_{vn,m}^{q-1}}{V_{IPqi,j}^{q-1}}\right)-\sum_{j=k+1}^{N}\frac{B\overline{V}_{ni,j}^{2}}{\prod_{m=1}^{j-1}A_{vn,m}^{2}}\right)
$$

(4.30)

The second-order sensitivity of NPD to block-level gain is proportional to the second-order derivative of NPD with respect to block-level gains as described in (4.31) and (4.32):

$$\left\{ \frac{\partial^2 NPD}{\partial A_{vn,k}\, \partial A_{vn,p}} = \frac{2R_{eq}^{-1}}{A_{vn,k}A_{vn,p}} \times \left(\sum_{q=2}^{\infty} \left[(q-1)^2 V_{interferer}^{2q} \times \sum_{j=p+1}^{N} \frac{\prod\limits_{m=1}^{j-1} A_{vn,m}^{q-1}}{V_{IPqi,j}^{q-1}} \times \left(\frac{1}{V_{IPqi,tot}^{q}} + \sum_{j=k+1}^{N} \frac{\prod\limits_{m=1}^{j-1} A_{vn,m}^{q-1}}{V_{IPqi,j}^{q-1}} \right) \right] + 2 \times \sum_{j=p+1}^{N} \frac{B\overline{V}_{ni,j}^2}{\prod\limits_{m=1}^{j-1} A_{vn,m}^2} \right) \right. \tag{4.31}$$
$$\left. (p > k) \right.$$

$$\frac{\partial^2 NPD}{\partial A_{vn,k}^2} = \frac{2R_{eq}^{-1}}{A_{vn,k}^2} \times \left(\sum_{q=2}^{\infty} V_{interferer}^{2q} \left[(q-1)^2 \left(\sum_{j=k+1}^{N} \frac{\prod\limits_{m=1}^{j-1} A_{vn,m}^{q-1}}{V_{IPqi,j}^{q-1}} \right)^2 + \frac{(q-1)(q-2)}{V_{IPqi,tot}^{q-1}} \times \sum_{j=k+1}^{N} \frac{\prod\limits_{m=1}^{j-1} A_{vn,m}^{q-1}}{V_{IPqi,j}^{q-1}} \right] + 3 \sum_{j=k+1}^{N} \frac{B\overline{V}_{ni,j}^2}{\prod\limits_{m=1}^{j-1} A_{vn,m}^2} \right) \tag{4.32}$$

These are valid for $(1 \le k < N)$.

An inspection of (4.19) and (4.28) and, in the general case, (4.12) and (4.25) shows that the sensitivity to the noise performance of each stage is proportional to its contribution to the total noise of the receiver. A similar inspection of (4.21) and (4.29) reveals that the sensitivity to the qth order nonlinearity of each component is proportional to its contribution to the total qth order nonlinearity. Therefore, for a given total noise and total nonlinearity, reducing the sensitivity to noise/nonlinearity of one stage results in increased sensitivity to noise/nonlinearity of other stages. On the other hand, the sensitivity to gain of each block can be set to zero as implied by (4.27) or (4.30). Furthermore, according to (4.31) and (4.32), the second-order sensitivity to block-level gains cannot be nullified, but can be reduced by lowering the contribution of the rear stages to the total noise and nonlinearity.

One solution for zeroing the first order sensitivity of NPD to gains in (4.30) is

$$\frac{\overline{V}_{ni,k}^2}{\overline{V}_{ni,k-1}^2} = \alpha_{k-1} A_{vn,k-1}^2 \tag{4.33}$$

$$\frac{V_{IPqi,k}^{q-1}}{V_{IPqi,k-1}^{q-1}} = \frac{1}{\alpha_{k-1}} A_{vn,k-1}^{q-1} \tag{4.34}$$

$$B\overline{V}_{ni,tot}^2 = \sum_{q=2}^{\infty} \frac{(q-1)V_{interferer}^{2q}}{V_{IPqi,tot}^{2(q-1)}} \tag{4.35}$$

where α_k $(1 \le k < N)$ is a parameter that must be chosen by the designer; we call it contribution factor of a stage, because it determines the ratio of the noise and non-linearity distortion contribution of each stage to that of its following stage. To nullify (4.30) for every stage, (4.33)–(4.35) must be satisfied for $(1 < k \le N)$. In case of dominance of third-order nonlinearity, (4.35) simplifies to

$$B\overline{V}_{ni,tot}^2 = \frac{2V_{interferer}^6}{V_{IP3i,tot}^4} = 2V_{IMD3i,tot}^2 . \tag{4.36}$$

which means that the third-order intermodulation distortion must be 3 dB below the level of the noise coming from the receiver itself.

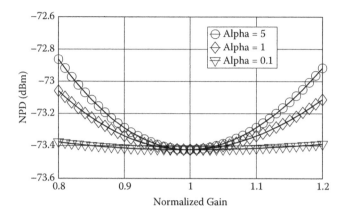

FIGURE 4.3 NPD versus normalized variation of the gain of the first stage for different values of α.

According to (4.31) and (4.32), the second-order sensitivity of NPD to the gain of front stages is bigger than that of rear stages. For a system that satisfies (4.33)–(4.35), the reduction of the second-order sensitivity to the gain of front stages requires their α to be smaller than unity, resulting in a lower contribution of rear stages to total noise and nonlinearity. In fact, if α is smaller than unity for every stage, the contribution of each stage to the total noise and nonlinearity, as quantified by (4.12), (4.19), and (4.21), will be smaller than that of its previous stage. It is worth mentioning that a higher contribution of a stage to the total noise (or nonlinearity distortion) does not necessarily mean that it is noisier (or more nonlinear), because the contribution of one stage to total noise (or total nonlinearity distortion) is not only a function of the noise (or nonlinearity) of the stage itself but also a function of the gain of its previous stages, as described by (4.12), (4.19), and (4.21). Figure 4.3 shows the NPD of a receiver as a function of the gain of its first stage (normalized to its nominal value) for different values of α. It shows the significant impact of α on the second-order sensitivity, with smaller values of α yielding smaller second-order sensitivity to the gain of the LNA.

4.5 DESIGN FOR ROBUSTNESS

The equations derived in Section 4.4 are insightful for both analysis and synthesis of a receiver. From the analysis perspective, they can determine the most sensitive building blocks of the receiver. From the synthesis perspective, they can be used to develop design approaches for minimum sensitivity to variability of building blocks. In this section, three design approaches are investigated. In all of them, the first-order sensitivity of the NPD to the gain of each block is set to zero. The difference between the three methods is in the different values chosen for α.

In the first approach, the sensitivities of the NPD to the noise and nonlinearity of the individual stages are all the same, while the sensitivity of the NPD to the gain of each block is set to zero by setting α to one. In the second approach, the second-order sensitivity to the gain of building blocks is reduced, while the first-order sensitivity to gain is kept at zero, by keeping α smaller than one. In the third approach, an optimum-power design method is described [3] and the first-order sensitivity to the gain of building blocks is set to zero by adding a new condition to the method; in this method, α is a function of the power consumption of the components. All the approaches are presented for the special case of the zero-IF receiver with a worst-case nonlinearity scenario, although they have the potential to be generalized to other receiver architectures and scenarios.

4.5.1 CONSTANT-SENSITIVITY APPROACH

For a given total noise and total nonlinearity, reducing the sensitivity to noise/non-linearity of one stage results in increased sensitivity to noise/nonlinearity of other stages, because reducing the contribution of one stage to total noise or nonlinearity, (4.19) and (4.21), increases the contribution of other stages. As a result, by keeping sensitivities equal for all the stages, one can avoid extra-sensitive nodes in the receiver. Forcing the sensitivity to noise in (4.28) to be equal for all k ($1 \leq k \leq N$) and using (4.17) and (4.19) yields

$$C_{Noise}(k) = \frac{\overline{V}^2_{ni,k}}{\prod\limits_{j=0}^{k-1} A^2_{vn,j}} = \frac{\overline{V}^2_{ni,tot}}{N} \qquad (4.37)$$

This means that the input-referred noise voltage of each stage must be better than that of its following stage by a factor of its voltage gain:

$$\frac{\overline{V}^2_{ni,k}}{\overline{V}^2_{ni,k-1}} = A^2_{vn,k-1} \qquad (4.38)$$

Forcing the sensitivity to nonlinearity in (4.29) to be constant for all k ($1 \leq k \leq N$) and using (4.20) and (4.21) yields

$$C_{Distortion,q}(k) = \frac{\prod\limits_{j=0}^{k-1} A^{q-1}_{vn,j}}{V^{q-1}_{IPqi,k}} = \frac{1}{N \times V^{q-1}_{IPqi,tot}} \qquad (4.39)$$

This means that the qth order nonlinearity performance of each stage must be better than that of its preceding stage by a factor of voltage gain of its preceding stage:

$$\frac{V^{q-1}_{IPqi,k}}{V^{q-1}_{IPqi,k-1}} = A^{q-1}_{vn,k-1} \qquad (4.40)$$

In the constant-sensitivity approach, (4.37) and (4.39) guarantee that the sensitivity to noise and nonlinearity of all the stages is the same and the contribution of all the stages to total noise and nonlinearity is equal.

According to (4.38) and (4.40) two conditions for zeroing the first-order sensitivity of NPD to the gain, (4.33) and (4.34), are automatically satisfied in this approach with α equal to unity for all the stages. Consequently, the designer has to satisfy (4.35) to nullify the first-order sensitivity to the gain.

Therefore, in this approach, first-order sensitivity to gain of each stage is set to zero and the sensitivity to noise and nonlinearity of all stages is equal.

4.5.2 REDUCED SECOND-ORDER SENSITIVITY APPROACH

In this approach, in addition to nullifying the first-order sensitivity to block-level gains in (4.30), the second-order sensitivity to the gain of front stages in (4.31) and (4.32) is reduced. To achieve that, α must be smaller than unity for every stage, resulting in

$$\frac{\overline{V}_{ni,k}^2}{\overline{V}_{ni,k-1}^2} < A_{vn,k-1}^2 \tag{4.41}$$

$$\frac{V_{IPqi,k}^{q-1}}{V_{IPqi,k-1}^{q-1}} > A_{vn,k-1}^{q-1}. \tag{4.42}$$

In contrast to the constant-sensitivity approach, in the reduced second-order sensitivity approach there is no unique solution for zeroing (4.30); that is, choosing any α smaller than unity is sufficient. In fact, making α smaller for each stage can further reduce the value of the second-order sensitivity to the gain of that stage. Theoretically, there is no lower limit for α and setting α to zero yields the lowest second-order sensitivity; however, in practice, obtaining very low values of α is difficult because it demands highly linear and low noise components, which are either impractical to implement or entail extremely high power consumptions.

In this approach, more attention is paid to the sensitivity of the NPD to the gain of individual blocks and no constraints are defined for sensitivity to the noise and nonlinearity of building blocks, because reducing the sensitivity of the NPD to the noise or nonlinearity of one stage increases the sensitivity of the NPD to the noise and nonlinearity of other stages, as described in the preceding subsection. Therefore, the overall impact of adjusting the sensitivities to block-level noise and nonlinearity on the robustness is not expected to be significant. Consequently, the reduced second-order sensitivity approach is expected to provide more robustness to process variations as compared to the constant-sensitivity approach.

As a side effect of reducing the second-order sensitivity of the NPD to the gain of the stages, the sensitivity to noise and nonlinearity of front stages (such as LNA) is increased in this approach, whereas the sensitivity to noise and nonlinearity of rear stages (like baseband filters) is reduced.

4.5.3 OPTIMUM-POWER DESIGN

In optimum-power design, as suggested in reference 3, a linear relationship is assumed between power consumption and dynamic range of the components:

$$P = \frac{V_{IP3i}^2}{V_{ni}^2} P_C \qquad (4.43)$$

where P is the power consumption of the component and P_C is the proportionality constant and is called power coefficient.

This assumption is valid in many RF components in specific operating regions [3]. Therefore, the resulting method is applicable solely in those regions. In this thesis, the optimum-power method is used as a means of comparison with the methods developed in this work so that we can verify that the methods aiming at robustness may also satisfy the requirements of optimum-power design. Optimum-power methodology proposes a block-level budgeting plan, which results in minimum power consumption for a given total noise and nonlinearity requirement [3]:

$$C_{Noise}(k) = \frac{\overline{V}_{ni,k}^2}{\prod_{j=0}^{k-1} A_{vn,j}^2} = \overline{V}_{ni,tot}^2 \times \frac{\sqrt[3]{P_{Ck}}}{\sum_{m=1}^{N} \sqrt[3]{P_{Cm}}} \qquad (4.44)$$

$$C_{Distortion,3}(k) = \frac{\prod_{j=0}^{k-1} A_{vn,j}^2}{V_{IP3i,k}^2} = \frac{1}{V_{IP3i,tot}^2} \times \frac{\sqrt[3]{P_{Ck}}}{\sum_{m=1}^{N} \sqrt[3]{P_{Cm}}} \qquad (4.45)$$

where P_{Ck} is the power coefficient of the kth stage and the total power consumption after optimization is obtained from (4.46). In this approach, third-order nonlinearity is assumed to be the dominant source of distortion and the other orders of nonlinearity are neglected.

$$P_{tot} = \frac{V_{IP3i,tot}^2}{V_{ni,tot}^2} \times \left(\sum_{m=1}^{N} \sqrt[3]{P_{Cm}} \right)^3 \qquad (4.46)$$

In the special case that the power coefficients of all blocks are equal, (4.44) and (4.45) simplify to (4.37) and (4.39), respectively, and the optimum-power design gives the same results as the constant-sensitivity design.

However, the guideline proposed in reference 3 does not determine the total noise and total nonlinearity of the receiver. As was mentioned in Section 4.1, the required BER determines the required noise plus distortion (NPD) but does not specify the contribution of the noise or nonlinearity to the total NPD. Thus, the designer needs to specify $V_{IP3i,tot}$ and $V_{ni,tot}$ in such a way that the total power consumption in (4.46) is minimized and the required NPD in (4.3) is satisfied. It can be proven that minimum power is achieved when the condition of (4.36) is met.

An inspection of (4.44)–(4.45) reveals that in this approach the noise and nonlinearity of consecutive stages have the following relationships:

$$\frac{\overline{V_{ni,k}^2}}{\overline{V_{ni,k-1}^2}} = A_{vn,k-1}^2 \times \frac{\sqrt[3]{P_{Ck}}}{\sqrt[3]{P_{Ck-1}}} \tag{4.47}$$

$$\frac{V_{IPqi,k}^{q-1}}{V_{IPqi,k-1}^{q-1}} = A_{vn,k-1}^{q-1} \times \frac{\sqrt[3]{P_{Ck-1}}}{\sqrt[3]{P_{Ck}}}. \tag{4.48}$$

This means that this approach also satisfies the first two conditions for zeroing the first-order sensitivity of NPD to the gain of stages as described in (4.33)–(4.34) and α is equal to the third root of the ratio of the power coefficients. Therefore, satisfying (4.36) is also necessary for zeroing the first-order sensitivity to the gains. Thus, selecting the total noise and nonlinearity of the receiver for minimum power leads to the nullification of the first-order sensitivity of NPD to block-level gains. This is an important advantage of the optimum-power design.

Based on (4.44) and (4.45), and (4.28) and (4.29), the sensitivity to the noise and nonlinearity of each block is directly proportional to the third root of its power coefficient—meaning that the overall system performance is more sensitive to power-hungry blocks.

In the rest of the chapter, the optimum-power approach refers to the herewith modified version of the method introduced in reference 3, which also satisfies the condition of (4.36).

In the course of development of this method, it is assumed that the noise and non-linearity performance of an RF circuit can be improved linearly and indefinitely by just increasing the power consumption and that, for constant power consumption, the noise and linearity performance can be traded for each other [3]. Due to the limited range of validity of these assumptions, the method should be used with caution.

4.5.4 SUMMARY AND DISCUSSION

The three approaches studied in this section are summarized in Table 4.3. Clearly, these approaches should be compared when their specified NPDs are equal. According to the second and last row of the table, the optimum-power design only satisfies the necessary conditions for reduced second-order sensitivity design if the power coefficients of the rear stages are smaller than those of the front stages. Alternatively, the optimum-power design only gives the same specifications of the constant-sensitivity approach if the power coefficients of all the stages are equal.

In the special case that the power coefficient of each block is smaller than that of its previous stage, the α of the optimum-power design is smaller than unity and the optimum-power design meets the conditions of the reduced second-order sensitivity approach. Accordingly, the second-order sensitivity of the NPD to the gain of a stage is higher if its following stages have larger power coefficients because, in optimum-power design, the blocks with larger power coefficients have higher contribution to the total noise (C_{Noise}) and nonlinearity distortion ($C_{Distortion}$); this can be verified by inspection of (4.19) and (4.21) and (4.47)–(4.48). In other words, for a stage followed by another one with a larger power coefficient, the value of α is larger

TABLE 4.3
Summary of the Three Explained Approaches

	Constant sensitivity	Reduced second-order sensitivity	Optimum power
Contribution factor (α_k)	= 1	<1	$= \dfrac{\sqrt[3]{P_{Ck+1}}}{\sqrt[3]{P_{Ck}}}$
First-order sensitivity to the gain of stages	0	0	0
Second-order sensitivity to the gain of front stages	High	Low	High if rear stages are power hungry; low if rear stages are low power
Sensitivity to noise and nonlinearity of each stage	Constant	High for front stages; low for rear stages	Proportional to third root of power coefficient

than unity, contradicting the requirements of the reduced second-order sensitivity approach. In fact, optimum-power design does not put any constraint on the position of large-power coefficient blocks, but having this sort of block at the front stages of the receiver would let the optimum-power design produce the same specifications as suggested by the reduced second-order sensitivity approach.

As a conclusion, the reduced second-order sensitivity approach is the ideal solution in terms of improving the robustness of a receiver to process variations. However, it is not always a practical solution, considering its implications for the power consumption. If the large-power coefficient blocks are located in the front stages of the receiver, the reduced second-order sensitivity approach can be used without excessive increase in the power consumption. However, if the large-power coefficient blocks are in the rear stages of the receiver, the optimum-power method is preferred as it meets the requirement for zeroing the first-order sensitivity to the gain of all the stages while yielding the minimum power consumption.

4.6   CASE STUDY

In this section we apply the design approaches described in Section 4.2.5 to 60 GHz zero-IF receivers. First, we investigate the case of a receiver without analog-to-digital converter (ADC) and then we study another receiver that includes an ADC.

4.6.1   RECEIVER WITHOUT ADC

The ADC-less 60 GHz zero-IF receiver shown in Figure 4.4 is considered to perform a comparison between different design approaches. A nonoptimum design based on a tested 60 GHz LNA and mixer is also included in the comparison. The power coefficients, used to estimate the power consumption in each case, and the block-level gains are extracted from simulated and measured circuits [7,8]. The requirements of the overall system are an SNDR of 12 dB and an MDS of −61.4 dBm, which result in an NPD of −73.4 dBm. The tolerable interference power at the input of the

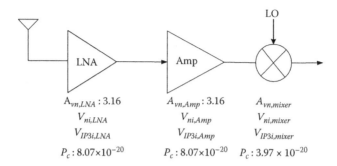

$A_{vn,LNA} : 3.16$ $A_{vn,Amp} : 3.16$ $A_{vn,mixer}$

$V_{ni,LNA}$ $V_{ni,Amp}$ $V_{ni,mixer}$

$V_{IP3i,LNA}$ $V_{IP3i,Amp}$ $V_{IP3i,mixer}$

$P_c : 8.07{\times}10^{-20}$ $P_c : 8.07{\times}10^{-20}$ $P_c : 3.97 \times 10^{-20}$

FIGURE 4.4   Three-stage ADC-less receiver used in the case study.

receiver is −33 dBm and the RF bandwidth is 2 GHz. For the sake of practicality, these requirements are based on state-of-the-art 60 GHz receivers [9]. Assuming that third-order intermodulation distortions are the dominant part of nonlinearities, (4.36), (4.3), (4.4), and (4.5) can be used to find the required noise figure and IIP3, which are 6 dB and −10 dBm, respectively. Block-level specifications obtained from each of the four design approaches are listed in Table 4.4. The input impedance of the receiver is assumed to be 50 Ω for all cases. All four designs satisfy the noise figure and IP3 requirements. The power consumption is estimated by [3]

$$P_{tot} = \sum_{m=1}^{N} \frac{V_{IP3i,m}^2}{\overline{V}_{ni,m}^2} P_{Cm}. \tag{4.49}$$

The variation of NPD as a function of the variations in the gain of the LNA, for the four approaches, is illustrated in Figure 4.5. The gain of the LNA is varied by ±20% around its nominal value.

Except for the nonoptimum design, all the approaches result in zero first-order sensitivity of the NPD to the gain of the LNA, whereas the reduced second-order sensitivity design shows minimum variation of NPD with respect to gain deviations

---

TABLE 4.4

Block-Level Specifications of the Four Design Approaches

Parameter	Constant sensitivity	Reduced second-order sensitivity	Optimum power	Nonoptimum
$V^2_{ni,LNA}$ (V²/Hz)	2.06×10^{-19}	3.599×10^{-19}	2.6×10^{-19}	2.896×10^{-19}
$V^2_{ni,Amp}$ (V²/Hz)	2.06×10^{-18}	1.542×10^{-18}	2.6×10^{-18}	1.274×10^{-18}
$V^2_{ni,Mixer}$ (V²/Hz)	2.06×10^{-17}	1.028×10^{-17}	9.56×10^{-18}	1.995×10^{-17}
$V^2_{IP3i,LNA}$ (V²)	0.015	0.0086	0.0118	0.1256
$V^2_{IP3i,Amp}$ (V²)	0.15	0.2000	0.118	0.32
$V^2_{IP3i,Mixer}$ (V²)	1.5	3.0001	3.2	0.621
Estimated power (mW)	12	13.6	8.6	55

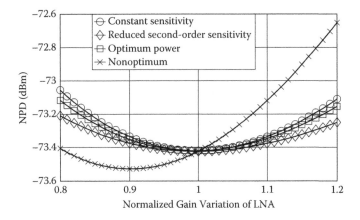

FIGURE 4.5 NPD versus normalized variation of LNA gain for the four different designs.

in the LNA. In this example the second-order sensitivity to gain in optimum-power design is not much higher than the one in the reduced second-order sensitivity approach because the power coefficient of the final stage is very low and therefore its noise and nonlinearity contribution are low. Please note that because only the gain of one stage is varied in Figure 4.6, the variations of NPD are small. The value of α in the case of reduced second-order sensitivity approach is 0.43. If smaller values of α could be obtained, smaller second-order sensitivities would be possible.

    In Figure 4.7 and Figure 4.8 the statistical behavior of the four designs of Table 4.4 is compared. Noise, linearity, and gain of all the stages are randomly changed in an interval of ±20% from their nominal values. Using MATLAB, half a million samples are made in this way for each design and the NPD of each sample is calculated.

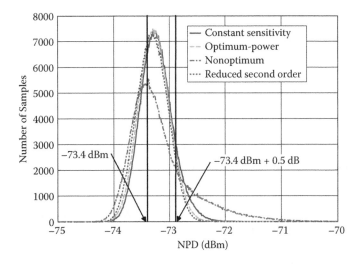

FIGURE 4.6 Comparing the four different designs in terms of number of random samples versus NPD.

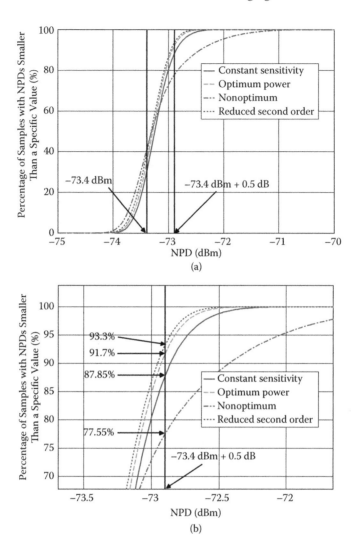

FIGURE 4.7 Comparing the four different designs in terms of (a) percentage of random samples with smaller than a specific NPD; (b) zoomed view of (a).

The range of NPDs between −72 and −68 dBm is divided into intervals of 0.01 dB and the number of samples falling in each interval is shown in Figure 4.7. The non-optimum design clearly shows the highest number of out-of-specification samples, whereas reduced second-order sensitivity design shows the lowest number of out-of-specification samples. Figure 4.7 shows the percentage of samples with NPDs smaller than a specific value. For instance the percentage of samples with NPDs more than 0.5 dB higher than nominal is 12.15%, 8.3%, 22.45%, and 6.7% for constant sensitivity, optimum-power, nonoptimum, and reduced second-order sensitivity designs, respectively. The percentage of samples with NPDs more than 1 dB higher than nominal is 0.59%, 0.09%, 9.58%, and 0.03% for the aforementioned designs

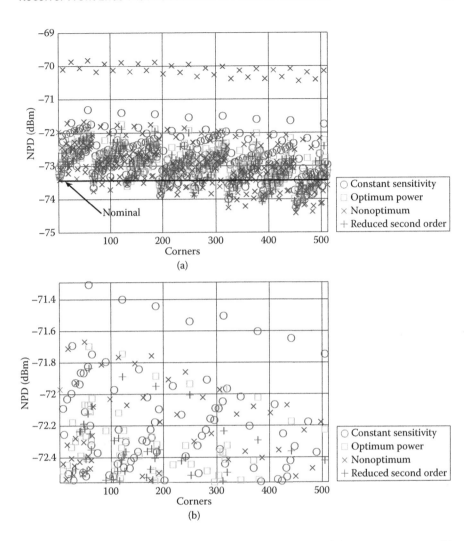

FIGURE 4.8 (a) NPDs of designs of Table 4.3 on the corners of the parameter space; (b) zoomed view.

respectively. The reduced second-order design achieves 99.9% yield if the acceptable limit for NPD is set to −72.5 dBm, whereas the nonoptimum design achieves such yield if that limit is set to −70.7 dBm. If one decides to compensate for this difference by overdesigning the nonoptimum case, one has to shift the nominal NPD by 1.8 dB, which means that $V^2_{ni,tot}$ must be improved by at least a factor of 1.5 and $1/V^2_{IP3i,tot}$ by at least a factor of 1.22, which in turn increases the power consumption by at least a factor of 1.84 (84%).

The receiver has three stages and each stage has three parameters: noise, linearity, and gain. If each parameter can vary by ±20% around its nominal value, a nine-dimensional parameter space is formed with 2^9 corners. Figure 4.9 shows the NPD of each of the four designs of Table 4.4 on the corners of this parameter space.

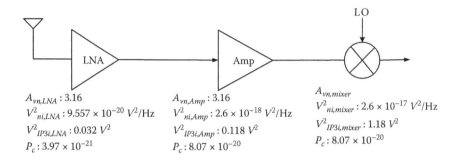

$A_{vn,LNA}$: 3.16

$V^2_{ni,LNA}$: 9.557×10^{-20} V^2/Hz

$V^2_{IP3i,LNA}$: 0.032 V^2

P_c : 3.97×10^{-21}

$A_{vn,Amp}$: 3.16

$V^2_{ni,Amp}$: 2.6×10^{-18} V^2/Hz

$V^2_{IP3i,Amp}$: 0.118 V^2

P_c : 8.07×10^{-20}

$A_{vn,mixer}$

$V^2_{ni,mixer}$: 2.6×10^{-17} V^2/Hz

$V^2_{IP3i,mixer}$: 1.18 V^2

P_c : 8.07×10^{-20}

FIGURE 4.9   Optimum-power three-stage receiver with different power coefficients.

Apparently, some corners yield better performance than the others. However, on the corners that give the worst performance, the nonoptimum design has the poorest performance level. It can be seen in Figure 4.8 that the nonoptimum design shows at least 2 dB more degradation of NPD at some corners than the reduced second-order sensitivity design, which achieves the best corner performance. This means that, to achieve the same corner performance from the nonoptimum and the reduced second-order design, one has to improve the NPD of the nonoptimum design by 2 dB. This means that $V^2_{ni,tot}$ must be improved by at least a factor of 1.58 and $1/V^2_{IP3i,tot}$ by at least a factor of 1.26, which in turn increases the power consumption by at least a factor of 2 (100%).

As mentioned before, the reduced second-order sensitivity approach is more focused on reducing the sensitivity of the NPD to the gains, whereas the constant-sensitivity approach puts constraints on the sensitivity of the NPD to the noise and nonlinearity. Comparing these two approaches in Figure 4.6, Figure 4.7, and Figure 4.8 reveals that the sensitivity of the NPD to gain variations of individual stages is more influential on the corner and statistical performance than the sensitivity of the NPD to the noise and nonlinearity of individual stages. This is because, for a constant total noise and nonlinearity, reducing the sensitivity of the NPD to the noise or nonlinearity of one stage increases the sensitivity to the noise or nonlinearity of other stages, whereas the sensitivity to the gain of all the stages can be reduced simultaneously.

The reason that the optimum-power design is showing a better corner performance in the preceding example than the constant-sensitivity design is that the NPD has a lower second-order sensitivity to the gain, which in turn is caused by the low power coefficient of the last stage (mixer). However, if the power coefficients are modified as in Figure 4.9, the optimum-power design will be more sensitive to block-level gains. The NPD histograms for both the optimum-power design of Figure 4.4 and that of Figure 4.9 are shown in Figure 4.10. The percentage of samples with NPDs more than 0.5 dB higher than nominal has increased from 8.3% to 18.57%.

Therefore, simultaneous optimum robustness and optimum power can be achieved only if the power coefficients of the rear stages are small compared to those of the front stages.

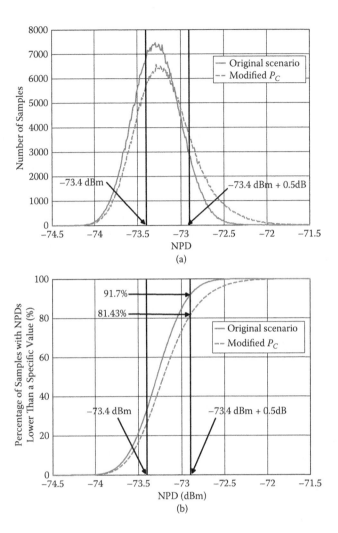

FIGURE 4.10   (a) Number of random samples versus NPD for two optimum-power designs; (b) percentage of random samples with smaller than a specific NPD.

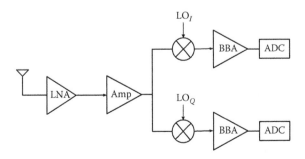

FIGURE 4.11   60 GHz zero-IF receiver including ADC.

4.6.2 RECEIVER WITH ADC

Now we investigate the possibility of applying the design approaches described in Section 4.5 to a 60 GHz receiver including ADC (shown in Figure 4.11). Considering the performance of state-of-the-art components in the 60 GHz band, the system requirements of a receiver including an ADC must be much more relaxed than a system without an ADC. Due to the inefficiency of traditional ADC/DSP approaches at gigahertz bandwidths, many designers tend to eliminate the ADC from the system and replace it with an analog demodulator or a mixed-signal circuit [9,10]. In this system-level study, a state-of-the-art ADC reported in reference 11 is used. The ADC has a sampling rate of 2.5 GS/s, ENOB of 5.4 bits, VDD of 1.1 V, SFDR of −43 dBc, and power consumption of 50 mW. Translating these parameters into the RF domain [12] results in a V^2_{ni} of 4.53×10^{-14} V²/Hz and V^2_{IP3i} of 5. The parameters of the other components are listed in Table 4.5. These parameters are realistic and consistent with state-of-the-art 60 GHz receiver components [7,8,11].

Using these components, a total noise figure of 12.9 dB and a total IIP3 of −22.8 dBm can be achieved from the receiver. The total power consumption of the receiver will be 155 mW. Assuming an SNDR of 12 dB and a P_{IMD3} 3 dB below the noise level, the minimum detectable signal for this receiver would be −54.4 dBm. Assuming an output power of 10 dBm for the desired transmitter and an antenna gain of 10 dBi for the transmitter and receiver, this MDS associates with a transmission distance of 6.6 m. Considering the total IP3 of the receiver, the maximum tolerable interferer power at the input of the receiver is −38.9 dBm. Assuming an interfering transmitter with an output power of 10 dBm and residing in a 60° direction, making the gain obtained from antenna 7 dBm, the tolerable interference distance will be 56 cm.

Doing an optimum-power design according to Table 4.6 requires very high performance components, which have not been reported in the literature yet. For example, the required noise figure of the LNA is 2.3 dB, which is well below the best reported figures in the literature [7]. This is due to the fact that the power coefficient of the ADC is much higher than that of the LNA, which results in relaxed specifications for the ADC and demanding specifications for the LNA by the optimum-power

---

TABLE 4.5
Component Parameters for a Nonoptimum Design Including ADC
of Reference 11

Parameter	LNA	Amp	Mixer	BBA	ADC
V^2_{ni} ((nV)²/Hz)	0.2896	1	20	100	4.53×10^4
V^2_{IP3i} (V²)	0.1256	0.1256	0.621	1	5
A_{vn}	3.16	3.16	1.3	10	—
NF (dB)	3.8	7.7	19.9	26.8	53.4
IP3 (dBm)	4	4	10.9	13	20
P_c (aW/Hz)	0.0807	0.0807	0.000794	1	906
P (mW)	35	10	0.25	10	100

TABLE 4.6

Component Parameters for an Optimum-Power Design Including ADC of Reference 11

Parameter	LNA	Amp	Mixer	BBA	ADC
V^2_{ni} ((nV)^2/Hz)	0.1421	1.421	6.56	55.6	5.38×10^4
V^2_{IP3i} (V^2)	0.00707	0.0707	1.53	0.52	5.3
A_{vn}	3.16	3.16	1.3	10	—
NF (dB)	2.3	9	15	24.3	54
IP3 (dBm)	−8.5	1.5	14.9	10	20
P_c (aW/Hz)	0.0807	0.0807	0.000794	1	906
P (mW)	4	4	0.19	9.4	89

approach. However, the resulting specifications for the LNA are not realistic. Therefore, even with such a relaxed overall requirement, optimum-power design is not yet feasible with state-of-the-art CMOS building blocks. Nevertheless, the total power consumption of a receiver designed based on Table 4.6 would be 106.6 mW, which is 31% lower than the receiver in Table 4.5.

This unrealistic prediction by the optimum-power method is due to some assumptions made in the course of development of the method that are not always valid. The assumptions are that the noise and nonlinearity performance of an RF circuit can be improved linearly and indefinitely by just increasing the power consumption and that, for constant power consumption, the noise and linearity performance can be traded for each other, resulting in the prediction of an LNA with 4 mW power consumption and 2.3 dB noise figure. Since these figures are not realistic, we can conclude that these assumptions are not valid in this case or that the case is beyond the validity region of the assumptions.

The optimum power design shown in Table 4.6 has zero first-order sensitivity to the block-level gains, provided that the specified worst-case interferer is −38.9 dBm, satisfying the requirement of (4.36). Designing for low second-order sensitivity to block-level gain variations requires the rear stages of the receiver to have lower contribution to the total noise and nonlinearity distortion. However, since in this case the power coefficient of the ADC is much higher than that of the other components, demanding better noise and linearity performance from the ADC dramatically increases the total power consumption of the receiver.

Although it does not meet the required sampling rate, another ADC reported in the literature has a much better power consumption [13]. With 6.7 mW power consumption, 1 GS/s sampling rate, ENOB of 5, and SNDR of 31.5 dB, this ADC can be considered for a 60 GHz receiver with only half the nominal IF bandwidth (500 MHz). The specifications converted to RF domain are shown in Table 4.7. Using the ADC with the same RF components of Table 4.5 results in a receiver with a power consumption of 68.7 mW, which is less than half of the power consumption of the receiver of Table 4.5. The resulting total NF and IP3 are 18 and −22 dBm, respectively. Although the power consumption is much better, the noise performance is inferior to that of the receiver of Table 4.5. This is not surprising because, although

TABLE 4.7

Component Parameters for a Nonoptimum Design Including ADC of Reference 13

Parameter	LNA	Amp	Mixer	BBA	ADC
V^2_{ni} ((nV)^2/Hz)	0.2896	1	20	100	2.07×10^5
V^2_{IP3i} (V^2)	0.1256	0.1256	0.621	1	5.7
A_{vn}	3.16	3.16	1.3	10	—
NF (dB)	3.8	7.7	19.9	26.8	60
IP3 (dBm)	4	4	10.9	13	20.6
P_c (aW/Hz)	0.0807	0.0807	0.000794	1	487
P (mW)	35	10	0.25	10	13.4

TABLE 4.8

Component Parameters for an Optimum-Power Design Including ADC of Reference 13

Parameter	LNA	Amp	Mixer	BBA	ADC
V^2_{ni} ((nV)^2/Hz)	0.5897	5.897	12.63	231	1.81×10^5
V^2_{IP3i} (V^2)	0.00682	0.0682	3.181	0.5	6.33
A_{vn}	3.16	3.16	1.3	10	—
NF (dB)	3.85	14.7	17.9	30.5	59.4
IP3 (dBm)	−8.6	1.35	18	10	21
P_c (aW/Hz)	0.0807	0.0807	0.000794	1	487
P (mW)	1	1	0.2	2	17

the power consumption of the ADC of reference 13 is 7.5 times smaller than that of reference 20, its power coefficient is only 0.54 of that of reference 11. An optimum-power design with the ADC of reference 13 results in the specifications given by Table 4.8, which are feasible as compared to the specifications of Table 4.6. Please note that the overall performance parameters are still 18 dB of NF and −22 dBm of IP3, which are relaxing compared to those of Table 4.6.

The field of ADC design for high speed communication is experiencing rapid progress and new high performance ADCs are reported in the literature that may soon make the ADC/DSP approach possible for millimeter-wave communication links [14].

4.7   ADJUSTMENT OF PROCESS VARIATION IMPACT

According to the presented system-level analysis, the overall performance of a receiver is more sensitive to the noise and nonlinearity of building blocks, which contribute more to the total noise and nonlinearity. Therefore, accumulating the noise and nonlinearity contributions in one stage has the advantage of accumulating the sensitivities in that stage (i.e., the overall performance of the receiver becomes more sensitive to the noise and nonlinearity of that single stage and less sensitive to the noise and nonlinearity of the other stages). This way, the performance degradations resulting from process variations can be compensated mostly by tuning the performance of that single stage. The ideal location for such a stage is at the input of the receiver (i.e., the LNA), resulting in lowered second-order sensitivity to the gains of the stages. In addition, an LNA with tuneable parameters can provide the possibility of input impedance corrections. However, implementing a tuneable LNA is not a trivial task. Furthermore, it is not always easy to accumulate the contributions in the first stage. In the following we compare two strategies of accumulating the contributions in the LNA or in the mixer in terms of feasibility, costs, and benefits.

4.7.1 TUNEABILITY IN DIFFERENT LNA TOPOLOGIES

Figure 4.12 shows three different LNA topologies: a common-gate LNA, an inductively degenerated common-source LNA, and a voltage–voltage transformer-feedback LNA. An inspection of these schematics suggests that the performance of these circuits is determined not only by the operating point and the parasitic of the

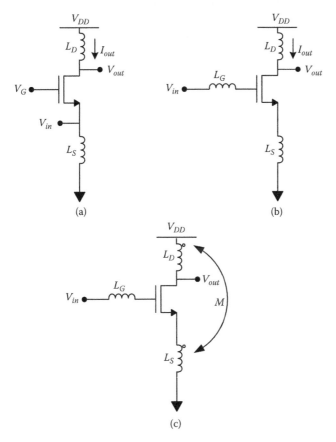

FIGURE 4.12 (a) A common-gate LNA, (b) an inductively degenerated common-source LNA, and (c) a voltage–voltage transformer-feedback LNA.

transistor but also by the value of the inductors. Therefore, in general, one needs to implement tuneable passives, as well as variable biasing for transistors, to deal with the resulting performance degradations. For instance, the input-referred noise voltage of an inductively degenerated common-source LNA is derived from the following:

$$\overline{v_{ni}^2} = \frac{4KT\gamma g_{d0}}{g_m^2} \frac{\left(R_S^2 + \left((L_G + L_S)\omega - \frac{1}{c_{gs}\omega}\right)^2\right)\left(\frac{g_m^2 L_S^2}{c_{gs}^2} + \left((L_G + L_S)\omega - \frac{1}{c_{gs}\omega}\right)^2\right)c_{gs}^2\omega^2}{\left((L_G + L_S)\omega - \frac{1}{c_{gs}\omega}\right)^2 + \left(\frac{g_m L_S}{c_{gs}} + R_S\right)^2} \tag{4.50}$$

According to (4.50), correcting the performance degradations resulting from the variability of c_{gs} and g_m (optimistically not including L_G and L_S) requires using variable biasing and a varactor. Due to the low quality factor of varactors at high frequencies, the noise figure of the LNA would be degraded substantially by using varactors. If the inductors are also prone to variations due to process spreading, variable inductors turn into a necessity. Variable inductors can be implemented using active inductors [15]. However, such circuits fail to behave inductively at millimeter-wave frequencies. In addition, even at low frequencies they tend to deteriorate the noise performance due to the additional active circuitry. Using varactors in parallel with the inductors can also change the effective value of the inductance. However, as mentioned before, this method suffers from the low quality factor of the varactors at high frequencies, which results in unacceptable levels of noise performance.

4.7.2 Tuneability in Other Stages

If, for some reason, we fail in providing tuneability to the LNA or rendering the LNA the dominant stage in terms of noise and nonlinearity contribution, it is still possible to design other stages, such as mixer, with such properties. In general, as explained before, the overall performance of the receiver is more sensitive to the performance of the stages with highest contribution to the total noise and nonlinearity. Therefore, if the noise and nonlinearity contribution of any stage is dominant compared to that of the other stages, it is practically possible to compensate the effect of process variations by just tuning the performance of that stage. Nevertheless, we know from the system-level analysis that the ideal situation is to accumulate all the contributions in the LNA; only if achieving this ideal becomes infeasible can we opt for accumulation of the contribution, and hence the sensitivities, in other stages.

Furthermore, it is worth noting that achieving the best performance is the first priority for every designer. Therefore, the performance should never be compromised for accumulating the sensitivities in one stage. In other words, any design that provides such property and does not give the required performance is obviously not acceptable.

Choosing the mixer or another IF stage for tuning obviates the need of dealing with delicate, high frequency components. In addition, as we will see later, it is much easier to accumulate the noise and nonlinearity distortion contributions in the mixer.

4.7.3 Overall Design Considerations

In the previous subsections, we compared the difficulties of implementing tuneability in the LNA as a high frequency component, and mixer, as a circuit partly located at the low frequency side. In this subsection, we investigate the complexities of accumulating the contributions in the LNA and mixer. A three-stage direct-conversion 60 GHz receiver, shown in Figure 4.13, is considered as a test vehicle for this study.

Two different design strategies are followed and compared. Ideally, as explained in Section 4.5, the ratio between the noise contribution of each stage to that of its following stage must be equal to the ratio between the distortion contribution of the stage to that of its following stage. This, along with an additional condition expressed by (4.35), assures a zero first-order sensitivity of the overall performance to the gains of the individual stages. In the first strategy, represented by receiver 1, the noise and nonlinearity contributions are accumulated in the last stage (mixer). On the other hand, as explained in Section 4.5.2, to reduce the second-order sensitivity of the overall performance to the individual gains, the noise and distortion contribution of the stages must be reduced in a step-wise manner as we move from the front stages to the rear ones. An attempt is made toward reaching this ideal situation in the second strategy. In this strategy, represented by receiver 2, the focus is toward accumulating the noise and nonlinearity contributions in the first stage (LNA).

An inductively degenerated common-source LNA, shown in Figure 4.12(a), and a single-balanced mixer, shown in Figure 4.14, are used in both designs as the first

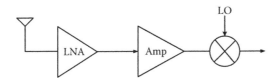

FIGURE 4.13 Three-stage receiver used as the test vehicle in the study.

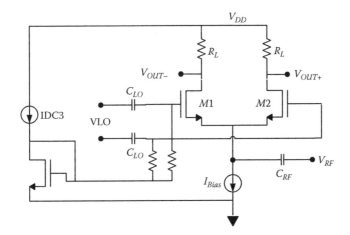

FIGURE 4.14 Single-balanced mixer used in receiver 1 and receiver 2.

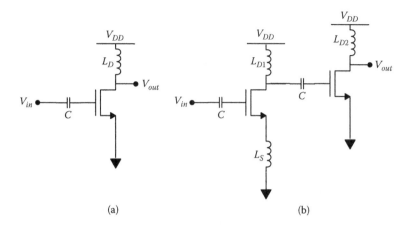

FIGURE 4.15   The second stage in (a) receiver 1, and (b) receiver 2.

and third stages, respectively. Apart from different component parameters in the first and third stages of receiver 1 and receiver 2, the two receivers use two different circuit topologies as the second stage. A single-stage common-source amplifier is used as the second stage of receiver 1, whereas a two-stage amplifier consisting of an inductively degenerated common-source stage cascaded with a common-source stage is used as the second stage of receiver 2, as shown in Figure 4.15. As a result, the second stage of receiver 2 provides a higher gain.

Figure 4.16 shows the noise and nonlinearity distortion contribution of each stage in receiver 1. According to Figure 4.16(b), the mixer has the highest contribution to the total nonlinearity distortion. Figure 4.16(a) shows that the mixer and LNA contribute to the total noise almost equally, whereas the noise contribution made by the second stage is smaller. This means that the noise contribution and nonlinearity distortion contribution are mostly accumulated in the mixer. One way to shift this accumulation to the first stage is to design a much more linear and low noise mixer, which is not realistic. Increasing the gain of the second stage can reduce the noise contribution of the mixer. However, it increases the nonlinearity distortion contribution of the mixer.

Figure 4.17 shows the noise and nonlinearity distortion contribution of each stage in receiver 2. The noise contribution of the mixer is well diminished by increasing the gain of the second stage, while its nonlinearity distortion contribution is further increased. Therefore, only receiver 1 has the desired characteristic of accumulating the noise and nonlinearity distortion contribution in one stage. The main reason for not being able to accumulate the contributions in the LNA stems from the very high nonlinearity contribution of the mixer, which cannot be reduced below the nonlinearity contribution of the LNA. However, this is not always true and the possibility of simultaneous accumulation of the noise and nonlinearity contribution in the LNA depends on many factors, such as the utilized circuit topologies, the application, the specifications, and the frequency of operation.

Based on Figure 4.16 and Figure 4.17, neither of the two receivers conforms to the aforementioned required conditions for zero sensitivity to the gains of individual

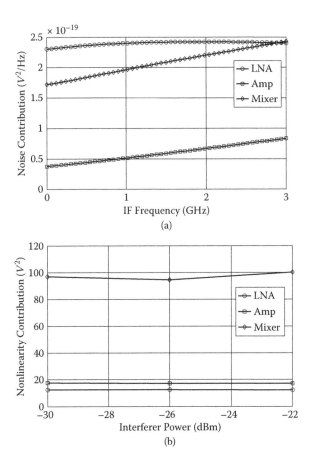

FIGURE 4.16 Contribution of each component in receiver 1 to (a) total noise, and (b) total nonlinearity.

stages. For instance, in receiver 1, the LNA and the mixer have the highest noise contribution, while the LNA shows the lowest distortion contribution. Also, in receiver 2, the LNA has the highest noise contribution and it has the lowest distortion contribution. It appears that the combination of noise and nonlinearity for the blocks, as proposed by the system-level design for zero sensitivity to the gains, is not easily achievable at the circuit level. However, our attempt is to get as close as possible to the ideal situation sketched by the system-level design. Consequently, the desired property of accumulating the noise and nonlinearity distortion contribution in one stage, which is also proposed by the system-level method for facilitating the adaptability, is achieved in receiver 1.

The noise and nonlinearity performances of the two designs are simulated on the process corners and the results are listed in Table 4.9. The power consumption is kept constant on all process corners by keeping the biasing current constant. The

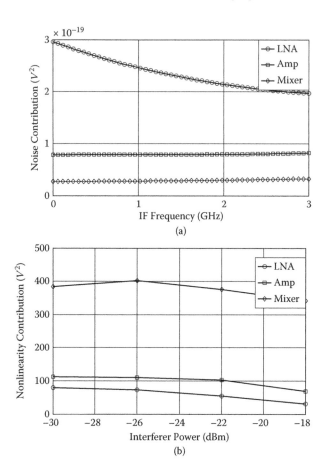

FIGURE 4.17   Contribution of each component in receiver 2 to (a) total noise, and (b) total nonlinearity.

simulations are performed with SPECTRE RF periodic steady state, periodic noise, and periodic AC analysis. The VDD is 1.2 V. The LO amplitude is 0 dBm and the LO frequency is 60 GHz. In the next subsection, we investigate the possibility of correcting the performance of receiver 1 on its corners by tuning the performance of the mixer as the stage with accumulated noise and distortion contributions.

4.7.4   CORRECTING THE CORNER PERFORMANCE

Now that we have a receiver with accumulated noise and nonlinearity distortion contribution in the mixer (receiver 1), the receiver performance degradations resulting from the process variations are expected to be (at least partly) correctable by just tuning the performance of the mixer. This idea is validated by further simulations performed on receiver 1. According to Table 4.9, the highest degradation of the IP3 of receiver 1 occurs at the slow–fast (sf) corner, while the NF is slightly improved at this corner. Choosing IDC3 (used in the biasing of the mixer) as the tuning parameter,

TABLE 4.9

Simulation Results of the Two Designs on Process Corners

	Receiver 1					Receiver 2				
	tt	ss	ff	sf	fs	tt	ss	ff	sf	fs
NF (dB)	6.54	6.61	6.45	6.46	6.47	5.55	5.6	5.45	5.55	5.56
IP3 (dBm)	−8.6	−8.1	−8.9	−13.5	−10.2	−15.8	−15.1	−15.8	−18.9	−17.5
NPD (dBm)	−74.1	−74.2	−74.1	−73.5	−74.1	−73.2	−73.6	−73.2	−69.8	−71.6
P_{DC}		12.3 mA × 1.2 V					18.4 mA × 1.2 V			

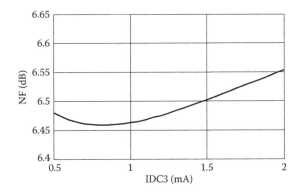

FIGURE 4.18 Noise figure of receiver 1 at sf corner versus IDC3: NF is back to its typical value when IDC3 is 1.87 mA, whereas the typical value of IDC3 is 1 mA.

the corresponding performance variations are simulated. Figure 4.18 shows the variations of the receiver noise figure as a function of IDC3 at the slow–fast corner. The typical value of IDC3 is 1 mA. According to Figure 4.18, the noise figure has its typical value when IDC3 is 1.87 mA. Then Figure 4.19(b) shows that the IP3 is corrected to above its typical value when IDC3 is 1.87 mA.

Figure 4.20 shows the variations of the receiver noise figure as a function of IDC3 at the fast–slow corner. According to Figure 4.20, the noise figure has its typical

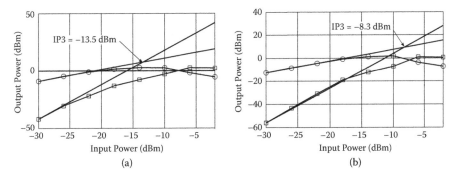

FIGURE 4.19 IP3 of receiver 1 at sf corner for (a) IDC3 of 1 mA, and (b) IDC3 of 1.87 mA.

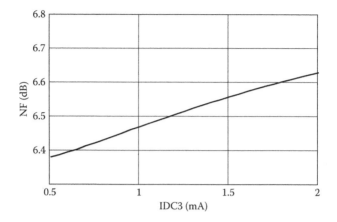

FIGURE 4.20 Noise figure of receiver 1 at fs corner versus IDC3: NF is back to its typical value when IDC3 is 1.5 mA, whereas the typical value of IDC3 is 1 mA.

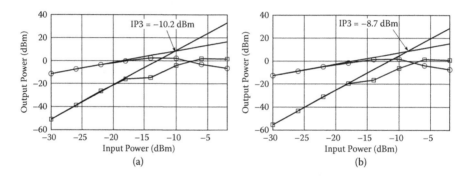

FIGURE 4.21 IP3 of receiver 1 at fs corner for (a) IDC3 of 1 mA, and (b) IDC3 of 1.5 mA.

value when IDC3 is 1.5 mA. Then Figure 4.21(b) shows that the IP3 is corrected to slightly less than its typical value when IDC3 is 1.5 mA.

As a conclusion, these tests validate the possibility of correcting the performance of the whole receiver on the process corners by tuning the performance of only a single stage. This is possible because the noise and distortion contributions are accumulated in that single stage (in this case, the mixer). This possibility can greatly facilitate the correction of process-induced performance degradations in a smart receiver, as it confines the required number of tuneable parameters.

4.7.5 LAYOUT IMPACT APPROXIMATED

To obtain a more accurate estimate of the receiver performance, the impact of the layout is approximated by adding more parasitic resistance and reactance at some specific points. The values of the additional parasitic are calculated based on prior experience with millimeter-wave layout and postlayout simulations. The RF lines, ground lines, and some of the DC lines are modeled by RLC π-networks. The

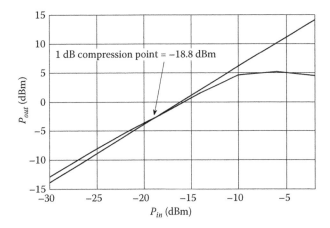

FIGURE 4.22   One-decibel compression point of receiver 1 after approximating the layout impact.

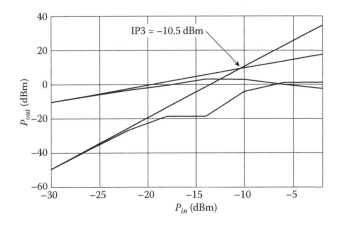

FIGURE 4.23   IP3 of receiver 1 after approximating the layout impact.

resulting typical performance parameters are shown in Figure 4.22, Figure 4.23, and Figure 4.24.

Comparing these results with Table 4.9 reveals that the noise performance is hardly degraded after adding the parasitic and the total IP3 is degraded by just 2 dB.

4.8   CONCLUSIONS

Based on a system-level sensitivity analysis performed on a generic RF receiver, it has been shown that the first-order sensitivities of the overall performance, represented by NPD, to the individual gains of the blocks can all be made zero. Applying the analysis to a zero-IF three-stage 60 GHz receiver shows a significant improvement in the design yield. A quantity called contribution factor is defined as the noise

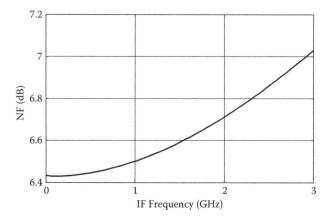

FIGURE 4.24   NF of receiver 1 after approximating the layout impact.

and nonlinearity distortion contribution of each stage with respect to that of its previous stage. Reduction of the second-order sensitivity of the NPD to the gain of individual stages, by keeping the contribution factor of all the stages below one, results in further improvements in the design yield. The conventional optimum-power design methodology has been modified in a way that it nullifies the first-order sensitivities of NPD to the individual gains of all the stages. It has been shown that simultaneous optimum power and optimum robustness can be achieved by using less power-hungry components at the rear stages of the receiver. Applying the analysis to a 60 GHz receiver including ADC shows that state-of-the-art ADCs are not adequate for optimum-power or optimum-robustness receiver design at 60 GHz.

A receiver is designed with good noise and nonlinearity performance and with accumulated noise and nonlinearity distortion contribution in its last stage (mixer). As a result, the overall performance of the receiver is more sensitive to the performance variations of the mixer. Simulations show that it is possible to correct the overall receiver performance degradations resulting from process variations by just tuning the performance of the mixer. In fact, these simulation tests validate the possibility of correcting the performance of the whole receiver on the process corners by tuning the performance of only a single stage (in this case, the mixer and only one parameter of the mixer). This possibility can greatly facilitate the correction of process-induced performance degradations in a smart receiver, as it confines the required number of tuneable parameters. It can also facilitate the performance trimming of the fabricated chips in the production line. The circuits are also simulated with additional parasitics calculated based on previous design experience to approximate the impact of interconnects added during the layout.

REFERENCES

1. L.-T. Pang, K. Qian, C. J. Spanos, and B. Nikolic, Measurement and analysis of variability in 45 nm strained-Si CMOS technology. *IEEE Journal of Solid State Circuits,* vol. 44, no. 8, pp. 2233–2243, Aug. 2009.

2. M. D. Meehan and J. Purviance, *Yield and reliability in microwave circuit and system design*. Boston: Artech House, 1993.

3. W. Sheng, A. Emira, and E. Sanchez-Sinencio, CMOS RF receiver system design: A systematic approach. *IEEE Transactions on Circuits and Systems-I: Regular Papers,* vol. 53, no. 5, pp. 1023–1034, May 2006.

4. P. Baltus, Minimum power design of RF front ends. PhD dissertation, Eindhoven University of Technology, 2004.

5. M. El-Nozahi, E. Sanchez-Sinencio, and K. Entesari, Power-aware multiband–multistandard CMOS receiver system-level budgeting. *IEEE Transactions on Circuits and Systems II: Express Briefs,* vol. 56, no. 7, pp. 570–574, July 2009.

6. Part 15.3: Wireless medium access control (MAC) and physical layer (PHY) specifications for high rate wireless personal area networks (WPANs): Amendment 2: Millimeterwave based alternative physical layer extension. IEEE 802.15.3c, Oct. 2009.

7. E. Janssen, R. Mahmoudi, E. van der Heijden, P. Sakian, A. de Graauw, R. Pijper, and A. van Roermund, Fully balanced 60 GHz LNA with 37% bandwidth, 3.8 dB NF, 10 dB gain and constant group delay over 6 GHz bandwidth. *10th Topical Meeting on Silicon Monolithic Integrated Circuits in RF Systems,* Jan. 2010.

8. P. Sakian, R. Mahmoudi, E. van der Heijden, A. de Graauw, and A. van Roermund, Wideband cancellation of second order intermodulation distortions in a 60 GHz zero-IF mixer. *11th Topical Meeting on Silicon Monolithic Integrated Circuits in RF Systems,* Jan. 2011.

9. A. Tomkins, R. A. Aroca, T. Yamamoto, S. T. Nicolson, Y. Doi, and S. P. Voinigescu, A zero-IF 60 GHz 65 nm CMOS transceiver with direct BPSK modulation demonstrating up to 6 Gb/s data rate over a 2 m wireless link. *IEEE Journal of Solid State Circuits,* vol. 44, no. 8, pp. 2085–2099, Aug. 2009.

10. C. Marcu, D. Chowdhury, C. Thakkar, J.-D. Park, L.-K. Kong, M. Tabesh, Y. Wang, et al., A 90 nm CMOS low-power 60 GHz transceiver with integrated baseband circuitry. *IEEE Journal of Solid State Circuits,* vol. 44, no. 12, pp. 3434–3447, Dec. 2009.

11. E. Alpman, H. Lakdawala, L. R. Carley, and K. Soumyanath, A 1.1V 50 mW 2.5 GS/s 7 b time-interleaved C-2C SAR ADC in 45 nm LP digital CMOS. *IEEE International Solid State Circuits Conference,* Feb. 2009.

12. W. Deng, R. Mahmoudi, P. Harpe, and A. van Roermund, An alternative design flow for receiver optimization through a trade-off between RF and ADC. *IEEE Radio Wireless Symposium,* Jan. 2008.

13. J. Yang, T. Lin Naing, and R. W. Brodersen, A 1 GS/s 6 bit 6.7 mW successive approximation ADC using asynchronous processing. *IEEE Journal of Solid State Circuits,* vol. 45, no. 8, pp. 1469–1478, Aug. 2010.

14. B. Verbruggen, J. Craninckx, M. Kuijk, P. Wambacq, and G. Van der Plas, A 2.6 mW 6 b 2.2 GS/s 4-times interleaved fully dynamic pipelined ADC in 40 nm digital CMOS. *IEEE International Solid-State Circuits Conference Digest of Technical Papers,* pp. 296–298, Feb. 2010.

15. C.-L. Ler, A. K. bin A'ain, and A. V. Kordesch, Compact, high-Q, and low-current dissipation CMOS differential active inductor. *IEEE Microwave and Wireless Components Letters,* vol. 18, no. 10, pp. 683–685, Oct. 2008.

5 Ultrawide Bandwidth Radio Frequency Identification and Localization

Davide Dardari

CONTENTS

5.1 INTRODUCTION

Radio-frequency identification (RFID) technology for use in real-time object identification is facing a rapid adoption in several fields, such as logistic, automotive, surveillance, and automation systems, etc. [1,2]. An RFID system consists of readers and tags applied to objects. The reader interrogates the tags via a wireless link to obtain the data stored on them. Tags equipped with a complete radio-frequency (RF) transmitter are denoted as *active*. The cheapest RFID tags with the largest commercial potential are *passive* or *semipassive,* where the energy necessary for tag-reader communication is harvested from the reader's signal or the surrounding environment. In current ultrahigh frequency (UHF) (semi)passive RFID technology, the reader sends an unmodulated continuous wave (CW) field that the tags modulate and scatter back to the reader. The backscatter modulation is realized by properly changing the tag's antenna load according to the data so that no active RF transmission happens [3].

It is expected that the global revenues coming from the RFID technology will amount to tens of billions of dollars in the near future. This includes many new markets that are being created, such as the market for real-time locating system (RTLS), which will itself be more than $6 billion in 2017 [4]. Therefore, future advanced RFID systems are expected to provide both reliable identification and high-definition localization of tags [5]. Accurate real-time localization at the submeter level, high security, and management of large numbers of tags, in addition to extremely low power consumption, small size, and low cost, will be new important requirements. Unfortunately, most of these requirements cannot be completely fulfilled by the current first- and second-generation RFID [2] or wireless sensor network (WSN) technologies such as those based on the ZigBee standard [6]. In fact, RFID systems using standard CW-oriented communication in the UHF band have an insufficient range resolution to achieve accurate localization, are affected by multipath signal cancellation (due to the extremely narrow bandwidth signal), are very sensitive to narrowband interference and multiuser interference, and have an intrinsic low security [1,7]. Although some of these limitations, such as security and signal cancellation due to multipath, are going to be reduced or overcome in future versions of UHF RFID systems [8,9], a technology change is required to satisfy new applications' requirements fully—especially those related to high-definition localization at the submeter level.

A promising wireless technique for future identification and localization systems is the ultrawide bandwith (UWB) technology characterized, in its impulse radio UWB (IR-UWB) implementation, by the transmission of subnanosecond duration pulses. The employment of wideband signals enables the resolution of multipath and extraordinary localization precision based on time of arrival (TOA) estimation signals. In addition, UWB allows for low power consumption at the transmitter side, extremely accurate ranging and positioning capability at the submeter level, robustness to multipath, low detection probability, efficient multiple channel access, and interference mitigation—thus leading to a potentially large number of devices operating and coexisting in small areas [7,10,11].

Thanks to their low power consumption, IR-UWB transmitters can be successfully adopted both for active and passive tags. UWB has been proposed to realize low consumption and low complexity active RF tags for precision asset location systems [12]. Recently, some commercial proprietary RTLSs have been introduced based on tags emitting UWB pulses with extremely low duty cycle to ensure high battery duration [13–15], and the IEEE 802.15.4f working group has been formed to define a dedicated standard to cover the RTLS segment [16].

As anticipated, passive tags solutions are of particular interest thank to their potential lower cost and power consumption. To this purpose, the idea of passive tags based on UWB backscatter signaling was introduced in Dardari et al. [5], Dardari [17], and Dardari and D'Errico [18], where tag architectures as well as backscatter signaling schemes robust to the presence of clutter (i.e., reflections coming from surrounding objects) are presented. Nevertheless, UWB RFID analysis based on backscatter modulation is at the beginning and different issues must still be investigated [5,19].

In this chapter, a survey of the UWB technology and its current application for RFID and RTLS systems is given. Particular emphasis will be paid to recent and promising solutions based on passive UWB RFID tags due to their potential low cost and hence wider market perspectives.

5.2   THE UWB TECHNOLOGY

5.2.1   UWB: Definitions and Regulatory Issues

The UWB technology has been around since 1960, when it was mainly used for radar and military applications. Nowadays it is a very promising technology for advances in wireless communications, networking, radar, imaging, positioning systems, WSNs, and RFID [6,10,11,20,21].

The most widely accepted definition of a UWB signal is a signal with instantaneous spectral occupancy in excess of 500 MHz or a fractional bandwidth of more than 20% [22]. The fractional bandwidth is defined as B/f_c, where B denotes the -10 dB bandwidth and f_c is the center frequency. A way to generate such signals is by driving an antenna with very short electrical pulses with duration in the order of 1 ns or less.

In 2002, the US Federal Communications Commission (FCC) issued the First Report and Order (R&O), which permitted unlicensed UWB operation and commercial deployment of UWB devices. There are three classes of devices defined in the R&O document: (1) imaging systems (e.g., ground penetrating radar systems, wall imaging systems, through-wall imaging systems, surveillance systems, and medical systems), (2) vehicular radar systems, and (3) communications and measurement systems. The FCC allocated a block of unlicensed radio spectrum from 3.1 to 10.6 GHz at the maximal effective radiated isotropic power (EIRP) spectral density of -41.3 dBm/MHz for these applications, where each category was allocated a specific spectral mask, as described in reference 22, and UWB radios overlaying coexistent RF systems can operate. In Figure 5.1 an example of an FCC spectral mask for indoor commercial systems is reported. With similar regulatory processes currently under way in many countries worldwide, government agencies responded to this FCC ruling.

Regarding Europe, it is important to mention that on February 21, 2007, the Commission of the European Communities released a decision on allowing the use of the radio spectrum for equipment using UWB technology in a harmonized manner in the community [23]. The decision concerns the use of the radio spectrum on a noninterference and nonprotected basis by equipment using UWB technology, with the definition of maximum allowed EIRP densities in the absence and in the presence

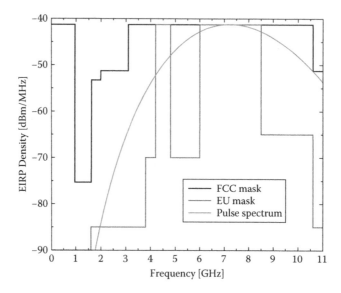

FIGURE 5.1 FCC and EU spectral masks, respectively, for indoor commercial systems in the absence of appropriate mitigation techniques. An example of sixth derivative of the Gaussian pulse spectrum with $t_p = 0.192$ ns, compliant with FCC mask, is also reported.

of appropriate interference mitigation techniques. In Figure 5.1 the maximum EIRP density in the absence of appropriate mitigation techniques is reported. In particular, the upper frequency band of 6–8.5 GHz has been identified in Europe for long-term UWB operation with a maximum mean EIRP spectral density of −41.3 dBm/MHz and a maximum peak EIRP of 0 dBm measured in a 50 MHz bandwidth without the requirement for additional mitigation [24]. Within the lower band of 3.1–4.8 GHz, low duty cycle (LDC) or interference detection and avoidance (DAA) UWB devices are permitted to operate with the same limits for the EIRP. In addition, dedicated standards have been amended for specific UWB applications such as RTLS [25,26]. Table 5.1 reports an overview of the current worldwide UWB emission masks.

5.2.2 IMPULSE RADIO UWB

UWB transmission systems can be realized through conventional modulation schemes by stretching the bandwidth to be larger than 500 MHz, for example, by adopting orthogonal frequency division multiplexing (OFDM) signaling. This approach has been followed by the WiMedia alliance standard for multimedia applications requiring high data rate transmissions [27]. One promising UWB technique, especially for RFID and WSN applications, is named IR-UWB. The IR-UWB technique relies on ultrashort (nanosecond scale) pulses that can be free of sine-wave carriers and do not require intermediate frequency processing because they can operate at baseband, thus drastically reducing the hardware complexity and power consumption. The IR-UWB technique has been selected as the physical (PHY) layer of the IEEE 802.15.4a Task Group for wireless personal area networks (WPAN) low rate

TABLE 5.1

Worldwide UWB Emission Masks

Band	EIRP	Mitigation technique
China		
4.2–4.8 GHz	−41.3 dBm/MHz	DAA
6.3–8.9 GHz	−41.3 dBm/MHz	No mitigation
Europe		
3.1–4.8 GHz	−41.3 dBm/MHz	LDC or DAA
2.7–3.4 GHz	−70 dBm/MHz	No mitigation
3.4–3.8 GHz	−80 dBm/MHz	No mitigation
3.4–6 GHz	−70 dBm/MHz	No mitigation
6–8.5 GHz	−41.3 dBm/MHz	No mitigation
8.5–9 GHz	−41.3 dBm/MHz	DAA
Japan		
3.4–4.8 GHz	−41.3 dBm/MHz	DAA
4.8–7.25 GHz	−70 dBm/MHz	No mitigation
7.25–10.25 GHz	−41.3 dBm/MHz	No mitigation
Korea		
3.1–4.8 GHz	−41.3 dBm/MHz	LDC or DAA
7.25–10.25 GHz	−41.3 dBm/MHz	No mitigation
United States		
3.1–10.6 GHz	−41.3 dBm/MHz	No mitigation

alternative PHY layer [20,28], and for the standard IEEE 802.15.4f related to RTLS [16], as will be detailed in Section 5.4.3.

As said, in IR-UWB the information is encoded using pulses. Typically, the adopted pulse $p(t)$ is derived by the Gaussian pulse $p_0(t) = \exp\left(-2\pi(t^2 / \tau_p^2)\right)$ and its derivatives due to its smallest possible time-bandwidth product, which maximizes range-rate resolution and is readily available from antenna pattern. Alternatively, the IEEE 802.15.4a standard suggests the following band-pass pulse with center frequency f_0 and root raised cosine (RRC) envelope [28]:

$$p(t) = \frac{4\,v\sqrt{2}}{\pi\sqrt{\tau_p}} \frac{\cos\left((1+v)\pi t / \tau_p\right) + \dfrac{\sin\left((1-v)\pi t / \tau_p\right)}{4vt / \tau_p}}{1-(4vt / \tau_p)^2} \cos(2\pi f_0 t) \qquad (5.1)$$

where parameter τ_p and roll-off factor v determine the bandwidth $W = (1 + v)/\tau_p$.[*]

In Figure 5.1 the sixth derivative of the Gaussian pulse with $\tau_p = 0.192$ ns is shown in the frequency domain. It can be noted that this pulse is compliant with the FCC

---

[*] Two different values of t_p are recommended [28]: $t_p = 1$ ns and $t_p = 3.2$ ns with $v = 0.6$, corresponding to two different bandwidths, $W = 1.6$ GHz and $W = 500$ MHz, respectively.

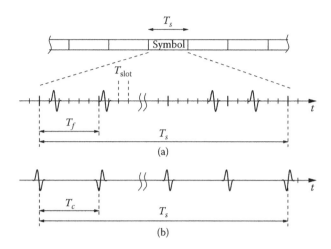

FIGURE 5.2 (a) UWB time-hopping frame structure; (b) UWB direct-sequence frame structure.

specifications. In general, due to the short pulse duration (typically less than 1 ns), the bandwidth of the transmitted signal can be on the order of one or more gigahertz.

To allow for multiuser communication in a typical IR-UWB communication system, each symbol (bit) is associated to multiple pulses.

In time-hopping (TH) schemes, symbols of duration T_s are divided into time intervals T_f called frames (see Figure 5.2a). The frames are further decomposed into smaller time slots T_{slot}. The UWB pulse $p(t)$, with duration $T_p < T_{\text{slot}}$, is transmitted in each frame in a slot position specified by a user-specific pseudorandom TH code $\{c_k\}$ having period N_s, where N_s is the number of pulses per symbol [10]. In direct sequence (DS) schemes (see Figure 5.2b), each pulse is modulated according to a pseudorandom binary code $\{c_k\}$ having period N_s and transmitted at regular intervals of T_c seconds—usually named, in this case, chips. The frame/chip time T_f/T_c is usually chosen to be greater than the maximum multipath delay to avoid intersymbol interference. The information can be associated to pulse polarity leading to pulse amplitude modulation (PAM) signaling or to pulse position, thus obtaining a pulse position modulation (PPM) signaling scheme.

Implementing an impulse-based radio allows for a simple circuit structure with low power dissipation since there is no need to up-convert a carrier signal [29]. The transmitter feeds these impulses to a very large bandwidth nonresonating antenna, or sometimes the antenna itself shapes the pulses to the required frequency of operation. In Figure 5.3 an example of a simple UWB pulse generator schematic proposed in Baghaei-Nejad et al. [29] is reported; it consists of a delay line, a NOR gate, and a pulse shaping circuit.

The basic UWB receiver is a correlation receiver [10] where the received signal is correlated with a local replica (template) of the transmitted pulse $p(t)$ or, equivalently, is filtered by a filter matched to $p(t)$ (matched filter [MF]). In a single-user additive white Gaussian noise (AWGN) scenario, the bit error probability (BEP) of a UWB link employing binary antipodal PAM is simply

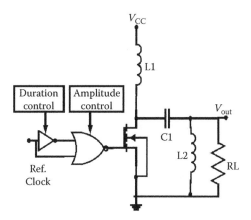

FIGURE 5.3 Example of UWB transmitter schematic [29].

$$P_b = \frac{1}{2} \text{erfc} \sqrt{\frac{E_r N_s}{N_0}} \qquad (5.2)$$

where E_r is the energy of the received pulse, N_0 is the thermal noise one-side power spectral density, and erfc($\cdot$) is the complementary error function. The received energy per symbol is $E_s = N_s E_r$.

Typical indoor environments often exhibit the presence of a dense multipath with delay spread much larger than the resolution capability of the signal being employed [30,31]. The transmission of ultrashort pulses can potentially resolve extremely large numbers of paths experienced by the received signal, especially in indoor environments, thus eliminating significant multipath fading [30]. This may considerably reduce fading margins in link budgets and may allow low transmission power operation. In addition, rich multipath diversity can be collected through the adoption of rake receivers, which combine the signals coming over resolvable propagation paths in a way that maximizes the signal-to-noise ratio (SNR) [32].

It can be shown that the bit error probability of a rake receiver is

$$P_b = \frac{1}{2} \text{erfc} \sqrt{\frac{E_s \eta_{cap}}{N_0}} \qquad (5.3)$$

where η_{cap} is called the *energy capture efficiency* and accounts for the ability of the RAKE receiver to collect the energy coming from different propagation paths. For more information on the fundamentals of UWB, we refer to references [10] ,[11], and [32] and references therein.

5.2.3 RANGING CAPABILITY OF UWB SIGNALS

Distance estimation (ranging) between tags and multiple readers represents the first step to localize the tag using, for example, multilateration algorithms [33].

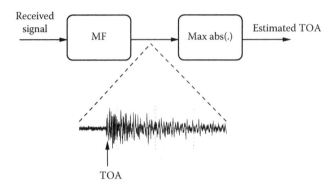

FIGURE 5.4   Classical MF-based TOA estimator.

Considering that the electromagnetic (e.m.) waves travel at the speed of light $c = 3 \cdot 10^8$ m/s, the distance estimate can be obtained from the measurement of the time of flight (TOF) τ_f of the signal and by observing that $\tau_f = d/c,$ where d is the actual distance between the tag and the reader. This requires an accurate estimation of the TOA of the received signal.

To understand which fundamental system parameters dominate ranging accuracy, we present an overview of the performance limits of TOA estimation in AWGN channels. We consider a scenario in which a unitary energy pulse $p(t)$ is transmitted (with duration T_p) through an AWGN channel.* In the absence of other error sources, the received signal can be written as

$$r(t) = \sqrt{E_r}\, p(t - \tau) + n(t) \tag{5.4}$$

where E_r is the received energy and $n(t)$ is AWGN with zero mean and two-sided power spectral density $N_0/2$.

The goal is to estimate the TOA τ, and hence the distance d, by observing the received signal $r(t)$. This task can be challenging due to the presence of thermal noise and multipath components. Under this simple model, TOA estimation is a classical nonlinear parameter estimation problem, with a solution based on an MF receiver [34]. As shown in Figure 5.4, the received signal is first processed by a filter matched to the pulse $p(t)$ or, equivalently, by a correlator with template $p(t)$. The TOA estimate is given by the time instant corresponding to the maximum absolute peak at the output of the MF over the observation interval. This scheme yields a maximum likelihood (ML) estimate, which is known to be asymptotically efficient; that is, the performance of the estimator achieves the Cramér-Rao bound (CRB) for large SNRs.†

The MSE of any unbiased estimate $\hat{d}$ of $d,$ derived from TOA estimation, satisfies the following inequality [34]:

---

* In general, $p(t)$ can be a part of a multiple access signaling such as DS or TH as explained in Section 5.2.2. For band-limited signals, we consider T_p as the interval duration containing most of the signal energy.

† The CRB gives the theoretical limit on the mean square error (MSE) of any unbiased estimator and hence it represents a useful benchmark to assess the performance of any practical estimator.

$$\text{Var}(\hat{d}) = \mathbb{E}\left\{\epsilon^2\right\} \geq \frac{c^2}{8\pi^2 B_{\text{eff}}^2 \text{SNR}} \tag{5.5}$$

where $\epsilon = \hat{d} - d$ is the ranging estimation error. The right-hand term in (5.5) represents the CRB. Here SNR $\triangleq E_r / N_0$ and parameter B_{eff}^2 represent the second moment of the Fourier transform $P(f)$ of $p(t)^*$; that is,

$$B_{\text{eff}}^2 \triangleq \frac{\int_{-\infty}^{\infty} f^2 |P(f)|^2 \, df}{\int_{-\infty}^{\infty} |P(f)|^2 \, df}. \tag{5.6}$$

Notice that the lower bound in (5.5) decreases with both SNR and the constant B_{eff}, which depends on the shape of the pulse. This reveals that signals with high power and/or wide transmission bandwidth are beneficial for ranging.

It is known that the ML estimation error tends asymptotically to the Gaussian distribution. Denoting d the true distance, the measured range $\hat{d}$ can be expressed as

$$\hat{d} = d + \epsilon \tag{5.7}$$

where the ranging estimation error ϵ can be modeled, as first approximation, as a Gaussian random variable (RV) with zero mean and variance σ^2, where σ^2 is bounded by (5.5).

Figure 5.5 shows the root mean square error (RMSE) for CRB using the RRC second and sixth-order Gaussian monocycle pulses. Note that higher derivative Gaussian monocycles or lower τ_p reduce the bound. It can be observed that centimeter level accuracy (e.g., RMSE on TOA less than 1 ns) is potentially feasible using UWB signals. Other improved but more complex bounds can be found in Dardari et al. [34].

In more realistic environments, numerous practical factors might affect ranging accuracy. Sources of error from wireless signal propagation include multipath, direct path excess delay and blockage incurred by propagation, respectively, of a partially obstructed or completely obstructed direct path component that travels through obstacles such as walls in buildings. For more information about ranging using the UWB signal, please refer to Dardari et al. [34] and references therein.

5.3   UWB PROPAGATION

5.3.1   UWB PROPAGATION CHARACTERISTICS

The performance limits of a communication or localization system are determined by the channel it operates in. UWB propagation channels show fundamental differences from conventional narrowband propagation in many respects. In

---

* Parameter B_{eff} is often called *effective bandwidth*.

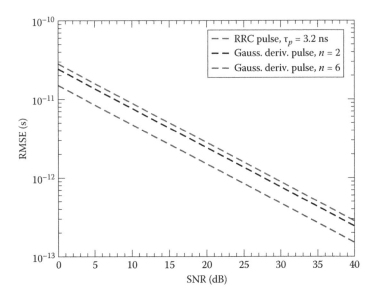

FIGURE 5.5 CRB on the TOA estimation RMSE as a function of SNR in an AWGN channel. The RRC pulse, the second-order Gaussian, and the sixth-order Gaussian monocycle pulses with, respectively, $t_p = 3.2$ ns, $t_p = 1$ ns, and $t_p = 0.192$ ns are considered.

this section a brief description of UWB propagation characteristics and models is presented.

In a free space environment the received signal power spectral density at distance d is

$$P_r(f) = P_t(f)\frac{G_t(f)G_r(f)c^2}{(4\pi fd)^2} \tag{5.8}$$

where $P_t(f)$ is the power density spectra of the transmitted signal; $G_t(f)$ and $G_r(f)$ are the transmit and receive antenna power gains, respectively; and c is the speed of light.

A number of UWB channel models have been proposed depending on the frequency range and operating conditions [30–32,35]. Widely adopted models valid for the 3–10 GHz frequency range in a number of different environments as well as for the frequency range below 1 GHz in office environments have been presented within the IEEE 802.15.4a Task Group and are briefly described in the following [31,36].

5.3.1.1 Path-Loss Model

In a UWB channel, the path loss is a function of frequency, distance, and the frequency characteristics of the antennas. In Molisch et al. [31], the following frequency-dependent path-loss model is given:

$$L_0(f,d) = 2\,k_0\frac{(f/f_c)^{2(\kappa+1)}\,d^\beta}{\eta_{TX}(f)\eta_{RX}(f)}, \tag{5.9}$$

where

k_0 is the isotropic path-loss at the reference distance of 1 m and center frequency f_c

the parameter κ is the frequency dependency decaying factor

$\eta_{TX}(f)$ and $\eta_{RX}(f)$ are, respectively, the transmit and the receive antenna efficiencies

f_c is the center frequency usually taken equal to 5 GHz.

Parameter β is the power path-loss exponent typically ranging between ≈ 1.8 and ≈ 4 [36], while the coefficient 2 accounts for the typical loss due to the presence of a person close to the antenna.

5.3.1.2 Multipath Characterization

Most of UWB channel models are based on an extended version of the classical Saleh-Valenzuela indoor channel model [37], where multipath components arrive at the receiver in groups (clusters) following the Poisson distribution. Specifically, the equivalent baseband channel impulse response related to the IEEE 802.15.4a models is [31]

$$h_0(t) = \sum_{l=1}^{L} \sum_{k=1}^{K_l} \alpha_{k,l} \exp(j\ \theta_{k,l}) \delta(t - T_l - \tau_{k,l}), \tag{5.10}$$

where

L represents the number of clusters

K_l is the number of paths in the lth cluster

T_l is the TOA of the lth cluster

$\alpha_{k,l}$, $\theta_{k,l}$, and $\tau_{k,l}$ represent, respectively, the amplitude, phase, and relative excess delay associated with the kth path within the lth cluster

$\delta(t)$ is the Dirac pseudofunction

The IEEE 802.15.4a working group defined several models based on (5.10), denoted with CM1-CM8, whose parameters' statistical characterization is specific to different reference environments (e.g., open office, industrial, etc.).

Another widely adopted model is the dense multipath model with a single cluster composed of L independent equally spaced paths and exponential power delay profile (PDP). The passband impulse response is given by

$$h(t) = \sum_{l=0}^{L} \alpha_l \delta(t - \tau_l) \tag{5.11}$$

where $\alpha_l = a_l p_l$, with a_l being the path amplitude and p_l an RV that takes, with equal probability, the values $\{-1,+1\}$ [30]. The average path power gains Λ_l are given by

$$\Lambda_l = \mathbb{E}\left\{|\alpha_l|^2\right\} = \frac{(e^{\Delta/\varepsilon} - 1)e^{-\Delta(l-1)/\varepsilon}}{e^{\Delta/\varepsilon}(1 - e^{L\Delta/\varepsilon})} \tag{5.12}$$

for $l = 1, 2, \ldots, L$. The parameter ε describes the multipath spread of the channel (decay time constant) and Δ is the resolvable time interval so that the lth path arrival time is $\tau_l = \tau_1 + (l - 1)\Delta$. In most of the previously mentioned models the small-scale fading, which characterizes the path amplitudes a_l or $\alpha_{k,l}$, follows a Nakagami-m distribution, with m being the severity parameter depending on the working conditions.

5.3.2 BACKSCATTER PROPAGATION

As already mentioned, passive RFID tags are based on backscatter modulation, where the antenna reflection properties are changed according to information data [3]. It is therefore fundamental to model the backscatter antenna characteristics. In general, when an e.m. wave encounters an antenna, it is partially reflected back depending on antenna configuration. An antenna scattering mechanism is composed of *structural* and *antenna mode* scattering [38,39].* Structural mode occurs owing to the antenna's given shape and material and it is independent of how the antenna is loaded. On the other hand, antenna mode scattering is a function of the antenna load; thus, data can be sent back to the reader through a proper variation of the antenna load characteristic without requiring a dedicated power source (*backscatter modulation*). This property is currently adopted in traditional passive UHF RFID tags based on CW signals to carry information from the tag to the reader [3].

While in UHF RFIDs an extensive literature exists dealing with this issue [3,39], the characterization of backscatter properties when operating with UWB signals is still not a well investigated and understood topic, especially in realistic environments [40–43]. In the following I give an overview of the UWB antenna backscatter properties.

5.3.2.1 UWB Antenna Backscattering

When a UWB pulse is transmitted and UWB antennas are employed, the reflected signal takes the form reported in Figure 5.6, where the structural and antenna modes' scattered components are plotted separately for convenience. The antenna mode scattered signal can be varied according to the antenna load Z_L, whereas the scattering of the structural mode will remain the same. Among the various possibilities, three particular choices are of interest for passive UWB RFID: $Z_L = 0$ (short circuit), $Z_L = \infty$ (open circuit), and $Z_L = Z_A^*$ (matched load), where Z_A^* is the conjugate antenna impedance. Ideally, antenna mode scattered waveforms have a phase difference of 180° between the case of open and short circuit loads, whereas no antenna mode scattering exists in the case of perfect matched load. In UWB antennas, the structural mode component takes a significant role in the total scattered signal; in fact, it is typically one or two orders of magnitude higher than that of the antenna mode [39,40]. In addition, signals scattered by the surrounding environment (clutter) are inevitably present and superimposed to the useful signal. In general, it is expected that the clutter and the antenna structural mode scattering have a significant impact at the reader's antenna, thus making the detection of the antenna mode scattered signal (which carries data) a main issue in passive UWB RFID systems.

---

* The structural mode scattered component is here conventionally defined as the signal scattered when the antenna load is matched. In the literature it is often defined with respect to the short circuit load [38]. In any case, whatever convention is adopted, the following analysis and the results are not affected.

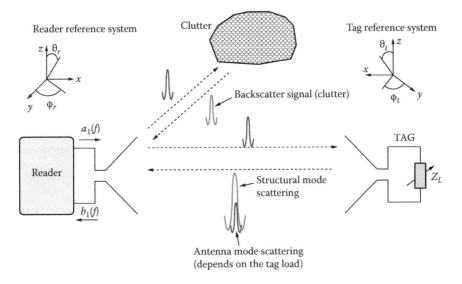

FIGURE 5.6 Example of backscatter mechanism of the transmitted pulse due to tag's antenna and the presence of scatterers [44].

This has not been widely addressed yet. To this purpose, ad hoc robust backscatter modulation schemes have to be designed. An example will be given in Section 5.5.

5.3.2.2 The Round-Trip Channel Transfer Function

To analyze the performance of backscatter modulation schemes, a proper model for the reader–tag–reader interaction is needed.

Consider a reference scenario, as shown in Figure 5.6, where a couple of UWB antennas, acting as tag and reader and located at distance d, are present. In the simple case of linear polarization of each antenna, the far field radiated from the reader and incident to the tag can be expressed as

$$E_{\text{inc}}(f; d, \Theta_r) = \frac{e^{-kd}}{d} \cdot \sqrt{\frac{\eta_0}{4\pi}} \cdot H_{\text{reader}}^{\text{T}}(f; \Theta_r) \cdot a_1(f), \qquad (5.13)$$

where
  λ is the wavelength
  $k = 2\pi/\lambda$
  η_0 is the free space impedance
  $a_1(f)$ is the incident wave at the reader's port
  $\Theta_r = (\theta_{\text{reader}}, \phi_{\text{reader}})$ is the reader orientation
  $H_{\text{reader}}^{\text{T}}(f; \Theta_r)$ is the antenna's transfer function in the transmitting mode of the reader [45]

The e.m. wave at the tag's antenna is partially backscattered according to the antenna's scattering characteristics, which depend on the antenna load (different load configurations will be referred to as tag status X) and reader–tag orientation in the three-dimensional space $\Theta = \{\Theta_r, \Theta_t\}$, with $\Theta_t = (\theta_{tag}, \phi_{tag})$ being the tag orientation. The antenna mode component of the received backscatter signal at the reader's antenna port is given by

$$b_1(f;d,X,\Theta) = H_{reader}^R(f;\Theta_r) \cdot \sqrt{\frac{4\pi}{\eta_0}} \cdot E_{SC}^{(a)}(f;d,X,\Theta), \qquad (5.14)$$

where $H_{reader}^R(f;\Theta_r)$ is the reader's antenna transfer function in receiving mode, and $E_{SC}^{(a)}(f;d,X,\Theta)$ is the component of the field scattered by the tag due only to the antenna mode. Therefore, we can express the tag's transfer function as

$$H_{tag}(f;\Theta,X) = \frac{4\pi d}{\lambda} \cdot \frac{E_{SC}^{(a)}(f;d,X,\Theta) \cdot e^{jkd}}{E_{inc}(f;d,\Theta_r)}. \qquad (5.15)$$

In particular, it can be shown that the amplitude characteristic of $H_{tag}(f;\Theta,X)$ can be expressed as

$$\left| H_{tag}(f;\Theta,X) \right| = \sqrt{\frac{4\pi\sigma(f;\Theta,X)}{\lambda^2}}, \qquad (5.16)$$

where $\sigma(f;\Theta,X)$ is the radar cross section of the tag. It turns out that the tag transfer function depends on the reader and local tag orientation Θ, but it is not dependent on the distance, since it relates the incoming plane wave complex amplitude at the tag to the far field radiated spherical wave. It is, for example, easy to prove that $H_{tag}(f;\Theta,X) = 1$ for a lossless tag reradiating isotropically. By considering the relationship between the transmitting and receiving modes of the reader

$$H_{reader}^R(f;\Theta_r) = -\jmath \, \frac{\lambda}{4\pi} H_{reader}^T(f;\Theta_r), \qquad (5.17)$$

we finally obtain the round-trip transfer function for linear polarized antennas:

$$H(f;d,X,\Theta) = \frac{b_1(f;d,X,\Theta)}{a_1(f)} = \left[H_{reader}^T(f;\Theta_r) \right]^2 \qquad (5.18)$$

$$\left(-\jmath \left(\frac{c}{f \cdot 4\pi \cdot d} \right)^2 e^{-\jmath \frac{2\pi f}{c} 2d} \right) \cdot H_{tag}(f;\Theta_t,X).$$

Note from (5.18) that, in free-space propagation conditions, the channel gain decreases with the distance d of an exponent factor of 2 instead of 1, as happens in conventional communication links. This typically reflects into a poor link budget when using the backscatter principle for data communication.

In a UWB RFID system the reader's antenna emits typically a very short pulse $p(t)$. We denote $w(t;d,X,\Theta)$ the backscattered signal, received back by the reader's

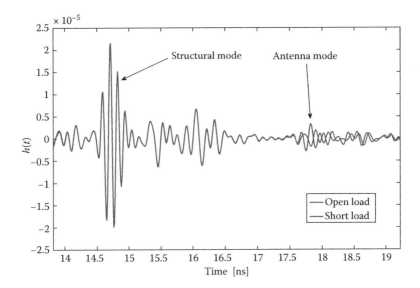

FIGURE 5.7 Example of backscatter impulse responses collected inside in the anechoic chamber at distance $d_{ref} = 2.15$ m for open and short circuit load conditions.

antenna, because of the tag's antenna mode, shape and energy of which are a function of the tag status X (open, short, loaded) as well as of Θ. In the frequency domain, it is

$$W(f;d,X,\Theta) = P(f)H(f;d,X,\Theta). \tag{5.19}$$

Figure 5.7 shows an example of backscattered signal measured in the anechoic chamber for open- and short-load conditions. In order to discriminate the structural mode from the antenna mode easily in the time domain, a delay line of electric length 40 cm was inserted between the tag antenna and the load. The structural and antenna modes can be clearly distinguished, where only the latter depends on the antenna load. In particular, the difference of about 180° for the antenna mode scattering between the two load conditions is clearly evident.

In a more realistic scenario, where several scatterers might be present, $H(f;d,X,\Theta)$ must also account for multipath components arising due to reflections. As of now, the only statistical model for a UWB backscatter round-trip channel with multipath available in the literature is that adopted in Sibille et al. [42] and Heiries et al. [46], where a double-convoluted IEEE 802.15.4a channel model based on (5.10) is proposed. However, this preliminary model has not been validated yet through extensive experimental campaigns in realistic scenarios.

5.3.2.3 Measurement Results

We present some experimental results collected in a typical laboratory scenario having dimensions 5.13 × 4.49 m² at ENSTA-ParisTech laboratory. A rectangular grid of nine test points as shown in Figure 5.8, spaced out of about 1 m in depth and 70 cm in width, was defined. The tag monopole dual feed stripline (DFMS) antenna was

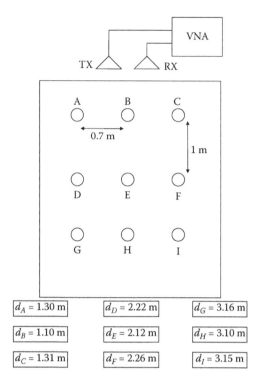

FIGURE 5.8 Indoor scenario considered for the measurement campaign at ENSTA-ParisTech. The distances between each point and the antennas connected to the vector network analyzer are also reported.

positioned alternatively in each point on a vertical support dressed with absorbers. Data postprocessing was performed to obtain the antenna backscattering response from the measured parameter $S_{21} = H(f;d,X,\Theta)$.

In Figure 5.9(a), an example of backscattered signal $h(t;d,X,\Theta)$ measured in the laboratory scenario from the tag placed at location I (distance 3.15 m) is reported. As can be noted, several clutter components (including the antenna structural mode) are present. Figure 5.9(b) shows the antenna mode backscattered signal (of interest) after clutter removal. Owing to its small amplitude, it turns out to be completely buried within the clutter component. The presence of an echo received after the first direct path is also clearly seen and can be ascribed to indirect paths between the tag and the reader. In most of the considered configurations, the normalized cross correlation between the backscattered signals in the case of open- and short-circuit loads is close to -1, as expected for antenna mode signals. This good symmetry property is useful in cases of signaling schemes employing antipodal pulses and justifies, in the following analysis, the approximation of perfect pulse symmetry—that is, $w(t) = w(t;d,0,\Theta) = -w(t;d,1,\Theta)$, where for notation compactness, we have hidden in $w(t)$ the explicit dependence on d and Θ.

In Sibille et al. [42] and Guidi et al. [47], extended measurement results in a realistic environment are presented. An important issue to be investigated is the

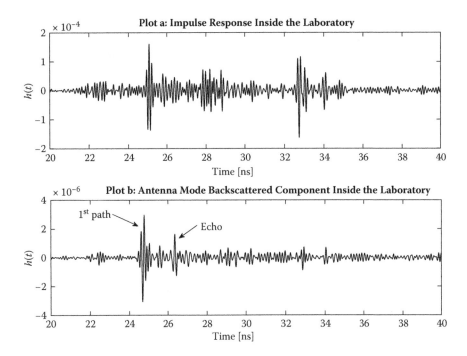

FIGURE 5.9 Example of backscatter responses collected inside the laboratory (plot a) grid location I at distance of 3.15 m, and of the only antenna mode contribution after clutter removal (plot b) in the same location I.

characterization of the effect of close-to-antenna objects. Some interesting preliminary results can be found in Guidi et al. [43], where it is shown how near metallic objects do not lead, in general, to a backscattered signal reduction—contrary to what happens in narrowband UHF RFID tags.

5.4 UWB RFID AND LOCALIZATION WITH ACTIVE TAGS

5.4.1 Low Duty Cycle UWB Tags

As already mentioned, an IR-UWB transmitter can be very simple and it is characterized by an extremely low duty cycle transmitted signal. Therefore, such a transmitter has low complexity and low consumption (less than 10 mW, depending on the pulse repetition rate) [49,50]. On the other hand, a UWB receiver, even a simple one, is typically characterized by higher values of consumed power that could be in the order of 100 mW [49]. Therefore, devices based on the IEEE 802.15.4a standard for WSNs, which allows for bidirectional communication, do not fit typical RFID applications requirements. For this reason most of proposed UWB RFID devices designed for RTLSs have only transmitting active tags operating with low duty cycle in order to extend battery operation up to a few years [12–15].

An alternative solution has been proposed [48] where the concept of *pseudorandom active reflector* is introduced. As shown in Figure 5.10, it consists of a simple device

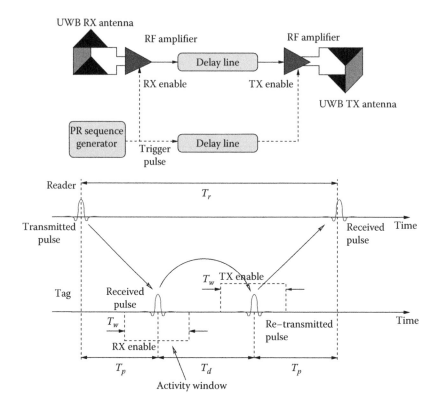

FIGURE 5.10 Pseudorandom active UWB reflector scheme and example of timing structure in case of single-pulse transmission [48].

that repeats a slightly delayed version of the received UWB signal only in certain time intervals according to a suitable pseudorandom (PR) TH sequence. In particular, the signal received by the antenna is amplified with gain G and delayed by a fixed quantity T_d of a few nanoseconds. The delayed version of the signal is used to drive the transmitter section composed of a power amplifier and a UWB antenna. The trigger pulse at the output of the PR generator enables the receiver amplifier for a certain time window T_w (activity window). A delayed version, by the quantity T_d, of the same trigger pulse is used to enable the transmitter. The transmission and receiving windows are not time overlapped, so no transmitter–receiver coupling occurs and the same antenna can be used for both the transmitter and receiver. The reader emits pulse trains with the same PR TH sequence used by the reflector it wants to communicate with. Each reflector has a unique PR sequence and reflects, with a delay T_d, the received UWB signal for a short time interval according to the TH sequence (see Figure 5.10). When synchronized, all transmitted pulses are reflected only by the tag adopting the same PR sequence of the reader, thus making the reader able to collect coherently the energy from that particular tag. As stated in Dardari [48], the advantages of this solution are in the hardware simplicity since only the analog section is present, in the low power consumption of the tag, and in the low timing constraint regarding the relative transmitter and reflector clock rates.

5.4.2 LOCALIZATION CAPABILITY

The first application of UWB technology in the RFID and RTLS fields was for precision asset location systems operating in indoor environments [12–15]. In these systems only UWB transmitting tags are employed; then, tag position estimation cannot rely on absolute distance estimate between tag and readers and localization schemes based on time difference of arrival (TDOA) are usually adopted. In TDOA-based localization systems, burst UWB signals are broadcast periodically by the tag and are received by several readers placed in known positions (see Figure 5.11a). The readers share their estimated TOA and compute the TDOA, provided that they have a common reference clock. Considering that time measurement accuracy should be in the order of 1 ns or less, readers must be kept tightly synchronized through a wired network connection. To calculate the position of the tag, at least three readers with known position and two TDOA measurements are required. Each TDOA measurement can be geometrically interpreted as a hyperbola formed by a set of points with constant range differences (time differences) from two readers [33].

Time Domain Plus RTLS is an example of a commercial proprietary system designed for locating personnel or mobile assets adopting the TDOA technique [13]. The active tag is a compact IR-UWB battery powered transmitter that is designed for long battery duration, up to 4 years (at 2 Hz update rate) or up to 1.5 years (at

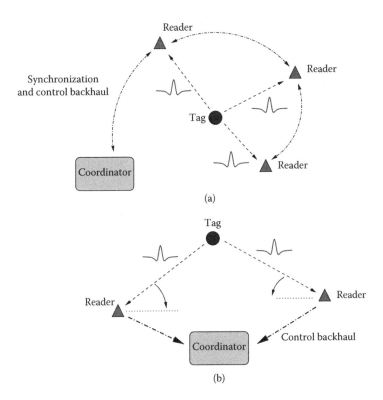

FIGURE 5.11 Active tag localization using (a) TDOA and (b) AOA estimation techniques.

4 Hz update rate). The system can locate and track thousands of tags with submeter location accuracy using an adequate number of tightly synchronized readers that perform the functions of receiving tag signals, demodulating the tag data, measuring the TOA for each tag, and computing the TDOA for location estimation.

Another possibility to localize the tag is to measure the angle of arrival (AOA) of the signal, thus obtaining the information about tag direction to neighboring readers (see Figure 5.11b). The AOA of an incoming radio signal can be estimated by using multiple antennas with known separation (antenna array) and by measuring the TOA of the signal at each antenna. Given the differences in arrival times and the array geometry, it is possible to estimate the direction of propagation of a radio-frequency wave incident on the antenna array. AOA does not require the precise time synchronization needed for TOA and TDOA techniques. Two angle measurements are required to determine node position (*triangulation*) [33]. In non-line-of-sight (NLOS) environments, the measured AOA might not correspond to the direct path component of the received signal and large angle estimation errors can occur. Due to the presence of multiple antenna elements, AOA techniques are in general more expensive in terms of cost and device dimensions than TDOA-based techniques.

The Ubisense platform is an example of a commercial precision localization system where active tags are localized using TDOA or AOA techniques [14]. When AOA localization is performed, no tight synchronization among readers is required, thus drastically reducing network requirements. However, the positioning accuracy is less than that obtainable using TDOA measurements.

Zebra Enterprise Solutions commercializes its RTLS, called Sapphire DART [15]. The location system is TOA- or TDOA based, the tag emits in the 6.5 GHz band, and the blink rate is similar to that of Ubisense at a few hertz.

5.4.3 THE IEEE 802.15.4F STANDARD

The IEEE 802.15.4f Active RFID System Task Group is chartered to define new wireless PHY layers and enhancements to the 802.15.4-2006 standard medium access control (MAC) layer [16]. This amendment defines a PHY layer, and only those MAC modifications required to support it, for active RTLS. It allows for efficient communications with active tags and sensor applications in an autonomous manner in a promiscuous network, using very low energy consumption (low duty cycle) and low transmitter power. The PHY layer parameters are flexible and configurable to provide optimized use in a variety of active tag operations including simplex and duplex transmission (reader to tags and tag to readers), multicast (reader to a select group of tags), unicast (reader to a single tag), tag-to-tag communication, and multihop capability.

At the time of writing, the group achieved a common proposal on the PHY layer based on three different physical sublayers, respectively: UWB and UHF at 433 MHz and 2.4 GHz. The UWB PHY is based on the on/off keying (OOK) modulation with 1–2 MHz pulse repetition frequency enabling high-accuracy TOA-based ranging.

The UHF interface is in the frequency range of 433.05–434.79 MHz and 1.74 MHz bandwidth. The modulation is minimum shift keying (MSK) with data rates of

250 Kb/s or 31.25 Kb/s. A received signal strength indicator is used as a low accuracy location determination mechanism.

The 2.4 GHz air interface is built on the IEEE 802.15.4 standard PHY layer. It could be used to stand alone for low precision RTLS or to provide assistance to UWB PHY layer. The operating bands have been chosen in order not to affect Wi-Fi and Zigbee nearby devices. Again, the MSK modulation is used with a bit rate of 250 Kbps found as a compromise between range, bandwidth, and power consumption.

The PHY layer specification supports a large tag population (hundreds of thousands) and basic functionalities such as read and write with authentication and accurate localization. The communication reliability of the system is expected to be high for applications such as active tag inventory counting or auditing, high-value asset location tracking, and personnel tracking. Typical requirements are operation in dense, metallic environments with submeter localization accuracy; real-time presence and location updates in seconds; and small tag sizes for easy placing on typical high-value assets. The active IEEE 802.15.4f PHY layer is capable of working in the presence of interference from other devices operating within the band of operation.

5.5   UWB RFID AND LOCALIZATION WITH (SEMI)PASSIVE TAGS

When tag cost, size, and power consumption requirements become particularly stringent, passive or semipassive tag solutions have to be taken into consideration. As already mentioned, communication with passive tags usually relies on backscatter modulation even though the tag's control logic and memory circuits must still be energized in order to have the tag working properly. Typically, passive RFID tags obtain the necessary power to operate from the RF signal sent by the reader. As a consequence, in conventional UHF RFID systems, the corresponding operating range is restricted to be no more than 7–8 m with a transmission power level of 2–4 W [3]. Unfortunately, due to regulatory constraints, the transmission power allowed for UWB devices is below 0 dBm. This means that no sufficient power can be derived from the received UWB signal to power up a remote tag at significative distance. A possible solution is represented by hybrid tag architectures, as will be illustrated in Section 5.5.1.

Besides the adoption of semipassive tags with battery-powered control logic, a promising possibility to retrieve the necessary energy is to use energy scavenging techniques that, in many cases, provide sufficient power (about 1–10 µW) for the control logic [51]. An interesting alternative is represented by chipless tags characterized by the absence of any control logic circuit, as will be illustrated in Section 5.6.

5.5.1   HYBRID TAGS BASED ON UHF AND UWB MODULATIONS

Hybrid tags solutions are proposed in references [29] and [52–55]. The main idea is to have an asymmetric link where in the downlink (reader–tag) a conventional transmission protocol at UHF is adopted to power up the tag and accumulate enough energy to allow an IR-UWB transmitter to send data for a short time interval at high data rate in the uplink (tag–reader).

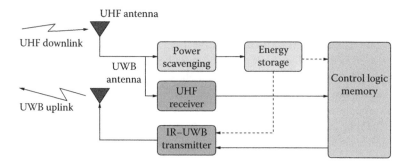

FIGURE 5.12   Hybrid UWB tag architecture proposed in Baghaei-Nejad et al. [29].

The block diagram of the module proposed in Baghaei-Nejad et al. [29] can be seen in Figure 5.12. It consists of a UHF receiving antenna, a power scavenging unit, an energy storage unit (basically a big surface-mounted capacitor), a UHF receiver, a low power IR-UWB transmitter, a UWB transmit antenna, and the control logic. Similarly to conventional passive tags, the incoming RF signal transmitted by the reader at UHF is used to provide power supply and receive the data. In particular, the reader radiates the RF signal with no data for at least 7 ms, which is the time for full charge of the storage capacitor. After enough energy has been collected, the tag goes to the receiving mode to receive commands from the reader at a low data rate (40 Kbps). The uplink transmission is performed using the IR-UWB transmitter. Thanks to the high transmission data rate (1 Mbps with $N_s = 10$ pulses per bit) and the low transmitter consumption (64 μW), the energy stored in the capacitor is sufficient to allow the transmission of packets containing more than 128 bits. Circuit implementation and simulations have shown that up to 10.7 m operating range with 4 W EIRP emission is feasible from the energy budget point of view. Unfortunately, no results are reported by the authors related to the uplink data transmission performance and the associated operating range.

Remote powering of a passive UWB tag by UHF has recently been achieved for high data rate exchanges (>50 Mbits/s at a few tens of centimeters) from a cell phone to a tag embedding a large memory [54].

5.5.2 Tags Based on Backscatter Modulation

Recently, some applications of the UWB technology in tags based on backscatter modulation have been proposed [5,17,18,44,56,57]. Due to its extremely low complexity, backscatter communication appears very promising, especially in the perspective of the adoption of efficient energy scavenging techniques for a tag's control logic alimentation. For this reason, in the following I describe more details to this solution.

In Figure 5.13 the architectures proposed in reference [17] and analyzed in references [18], [44], [47], and [57] for tag and reader are reported. The reader is composed of a transmitter and a receiver section.

During the interrogation cycle, the reader transmits a sequence of UWB pulses modulated by a periodic binary spreading sequence $\{d_n\}$ of period N_c with $d_n \in \{-1,1\}$

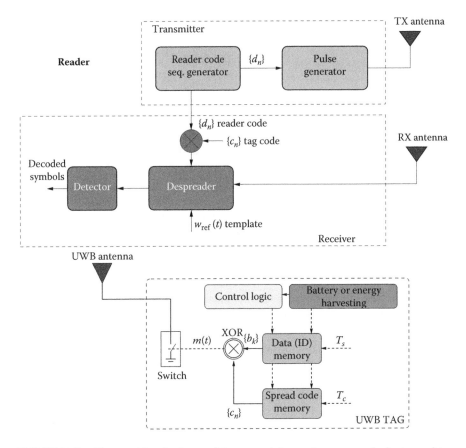

FIGURE 5.13 The considered scheme of the tag and the reader composed of a transmitter and a receiver section [18].

specific to that particular reader (reader's code). In general, N_{pc} pulses are associated to each code symbol (chip). To accommodate the signals backscattered by tags corresponding to an entire packet of N_r bits, the UWB interrogation contains $N_t = N_r N_s$ pulses, where $N_s = N_c N_{pc}$ is the number of pulses associated to each bit. Pulses are separated by T_p seconds. Then the UWB transmitted signal takes the form of

$$s_{reader}(t) = \sum_{k=0}^{N_r-1} s(t - kN_cT_c) \qquad (5.20)$$

This is characterized by a transmission power P_t, where $T_c = N_{pc}T_p$ is the chip time, $s(t)$ is the composite sequence

$$s(t) = \sum_{n=0}^{N_c-1} d_n g(t - nT_c), \qquad (5.21)$$

and

$$g(t) = \sum_{i=0}^{N_{pc}-1} p(t - iT_p) \tag{5.22}$$

is the waveform associated to each chip d_n composed of N_{pc} elementary UWB pulses $p(t)$. The pulse repetition period T_p is chosen so that all signals backscattered by the environment are received by the reader before the transmission of the successive pulse. In the indoor scenario, $T_p = 50\text{--}100$ ns is usually sufficient to this purpose.

Each pulse in (5.20) is backscattered by the tag's antenna as well as by all the surrounding scatterers present in the environment, which form the clutter component.

The main task of the receiver section of the reader is to detect the useful backscattered signal (i.e., that coming from the tag's antenna mode scattering, which depends on antenna load changes) from those backscattered by the antenna structural mode and other scatterers (clutter), which are, in general, dominant. To make the uplink communication robust to the presence of clutter and interference, and to allow multiple access, a proper backscatter modulation strategy is necessary at tag side.

In Figure 5.13, an example of tag architecture employing a binary backscatter modulator composed of a UWB switch is shown. The switch is controlled by a microcontroller; its purpose is to change the switch status X (short or open circuit) at each chip time T_c according to the data to be transmitted and a zero mean (balanced) periodic tag's code $\{c_n\}$, with $c_n \in \{-1,+1\}$, of period N_c [17,18,46]. Specifically, each tag information bit $b_k \in \{-1,+1\}$ is associated to N_s pulses; thus, the symbol time results $T_s = T_c N_c = T_p N_s$, as illustrated in Figure 5.14. In this way the polarity of the reflected signal changes according to the tag's code during a symbol time, whereas the information symbol affects the polarity of all pulses composing the sequence each symbol time. Therefore, the backscatter modulator signal, commanding the tag's switch, can be expressed as

$$m(t) = \sum_{k=0}^{N_r-1} b_k \sum_{n=0}^{N_c-1} c_n \, \Pi\left(\frac{1}{T_c}(t - nT_s - iT_c - \Delta) \right) \tag{5.23}$$

having defined $\Pi(t) \triangleq 1$ for $t \in [0,1]$ and 0 otherwise. Note that, in general, reader and tag have their own independent clock sources and hence they have to be treated as asynchronous. Then the tag's code $\{c_n\}$ is not in general time aligned to the reader's code $\{d_n\}$. The parameter Δ in (5.23) denotes the clock offset of the tag with respect to the reader's clock.

Powerful acquisition techniques [58] and the TOA estimator robust to clutter proposed in Xu and Law [59] can be adopted at a reader's side to compensate for clock offset. Once the tag and reader codes are aligned and the TOA is estimated, the reader can adjust its internal clock so that it becomes synchronous to that of the intended tag. A more general analysis considering asynchronous multiple tags can be found in Dardari et al. [44] and in [66]. For the sake of illustration, in the following I refer to a single tag scenario and consider perfect synchronization between reader

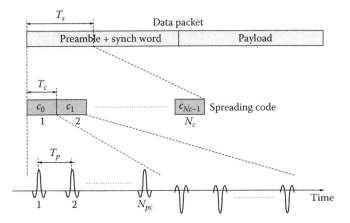

FIGURE 5.14 Example of backscattered signal structure. Only useful component reported.

and tag (i.e., $\Delta = 0$). In addition, the tag response due to the antenna mode is examined, whereas the antenna structural mode will be treated as part of the clutter, since it does not depend on data symbols. As a consequence, any clutter removal technique adopted will be effective on the antenna structural mode component as well.

Considering perfect backscattered pulse symmetry—that is, $w(t;0,d,\Theta) = -w(t;1,d,\Theta) \triangleq w(t)$—the received signal at the reader can be expressed as [44]

$$r(t) = \sum_{k=0}^{N_r-1}\sum_{n=0}^{N_c-1} d_n \sum_{i=0}^{N_{pc}-1} w(t - iT_p - nT_c - kT_s) \cdot m(t) \qquad (5.24)$$

$$+ \sum_{k=0}^{N_r-1}\sum_{n=0}^{N_c-1} d_n \sum_{i=0}^{N_{pc}-1} w^{(c)}(t - iT_p - nT_c - kT_s) + n(t),$$

$$= \sum_{k=0}^{N_r-1} b_k \sum_{n=0}^{N_c-1} d_n c_n \sum_{i=0}^{N_{pc}-1} w(t - iT_p - nT_c - kT_s)$$

$$+ \sum_{k=0}^{N_r-1}\sum_{n=0}^{N_c-1} d_n \sum_{i=0}^{N_{pc}-1} w^{(c)}(t - iT_p - nT_c - kT_s) + n(t)$$

where $n(t)$ is the AWGN with two-sided power spectra density $N_0/2$, and the signal $w(t)$ represents the backscattered version of the pulse $p(t)$ when short-circuit load is applied to the antenna, which is in general distorted due to antennas and channel frequency selectivity. The signal $w^{(c)}(t)$ represents the backscattered version of the pulse $p(t)$ due to the clutter component, which also accounts for pulse distortion and tag antenna structural mode.

Looking at (5.24), it can be noted that only the antenna mode scattered component results modulated by the combination of the tag's and reader's codes $\{c_n\}$ and

$\{d_n\}$, whereas all clutter signal components (including the antenna structural mode scattering) are received modulated only by the reader's code $\{d_n\}$. This property can be usefully exploited to remove the clutter component through a proper receiver and code design.

An example of a possible receiver scheme is that reported in Figure 5.13, where a correlator-based demodulator is adopted with a template:

$$s_{\text{ref}}(t) = \sum_{n=0}^{N_c-1} d_n\, c_n \sum_{i=0}^{N_{\text{pc}}-1} w_{\text{ref}}(t - iT_{\text{p}} - nT_{\text{c}}). \tag{5.25}$$

The optimum receiver can be obtained ideally by choosing the waveform $w_{\text{ref}}(t) = w(t)$ (perfect matched receiver) as pulses composing the local template in the correlator. Unfortunately, in practical situations, $w_{\text{ref}}(t) \neq w(t)$; we indicate with ρ the normalized cross correlation between the actual received pulse $w(t)$ and the template $w_{\text{ref}}(t)$, which accounts for the mismatch due to pulse distortion (note that $\rho = 1$ in cases of perfectly matched receivers). In substance, this scheme performs a de-spreading operation using the combined code $\{c_n \cdot d_n\}$, which identifies both the reader and the desired tag.[*]

Every T_s seconds, the output of the correlator is sampled, thus obtaining the decision variable for the kth bit b_k:

$$y_k = b_k\, E_r\, \rho\, N_s + \gamma^{(c)}(kT_s) \sum_{i=0}^{N_c-1} c_i + z_k, \tag{5.26}$$

where z_k is the thermal noise sample with variance

$$\sigma_z^2 = \frac{N_0}{2} N_s$$

and E_r represents the average received energy per pulse. Using (5.19) E_r is given by

$$E_r = \int_{-\infty}^{\infty} |W(f;0,d,\Theta)|^2\, df. \tag{5.27}$$

As can be deduced from (5.26), to remove the clutter component completely, it is sufficient that the tag's code $\{c_n\}$ has zero mean; that is,

$$\sum_{n=0}^{N_c-1} c_n = 0.$$

---

[*] Multiple readers can access the same tag by using different reader codes, provided that they are designed with good cross-correlation properties.

In such a case (5.26) can be further simplified, leading to

$$y_k = b_k \, E_r \, N_s \, \rho + z_k. \tag{5.28}$$

In general, from (5.28), it is easy to show that the BEP is given simply by

$$P_b = \frac{1}{2} \text{erfc}\left(\sqrt{\frac{E_r \, N_s \, \rho^2}{N_0}} \right). \tag{5.29}$$

For further convenience, we define G_{ref} the round-trip channel power gain at the reference distance d_{ref} and at the maximum direction of radiation $\Theta_{\max}$ in the AWGN scenario. In addition, we assume a typical exponential path-loss law where the power path-loss exponent β usually ranges between ≈ 1.8 and ≈ 4 [36]. The BEP can be rewritten as [18]

$$P_b = \frac{1}{2} \text{erfc}\left(\sqrt{\frac{P_t \, G_{\text{ref}} \, \rho^2 \left(\dfrac{d_{\text{ref}}}{d} \right)^{2\beta}}{N_0 \, R_b}} \right), \tag{5.30}$$

where $R_b = 1/(N_s T_p)$ is the data rate (bit rate). It is interesting to note that the exponent 2β is present in (5.30), instead of β, to account for the two-way link.

5.5.2.1 Performance Analysis in Anechoic Chamber

We now investigate the potential operating range of the passive UWB RFID system described in this section considering the following parameter values: $T_p = 100$ ns, $F = 4$ dB (receiver noise figure), and EIRP = -6.7 dBm. The FCC-compliant sixth derivative Gaussian monocycle has been considered for the transmitted pulse.

Figure 5.15 shows the achievable operating range as a function of the data rate R_b for a fixed target BEP $P_b = 10^{-3}$. A perfectly matched receiver in anechoic chamber scenario (AWGN channel) is considered using measurement data for different tags' antenna orientation offsets ϕ with respect to the maximum radiating angle. For each measured data set, the normalized cross-correlation coefficient ρ was calculated and used in (5.30). For the balanced antipodal Vivaldi (BAV) antenna considered in the measurements, $G_{\text{ref}} = -75$ dB, which accounts also for the reader's antenna gain $G_{\text{reader}} = 5$ dB. From the figure, it can be seen that, for example, for data rate $R_b = 10^3$ bits/s, an operating range larger than 20 m can be achieved. However, antenna radiation pattern and pulse distortion might determine a significant performance degradation when devices are not oriented to the maximum radiating direction, as can be noted in Figure 5.15 (see curves with $\phi \neq 0$). For comparison, the corresponding operating ranges in free-space conditions are reported for UHF-passive RFID tags operating at 868 MHz and 2.4 GHz, respectively, with a transmitted EIRP of 500 mW according to European regulations [60]. As can be noted, using the UWB technology, a significantly larger operating range is achievable with respect to that of

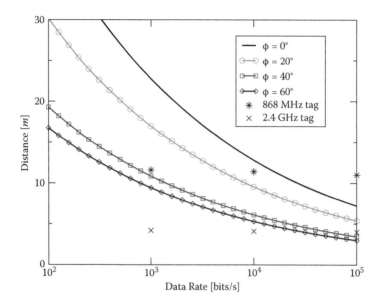

FIGURE 5.15 Tag–reader operating range in meters as a function of the data rate for $P_b = 10^{-3}$ and different tag orientations in the anechoic chamber scenario [44]. BAV antenna for the tag is considered. Cross and star dots refer to the corresponding performance of UHF tags operating at 868 MHz and 2.4 GHz, respectively. (From G. D. Vita and G. Iannaccone. *IEEE Transactions on Microwave Theory and Techniques,* vol. 59, no. 9, pp. 2978–2990, Sept. 2005.)

UHF-based RFID systems, especially for low data rates, with a dramatically reduced transmitted power level (≈ 0.09 vs. 500 mW).

5.5.2.2   Performance in Single-Tag Scenario Using Measured Data

Results related to the perfect matched receiver in every location in the grid inside the laboratory, shown in Figure 5.8, are reported in Figure 5.16 in terms of BEP calculated using (5.30). For the sake of comparison, the performance in AWGN is also reported for $d = 1.46$ m.

The performance obtained with the tag located at location B is better than that in AWGN condition because of the shorter distance (1.10 vs. 1.46 m). Note that in some cases tags placed at larger distances provide a better performance. This depends on the higher amount of energy that can be collected in some locations because of the presence of richer multipath components in the received signal. As a numerical example, with a target BEP $P_b = 10^{-3}$, data rates up to 200 kbit/s at a distance of 3.10 m with a transmitted power lower than 1 mW are feasible in a realistic environment.

5.5.2.3   Performance in Multitag Scenario: The Impact of Spreading Codes

The design of the spreading code $\{c_n\}$ of the tag must be oriented to obtain a good performance both in terms of multiuser interference (MUI) rejection and clutter mitigation. We have seen that to remove the clutter component completely and hence the

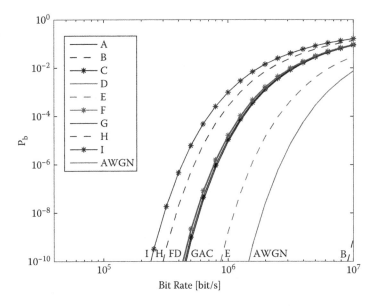

FIGURE 5.16   BEP as a function of the bit rate R_b in different tag locations (laboratory scenario) [44]. The perfect matched receiver is considered.

antenna structural mode component, it is sufficient that the tag's code $\{c_n\}$ has zero mean (balanced code).

Regarding the MUI, the situation is similar to what happens in conventional code division multiple access systems where the performance is strictly related to the partial cross-correlation properties of codes. Classical codes such as Gold codes or m-sequences offer good performance. Unfortunately, they are composed of an odd number of symbols, and hence there is no way to obtain a balanced code to remove the clutter completely. However, considering that m-sequences have a quasibalanced number of "+1" and "−1" (i.e., their number differs no more than 1), one option to achieve clutter removal is to lengthen the code by one symbol so that the resulting code becomes balanced. As a consequence, we expect a potential degradation in terms of multiple access performance, especially when short codes are adopted. This aspect will be investigated in the numerical example.

When the scenario is quasi-synchronous (i.e., synchronous at chip level), orthogonal codes such as Hadamard codes represent a good choice, as the interference can be completely removed, at least in ideal scenarios. This could be the situation where a wake-up signal is sent by the reader to switch the tag on and reset the code phase. Such a solution, where the UWB tag is supposed to be woken up by a dedicated control signal sent in the UHF band, is under investigation [19].

As numerical example to investigate the effectiveness of code design, we consider one reader and one useful tag placed at 7 m from the reader in the direction of its maximum radiation, with 59 interfering tags randomly distributed in a range of 1 m around the useful tag. An RRC waveform with roll-off factor of $\nu = 0.6$, $\tau_p = 0.95$ ns, and $f_0 = 3.95$ GHz is used as transmitted pulse. The system parameters are $T_p = 128$ ns,

$N_c = 1024$, receiver noise figure $F = 4$ dB, reader antenna gain $G_{reader} = 5$ dBi, tag's antenna gain $G_{tag} = 1$ dBi, and tag's switch loss = 2 dB. The receiver template $w_{ref}(t)$ is matched to the pulse that would have been received in free-space propagation in the direction of the tag's maximum radiation. Results have been obtained by Monte Carlo simulations, starting from channel responses drawn from the double-convoluted IEEE 802.15.4a CM1 channel model to account for the two-way link of the backscattered signal [46]. As far as the clutter is concerned, a uniform power delay profile in the overall interval T_p has been taken into account, with paths spaced apart of 0.95 ns; each path's amplitude is characterized by Nakagami-m fading, with $m = 3$, and a root mean squared value of 0.5 mV at the receiver.

In Figure 5.17 we report the performance terms of bit error rate (BER) as a function of the number of pulses per symbol N_s and different spreading codes, respectively; Hadamard; m-sequences; and extended m-sequences. In a quasisynchronous scenario, where tags' and reader's code generators are supposed to be synchronized (e.g., using a wake-up signal), the performance of orthogonal Hadamard codes results are not affected by MUI and clutter. On the other hand, when reader and tags are asynchronous, the performance of the system, when using orthogonal codes, drastically degrades due to the joint effect of multipath and MUI caused by the poor cross-correlation properties of Hadamard code words that are not aligned. By using m-sequences, MUI effects are well mitigated by the good cross-correlation properties of the code, even in the asynchronous scenario. However, it is evident that the clutter effect is dominant, leading to an error floor due to code unbalance. On the other hand, as can be noted in Figure 5.17, extended (balanced) m-sequences represent a good solution in terms of clutter suppression at the expense of a negligible loss in MUI mitigation.

For a complete characterization of passive UWB RFID technology, important related topics not yet sufficiently addressed in the literature need to be investigated,

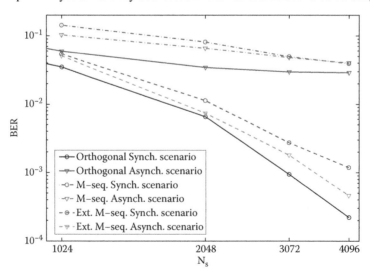

FIGURE 5.17 BER as a function of N_s and different spreading codes; 59 interfering tags. Useful tag at $d = 7$ m. Multipath 802.15.4a CM1 channel considered [57].

such as experimental clutter and round-trip multipath channel characterization, performance assessment in the presence of interference, design of codes robust to clutter, MUI, and jitter caused by the tag's oscillator drift [44] and [66].

5.5.3 LOCALIZATION CAPABILITY

We investigate the theoretical performance of a localization system based on UWB backscatter signals. Consider a scenario composed of N readers placed at known coordinates (x_i, y_i), with $i = 1, 2, \ldots N$, which interrogate a tag located in unknown coordinates $\mathbf{p} = (x, y)$ with the purpose to obtain an estimate of the reader-tag distance d_i (through the measurement of the round-trip time of the backscattered signal) and determine the tag's position by means, for example, of a classical multilateration localization algorithm [33].

Range measurement $\hat{d}_i$ between the tag and the ith reader is, in practice, affected by errors. We have seen, for example, that in AWGN the optimum distance estimator is characterized by an asymptotic MSE given by the CRB, as expressed by (5.5), which is a decreasing function of the SNR. According to (5.5) and assuming an exponential path-loss model, the dependence of the variance of the estimation error ϵ_i on distance d_i can be modeled as $\sigma_i^2 \equiv \sigma^2(d_i) = \sigma_0^2 d_i^\beta$, where β is the path-loss exponent and σ_0^2 is the variance at 1 m. The probability distribution function (PDF) of the ith range measurement, conditioned to the true position $\mathbf{p}$ of the tag, is therefore given by

$$f_i(\hat{d}_i|\mathbf{p}) = \frac{1}{\sqrt{2\pi}\sigma(d_i(\mathbf{p}))} \exp\left(-\frac{\left(\hat{d}_i - d_i(\mathbf{p})\right)^2}{2\sigma^2(d_i(\mathbf{p}))}\right). \tag{5.31}$$

Let $\hat{\mathbf{p}} = (\hat{x}, \hat{y})$ be any position estimation. The fundamental limit on the accuracy of any localization method is given by the CRB on the MSE of position estimation, called position error bound (PEB). Considering independent distance measurements, the PEB is [61]

$$\text{PEB}(x, y) = \sqrt{\frac{\sum_{i=1}^{N} A_i}{\left(\sum_{i=1}^{N} A_i c_i^2\right)\left(\sum_{i=1}^{N} A_i s_i^2\right) - \left(\sum_{i=1}^{N} A_i c_i s_i\right)^2}} \tag{5.32}$$

where $A_i \triangleq 1/\sigma_i^2$, $c_i \triangleq \cos\theta_i$, and $s_i \triangleq \sin\theta_i$, and θ_i is the angle between the ith reader and the tag with respect to the horizontal. Due to multipath and extra propagation delays (caused by obstacles such as walls), range measurements could be positively biased, thus affecting the achievable performance [34]. This effect has been accounted for in Jourdan, Dardari, and Win [61].

In Figure 5.18 an example of theoretical achievable localization accuracy evaluated using (5.5) and (5.32) in a square area of 20 × 20 m composed of four readers is

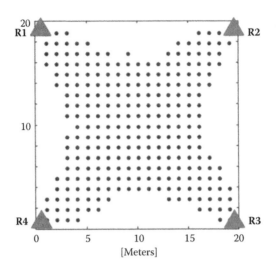

FIGURE 5.18 Coverage map using four readers (R1-R4) and passive UWB tags. Locations with position estimation RMSE less than 20 cm are marked [5].

given. Distance estimation between each reader and the tag is obtained by measuring the signal propagation round-trip time. Results are derived starting from measured data in AWGN and considering random tag orientation. The same system parameters used in Section 5.5.2.1 are considered with $N_s = 1000$. In particular, Figure 5.18 shows the covered locations in a predefined grid, where a location is defined to be covered if, for at least 70% of possible tag orientations, the localization estimation RMSE is less than 20 cm. As can be noted, even in a relatively big area, the number of locations covered with high-accuracy results can be quite large, thus making submeter accuracy localization using backscatter signaling feasible. More extended analysis starting from real measurements can be found in Heiries et al. [46] and Bartoletti and Conti [62].

5.6 ADVANCED ISSUES

5.6.1 CHIPLESS TAGS

An interesting technology exploiting backscatter signaling is represented by full passive or chipless tags. In chipless tags neither the microcontroller nor backscatter modulator is present, so then no energy source is required. They are based on passive antenna loads whose response is tag dependent. Therefore, the reader can identify different tags by analyzing the structure of the signal backscattered.

An example of chipless tag can be found in Zheng et al. [63], where the proposed tag is composed of a transmission line with mismatched reconfigurable impedances encoding the tag ID as shown in Figure 5.19. Very low cost inkjet printing is used to reconfigure the tag ID by creating intersections between the preprinted planar capacitors and the transmission line. When a UWB interrogation signal is received, the impedances cause reflected waves that travel backward along the transmission

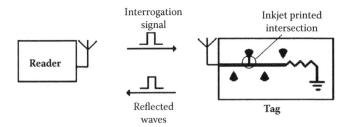

FIGURE 5.19   Example of UWB chipless tag proposed in Zheng et al. [63].

line. The encoded ID can be gained by analyzing the reflected response at the reader side. This technology is quite promising thanks to the extremely low cost implementation, even though current implementations suffer from several drawbacks in terms of maximum ID length (up to 8 bits) and dimensions (a 40 cm long transmission line is required).

Similar approaches are represented by multiresonator chipless tags, where tag ID is encoded into the spectral domain in both magnitude and phase of the spectrum [64].

5.6.2   HIGH-ACCURACY RADIO DETECTION, IDENTIFICATION, AND LOCALIZATION APPLICATIONS

The possibility of designing a cheap wireless network composed of several radio transmitting/receiving nodes able to detect, localize, and track with high precision moving objects—also discriminating between collaborative objects (objects equipped with tags) and noncollaborative objects (i.e., nonidentifiable [targets])—is particularly attractive. UWB is the only technology today that allows in principle the integration of object detection (through the concept of wireless sensor radar [WSR]) [21,62,65], object identification (RFID), and localization features (RTLS) to form a new concept of integrated system we name *hi*gh-accuracy *ra*dio *d*etection, *i*dentification, *a*nd *l*ocalization (Hi-RADIAL).

A wireless network integrating such capabilities would lead to a relevant improvement in development and reliability of a wide range of advanced applications including logistics (package tracking), security (localizing authorized persons in high-security areas as well as homeland), medical (monitoring of patients), family communications/supervision of children, search and rescue (communications with fire fighters or avalanche/earthquake victims), automotive safety, and military applications.

A possible application scenario example is reported in Figure 5.20, where an indoor area must be monitored and protected by a Hi-RADIAL network composed of several UWB radio transmitters/receivers placed in known positions. All authorized persons (or objects) are equipped with a UWB RFID tag. The system has the task to detect a noncollaborative moving target (intruder) entering the area and, at the same time, discriminate between authorized persons (by identifying their tags) and unauthorized persons. The position of all targets and tags is tracked in real time with submeter precision. Examples of specific applications falling in this scenario are represented by indoor/outdoor advanced security systems for personal safety,

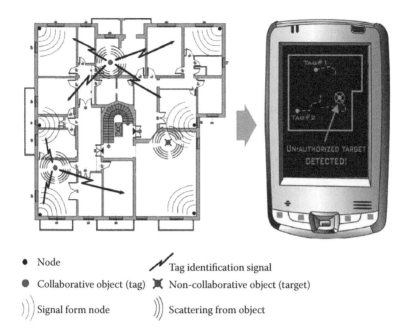

- • Node
- ● Collaborative object (tag)
-))) Signal form node

- ↗ Tag identification signal
- ✖ Non-collaborative object (target)
-))) Scattering from object

FIGURE 5.20   Example of advanced security application using the Hi-RADIAL concept.

package tracking in supply chain management, advanced systems for the surveillance of factory warehouses, and automotive applications.

The Hi-RADIAL concept opens new, challenging issues determined by spectrum use regulatory constraints; application requirements; wireless communication issues, such as the presence of dense multipath; clutter; NLOS conditions; interference; and the need of low cost and low consumption devices [19].

ACKNOWLEDGMENTS

The author would like to thank Andrea Conti, Nicoló Decarli, Raffaele D'Errico, Francesco Guidi, and Alain Sibille. This work has been performed within the framework FP7 European Project SELECT (grant agreement no. 257544).

REFERENCES

1. K. Finkenzeller, in *RFID handbook: Fundamentals and applications in contactless smart cards and identification,* 2nd ed. New York: Wiley, 2004.
2. E. Ngai, K. K. Moon, F. J. Riggins, and C. Y. Yi, RFID research: An academic literature review (1995–2005) and future research directions. *International Journal of Production Economics,* vol. 112, no. 2, pp. 510–520, 2008, Special Section on RFID: Technology, Applications, and Impact on Business Operations.
3. V. Chawla and D. S. Ha, An overview of passive RFID. *IEEE Applications & Practice,* pp. 11–17, Sept. 2007.
4. R. Das and P. Harrop, RFID forecast, players and opportunities 2007–2017 (available online at http://www.idtechex.com), 2007.

5. D. Dardari, R. D'Errico, C. Roblin, A. Sibille, and M. Z. Win, Ultrawide bandwidth RFID: The next generation? *Proceedings of the IEEE,* Special Issue on RFID—A Unique Radio Innovation for the 21st Century, vol. 98, no. 9, pp. 1570–1582, Sept. 2010.

6. R. Verdone, D. Dardari, G. Mazzini, and A. Conti, *Wireless sensor and actuator networks: Technologies, analysis and design.* London: Elsevier Ltd., 2008.

7. D. Ha and P. Schaumont, Replacing cryptography with ultra wideband (UWB) modulation in secure RFID. 2007 IEEE International Conference on RFID, Grapevine, TX, March 2007.

8. T. van Deursen and S. Radomirović, Security of RFID protocols—A case study. *Electron Notes in Theoretical Computer Science,* vol. 244, pp. 41–52, 2009.

9. J. Griffin and G. Durgin, Reduced fading for RFID tags with multiple antennas. *2007 IEEE Antennas and Propagation International Symposium,* pp. 1201–1204, June 2007.

10. M. Z. Win and R. A. Scholtz, Impulse radio: How it works. *IEEE Communications Letters,* vol. 2, no. 2, pp. 36–38, Feb. 1998.

11. *Proceedings of IEEE,* Special Issue on UWB Technology & Emerging Applications, Feb. 2009.

12. R. J. Fontana and S. J. Gunderson, Ultra-wideband precision asset location system. *Proceedings of IEEE Conference on Ultra Wideband Systems and Technologies (UWBST),* vol. 21, no. 1, pp. 147–150, May 2002.

13. Time domain corporation, www.timedomain.com

14. Ubisense limited, www.ubisense.net

15. Zebra enterprise solutions, zes.zebra.com

16. IEEE 802.15.4f Draft, IEEE standard for information technology—Telecommunications and information exchange between systems—Local and metropolitan area networks—Specific requirements—Part 15.4: Wireless medium access control (MAC) and physical layer (PHY) specifications for low rate wireless personal area networks (WPANs)—Amendment: Active RFID system PHY. 2011. Online. Available: http://www.ieee802.org/15/pub/TG4f.html

17. D. Dardari, Method and apparatus for communication in ultra-wide bandwidth RFID systems. International Patent Application PCT/IB2009/000 360, Feb. 25, 2009.

18. D. Dardari and R. D'Errico, Passive ultrawide bandwidth RFID. IEEE Global Communications Conference (GLOBECOM 2008), New Orleans, LA, Nov. 2008.

19. SELECT (smart and efficient location, identification, and cooperation techniques), FP7 European Project, http://www.selectwireless.eu

20. Z. Sahinoglu, S. Gezici, and I. Guvenc, in *Ultra-wideband positioning systems: Theoretical limits, ranging algorithms, and protocols.* New York: Cambridge University Press, 2008.

21. E. Paolini, A. Giorgetti, M. Chiani, R. Minutolo, and M. Montanari, Localization capability of cooperative anti-intruder radar systems. *EURASIP Journal on Advances in Signal Processing,* Special Issue on Cooperative Localization in Wireless Ad Hoc and Sensor Networks, vol. 2008, Article ID 726854, p. 14, 2008.

22. Federal Communications Commission, Revision of part 15 of the commission's rules regarding ultra-wideband transmission systems, first report and order (ET Docket 98-153), adopted Feb. 14, 2002, released April 22, 2002.

23. European Commission, Commission decision of 21 February 2007 on allowing the use of the radio spectrum for equipment using ultra-wideband technology in a harmonized manner in the community. *Official Journal of the European Union,* vol. C (2007) 522, Feb. 2007.

24. European Commission, Amendment to ECC decision to include DAA: Ecc/dec/(06)12 amended, October 2008.

25. European Commission, Draft ECC report 170 in ECC consultation ECC report on specific UWB applications in the bands 3.4–4.8 GHz and 6–8.5 GHz location tracking applications for emergency services (LAES), location tracking applications type 2 (LT2) and location tracking and sensor applications for automotive and transportation environments (LTA).

26. European Commission, Draft ECC recommendation (11)09 UWB location tracking systems type 2 (LT2). Recommendation adopted by the Working Group Frequency Management (WG FM), 2009.

27. WiMedia Alliance, Wimedia, 2005. Online. Available: http://www.wimedia.org

28. IEEE standard for information technology—Telecommunications and information exchange between systems—Local and metropolitan area networks—Specific requirement part 15.4: Wireless medium access control (MAC) and physical layer (PHY) specifications for low-rate wireless personal area networks (WPANs), IEEE Std 802.15.4a-2007 (amendment to IEEE Std 802.15.4-2006), pp. 1–203, 2007.

29. M. Baghaei-Nejad, Z. Zou, H. Tenhunen, and L.-R. Zheng, A novel passive tag with asymmetric wireless link for RFID and WSN applications. ISCAS 2007. *IEEE International Symposium on Circuits and Systems,* 2007, pp. 1593–1596, May 2007.

30. D. Cassioli, M. Z. Win, and A. F. Molisch, The ultra-wide bandwidth indoor channel: From statistical model to simulations. *IEEE Journal of Select Areas in Communication,* vol. 20, no. 6, pp. 1247–1257, Aug. 2002.

31. A. F. Molisch, D. Cassioli, C.-C. Chong, S. Emami, A. Fort, B. Kannan, J. Karedal, et al., A comprehensive standardized model for ultrawideband propagation channels. *IEEE Transactions Antennas Propagation,* vol. 54, no. 11, pp. 3151–3166, Nov. 2006, Special Issue on Wireless Communications.

32. M. Z. Win and R. A. Scholtz, Characterization of ultra-wide bandwidth wireless indoor communications channel: A communication theoretic view. *IEEE Journal of Select Areas Communication,* vol. 20, no. 9, pp. 1613–1627, Dec. 2002.

33. D. Dardari, E. Falletti, and M. Luise, in *Satellite and terrestrial radio positioning techniques—A signal processing perspective.* Elsevier Ltd, London, 2011.

34. D. Dardari, A. Conti, U. Ferner, A. Giorgetti, and M. Z. Win, Ranging with ultrawide bandwidth signals in multipath environments. *Proceedings of IEEE,* Special Issue on UWB Technology & Emerging Applications, vol. 97, no. 2, pp. 404–426, Feb. 2009.

35. C.-C. Chong and S. K. Yong, A generic statistical-based UWB channel model for highrise apartments. *IEEE Transactions Antennas Propagation,* vol. 53, pp. 2389–2399, 2005.

36. A. Molisch, K. Balakrishnan, D. Cassioli, C.-C. Chong, S. Emami, A. Fort, J. Karedal, et al., IEEE 802.15.4a channel model—Final report, 2005.

37. A. Saleh and R. A. Valenzuela, A statistical model for indoor multipath propagation. *IEEE Journal of Select Areas Communication,* vol. 5, no. 2, pp. 128–137, Feb. 1987.

38. R. C. Hansen, Relationship between antennas as scatters and as radiators. *Proceedings of the IEEE,* vol. 77, no. 5, 1989.

39. K. Penttila, M. Keskilammi, L. Sydanheimo, and M. Kivikoski, Radar cross-section analysis for passive RFID systems. *IEE Proceedings on Microwave and Antennas Propagation,* vol. 153, no. 1, pp. 103–109, Feb. 2006.

40. S. Hu, C. L. Law, Z. Shen, L. Zhu, W. Zhang, and W. Dou, Backscattering cross section of ultrawideband antennas. *IEEE Antennas Wireless Propagation Letters,* vol. 6, pp. 70–72, 2007.

41. S. Hu, Y. Zhou, C. L. Law, and W. Dou, Study of a uniplanar monopole antenna for passive chipless UWB-RFID localization system. *IEEE Transactions Antennas Propagation,* vol. 58, no. 2, pp. 271–278, 2010.

42. A. Sibille, M. Sacko, Z. Mhanna, F. Guidi, and C. Roblin, Joint antenna-channel statistical modeling of UWB backscattering RFID. *Proceedings of the 2011 IEEE International Conference on Ultra-Wideband (ICUWB 2011),* Sept. 2011, pp. 474–478.

43. F. Guidi, A. Sibille, D. Dardari, and C. Roblin, UWB RFID backscattered energy in the presence of nearby metallic reflectors. European Conference on Antennas and Propagation 2011, Rome, Italy, April 2011.

44. D. Dardari, F. Guidi, C. Roblin, and A. Sibille. Ultra-wide bandwidth backscatter modulation: Processing schemes and performance. *EURASIP Journal on Wireless Communications and Networking,* vol. 2011, no. 1, 2011.

45. C. Roblin, S. Bories, and A. Sibille, Characterization tools of antennas in the time domain. Proceedings of International Workshop on Ultra Wideband Band Systems (IWUWBS 2003).

46. V. Heiries, K. Belmkaddem, F. Dehmas, B. Denis, L. Ouvry, and R. D'Errico, UWB backscattering system for passive RFID tag ranging and tracking. *Proceedings of the 2011 IEEE International Conference on Ultra-Wideband (ICUWB 2011),* pp. 489–493, Sept. 2011.

47. F. Guidi, D. Dardari, C. Roblin, and A. Sibille, Backscatter communication using ultrawide bandwidth signals for RFID applications, in D. Giusto et al. (eds.), *The Internet of things: 20th Tyrrhenian Workshop on Digital Communications,* SpringerScience+BusinessMedia, Pula, Sardinia, Italy, pp. 251–262, Sept. 2009.

48. D. Dardari, Pseudo-random active UWB reflectors for accurate ranging. *IEEE Communications Letters,* vol. 8, no. 10, pp. 608–610, Oct. 2004.

49. L. Stoica, A. Rabbachin, H. O. Repo, T. S. Tiuraniemi, and I. Oppermann, An ultrawideband system architecture for tag based wireless sensor networks, *IEEE Transactions Vehicular Technolology,* vol. 54, no. 5, pp. 1632–1645, Sept. 2005.

50. Z. Zou, M. Baghaei-Nejad, H. Tenhunen, and L.-R. Zheng, An efficient passive RFID system for ubiquitous identification and sensing using impulse UWB radio, *e & i Elektrotechnik und Informationstechnik,* Springer Wien, vol. 124, no. 11, pp. 397–403, Nov. 2007.

51. J. A. Paradiso and T. Starner, Energy scavenging for mobile and wireless electronics, *IEEE Pervasive Computing,* pp. 18–26, Jan.–March 2005.

52. K. Pahlaven and H. Eskafi, Radio frequency tag and reader with asymmetric communication bandwidth, US Patent 7,180,421 B2, Feb. 20, 2007.

53. A. Muchkaev, Carrierless RFID system, US Patent 7,385,511 B2, June 10, 2008.

54. J. Jantunen, A. Lappetelainen, J. Arponen, A. Parssinen, M. Pelissier, B. Gomez, and J. Keignart, A new symmetric transceiver architecture for pulsed short-range communication. *IEEE Global Communications Conference* (GLOBECOM 2008), New Orleans, LA, Nov. 2008.

55. R. Vauche, E. Bergeret, J. Gaubert, S. Bourdel, O. Fourquin, and N. Dehaese, A remotely UHF powered UWB transmitter for high precision localization of RFID tag. *Proceedings of the 2011 IEEE International Conference on Ultra-Wideband (ICUWB 2011),* pp. 494–498, Sept. 2011.

56. J. Reunamaki, Ultra wideband radio frequency identification techniques, US Patent 7,154,396, Dec. 26, 2006.

57. F. Guidi, N. Decarli, D. Dardari, C. Roblin, and A. Sibille. Performance of UWB backscatter modulation in multi-tag RFID scenario using experimental data. *IEEE International Conference on Ultra-Wideband,* ICUWB 2011, Bologna, Italy, pp. 1–5, Sept. 2011.

58. W. Suwansantisuk and M. Z. Win, Multipath aided rapid acquisition: Optimal search strategies. *IEEE Transactions Information Theory,* vol. 52, no. 1, pp. 174–193, Jan. 2007.

59. C. Xu and C. L. Law, TOA estimator for UWB backscattering RFID system with clutter suppression capability. *EURASIP Journal on Wireless Communications and Networking,* vol. 2010, Article ID 753129, 14 pp., 2010.
60. G. D. Vita and G. Iannaccone, Design criteria for the RF section of UHF and microwave passive RFID transponders. *IEEE Transactions on Microwave Theory and Techniques,* vol. 59, no. 9, pp. 2978–2990, Sept. 2005.
61. D. B. Jourdan, D. Dardari, and M. Z. Win, Position error bound for UWB localization in dense cluttered environments. *IEEE Transactions Aerospace Electronic Systems,* vol. 44, no. 2, pp. 613–628, April 2008.
62. S. Bartoletti and A. Conti, Passive network localization via UWB wireless sensor radars: The impact of TOA estimation. *Proceedings of the 2011 IEEE International Conference on Ultra-Wideband (ICUWB 2011),* pp. 258–262, Sept. 2011.
63. L. Zheng, S. Rodriguez, L. Zhang, B. Shao, and L.-R. Zheng, Design and implementation of a fully reconfigurable chipless RFID tag using inkjet printing technology. *IEEE International Symposium on Circuits and Systems 2008* (ISCAS 2008), pp. 1524–1527, May 2008.
64. S. Preradovic, I. Balbin, N. Karmakar, and G. Swiegers, Multiresonator-based chipless RFID system for low-cost item tracking. *IEEE Transactions on Microwave Theory and Techniques,* vol. 57, no. 5, pp. 1411–1419, May 2009.
65. V. Casadei, N. Nanna, and D. Dardari, Experimental study in breath detection and human target ranging in the presence of obstacles using ultra-wideband signals. *International Journal of Ultra Wideband Communications and Systems,* Special Issue on Applications of Ultra Wideband Technology in Healthcare, 2011.
66. N. Decarli, F. Guidi, A. Conti, D. Dardari, Interference and clock drift in UWB RFID systems using backscatter modulation. *IEEE International Conference on Ultra-Wideband,* ICUWB 2012, Syracuse, NY September 2012, pp. 1–5.

6 Multifunctional Software-Defined Radar Sensors for Detection, Imaging, and Navigation

*Dmitriy Garmatyuk, Kyle Kauffman,
Y. T. (Jade) Morton, and John Raquet*

CONTENTS

6.1   INTRODUCTION

Multifunctional radio frequency (RF) sensors are a natural evolution of radar. When first deployed in World War II, the early radar systems were tasked with estimating the range to the potential targets of interest only, and even due to this single functionality it has been claimed that it "is obvious that radar transformed the nature of war more than has any other single invention" [1]. Less than two decades after this triumphant entrance, the radar functionalities were enriched with imaging [2]. This, too, signaled a fundamental change in a number of fields—from the military, which could now perform all-weather, day or night surveillance of obscured targets [3], to the civilian applications of radar, such as monitoring ice sheets or obtaining imagery of space bodies [4]. In parallel, the radar as a navigational tool was also explored, primarily for airborne and seaborne navigation [5]. Once these capabilities were established, the need to distinguish between the targets and objects of different shapes and dimensions quickly emerged and, with it, the automatic target recognition (ATR) field [6]. Thus, the functionalities of imaging and detection began to converge.

As radar technology has further developed, the desire for better, more high-resolution imagery has strengthened, as a better image would produce better ATR and detection results. A fundamental law of radar dictates that range resolution is inversely proportional to the radar signal's bandwidth [7], which, in turn, led to the development of wideband and ultra-wideband (UWB) radar systems [3,8]. The downside of this technology, however, is the vast amount of information that UWB signals can provide, which initially made their real-time processing a near impossibility, especially on board of airborne vehicles that carried the actual radar platform for, e.g., reconnaissance purposes, thus necessitating off-line image formation and processing in a ground-based facility. Furthermore, the growing list of onboard devices and their functionalities—detection, imaging, communications, location—combined with the desire to utilize them in ever smaller, autonomous systems, such as unmanned aerial vehicles (UAVs) started to present a challenge in terms of increased weight, size, power consumption, and interoperability. A potential solution to these problems can consist of two parts:

- Better (more computationally efficient) data processing algorithms run on ever faster computing platforms [9]
- Employing multifunctional, software-defined sensor architectures, which enable multiple tasks to be carried out via a single analog front end (AFE) and processed locally in real time

The focus of this chapter is on the latter approach.

The interest in multifunctional systems can be traced back to a 1978 paper on the Ku-band NASA Space Shuttle Orbiter subsystem, which combined the radar and communication functionalities [10]. The subsystem described in the paper could perform as a wideband data communication device, with a data rate up to 216 kb/s, or as frequency-hopping pulse-Doppler radar, providing range accuracy of 80 ft and velocity accuracy of 1 ft/s. The authors indicated that the radar and communication functions shared a wideband, multifunction transmitter, a two-channel receiver, and an antenna, while each function had its own signal processing mechanisms. This description provides a natural connection of the dual-use concept and a concept of software-defined radar sensor (SDRS) suggested in 2001 by Wiesbeck [11]. In SDRS and in software-defined synthetic aperture radar (SD-SAR) systems, the concept of fully software-configurable transmitters and modular digital receiver data processors with downloadable software for a variety of scenarios is envisioned. Conceptual design of a multifunctional RF system test bed developed for the Advanced Multifunction Radio Frequency Concept (AMRFC) Program is summarized in Tavik et al. [12].

In recent years there have been explorations of radar and communications functionalities' fusion via several different approaches; for example, in Saddik, Singh, and Brown [13], a system based on implementing UWB linear frequency modulated (LFM) signals of opposite chirp slopes to achieve orthogonality between the radar and communications channels is described. In Sturm and Wiesbeck [14], a conceptual approach to fusing ranging and communications for vehicular applications is presented. Furthermore, software-defined sensor systems have been recently researched and reported as well. For example, a UWB stepped-frequency radar sensor for subsurface probing is presented in Park and Nguyen [15] and a high-resolution multiband UWB short-pulse radar sensor is described in Han and Nguyen [16]. A software-defined radar system for high-precision range measurements was also recently described in Zhang, Li, and Wu [17].

In our work we merged the radar/communications functionality with the SDRS concept and UWB performance. Our system was designed as a UWB orthogonal frequency-division multiplexing (OFDM) architecture, which has been shown to possess high spectral efficiency, excellent spectral flexibility with the potential to coexist with narrowband systems and/or effectively combat narrowband interference, and good performance in multipath scenarios [18]. UWB-OFDM in communications has been researched and quickly accepted by the industry as a system architecture and coding method of choice for low power, broadband data communications [19]. In radar, UWB-OFDM signaling and system architecture has not been explored as much; OFDM coding structure was first used to create a multifrequency complementary phase-coded radar signal concept by Levanon [20,21]. OFDM was proposed to be used for interferometer implementation [22]. Then, in recent years the interest in OFDM in radar has intensified and resulted in several works on various aspects of it [23–27]. In our work, we suggested using UWB-OFDM SDRS for ranging, communications [28], adaptive imaging [29], detection [30], and navigation [31]. In the following sections we will summarize the current state of the art in multifunctional UWB-OFDM SDRS.

6.2 SOFTWARE-DEFINED RADAR SYSTEM ARCHITECTURE

6.2.1 WAVE FORMING VIA OFDM

OFDM lends itself very well to implementation in software—from spectrum allocation and signal construction to frequency-domain signal processing upon reception. A conceptual diagram showing the frequency composition of an OFDM pulse transmitted at a carrier frequency f_0 is shown in Figure 6.1. From it, it is easy to see that any OFDM signal is essentially a linear combination of RF pulses, each at the carrier frequency that is a multiple of

$$f_{scl} = \frac{F_s}{2N} \tag{6.1}$$

where F_s is the sampling frequency of the digital-to-analog converter (DAC)—a principal component of an SDRS—and N is the desired total number of such RF pulses,

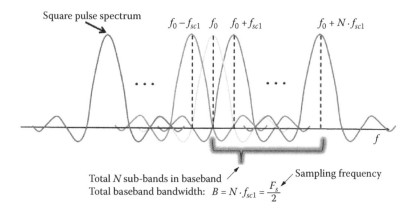

FIGURE 6.1   Conceptual representation of an OFDM pulse in frequency domain.

or sub-bands in frequency domain, which is arbitrary and can even be changed in an OFDM SDRS on a pulse-to-pulse basis.

It can be seen that the OFDM pulse can be designed using both the sub-band weighting (in which each RF pulse obtains an individual amplitude and phase characteristics) and sub-band turning on or off (in which some RF pulses are simply eliminated from the sum). Once the signal is created in frequency domain, it can be translated to time domain via a simple inverse fast Fourier transform (IFFT) operation, which also lends itself well to efficient software implementation. These operations are performed on signal samples—that is, frequency-domain wave forming amounts to creating a complex vector of amplitude and phase coefficients of OFDM signal values at the sub-carrier locations (multiples of f_{sc1}).

6.2.2   TRANSCEIVER DESIGN

The transceiver in our system consists of the two principal parts: the software-defined part and the hardware-implemented AFE. Naturally, the former will contain as much functionality as possible, as dictated by the SDRS concept. The conceptual representation of the software-enabled functions is shown in Figure 6.2. It is easy to see that, with such architecture, the following parameters of the transmit signal can be changed on a pulse-to-pulse basis: duration, spectral composition, bandwidth, envelope shape, and phase characteristic of individual frequency components.

After passing through the software-defined part of the SDRS, the signal is passed to the DAC in the form of discrete samples. The output from the DAC is sent to the AFE portion of the system, where it is up-converted to a 7.5 GHz carrier frequency and is transmitted as a 1 GHz bandwidth pulse using the horn antenna. Upon signal reception with a second horn antenna, the signal is amplified using ultralow noise amplifiers (ULNA) and is down-converted to baseband before being sent to the high-speed digitizer in the software-controlled part of the system. The block diagram of the AFE is illustrated in Figure 6.3.

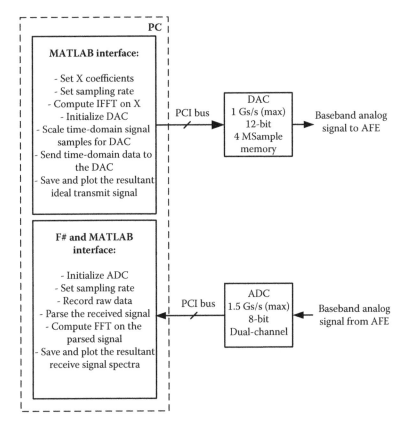

FIGURE 6.2 Conceptual representation of the software-defined part of the SDRS.

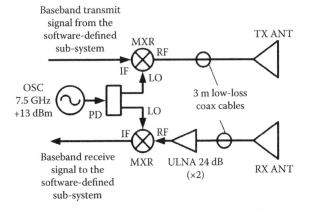

FIGURE 6.3 Conceptual representation of the AFE.

6.2.3   Sensor Signal Processing

Upon transmission, the receiver part of the AFE starts to receive the signals, which may consist of reflections from targets and noise, or contain noise only. Once the signal is received, it is sampled and passed through the FFT to arrive at the frequency-domain representation. We aim to reconstruct the target response (consisting of M targets), which is a part of the response vector:

$$\Theta = \sum_{q=1}^{M} (\alpha_q + \mathbf{w})^T \Psi_q \tag{6.2}$$

where α_k is frequency response of the qth target; $\mathbf{w}$ is cumulative disturbance vector consisting of clutter, noise, and system nonlinearity; and ψ_k is the vector:

$$\Psi_q = \left[e^{+j2\pi\frac{N \cdot f_s}{2N}\Delta t_q}, \ldots e^{+j2\pi\frac{(N-1)f_s}{2N}\Delta t_q}, \ldots, 0, \atop e^{-j2\pi\frac{f_s}{2N}\Delta t_q}, \ldots, e^{-j2\pi\frac{N \cdot f_s}{2N}\Delta t_q} \right]^T \tag{6.3}$$

where Δt_q is the round-trip time delay associated with the received signal reflected by the qth target and the exponential values are shown for the samples associated with the sub-carrier locations only. The rest of the values can be interpolated without loss of generality.

For range compression, the conventional technique in radar is to perform matched filtering of the received signal with the transmit signal replica stored locally. This is often done in frequency domain to take advantage of the fast and efficient FFT algorithm implementation. Since OFDM signals are converted into the frequency domain upon reception by default, this makes it a natural choice for our system as well. The received signal will contain K samples associated with K individual ranges, as determined by the system's sampling rate and the maximum range. Then, the original transmit signal is up-sampled (zero-padded) by a factor of $K/(2N + 1)$ to ensure dimension match. The resultant range profile can then be computed as

$$\mathbf{y} = \mathcal{F}^{-1}\left[\mathbf{s}^H \mathbf{X}^K \right] \tag{6.4}$$

where $\mathcal{F}^{-1}$ denotes inverse Fourier transform, H denotes Hermitian operator, and $\mathbf{X}^K$ denotes the original frequency-domain vector of the transmit OFDM signal's coefficients, which had been up-sampled to the length K via time-domain zero-padding. In the next section we will present some of the experimental results of this signal processing approach.

6.3   EXPERIMENTAL RESULTS

The experimental tests took place in an indoor laboratory setting with a multipath environment. Three types of experimental results are presented: (1) single-target

SAR imaging, (2) angle-frequency profile reconstruction, and (3) communication performance in radar setup.

6.3.1 SINGLE-TARGET IMAGING

We created a distributed aperture by moving the antenna platform along a straight path and taking radar measurements of the reflections from a single target at a number of positions, as shown in Figure 6.4, thus performing SAR imaging test.

The signals used in this experiment were constructed from $N = 129$ sub-bands, with the sub-bands weighted with vectors of random real coefficients $\mathbf{X_m}$, where $\mathbf{m} = 1 \dots 11$ is the index of radar cross-range position. Range compression was achieved, as in the case of ranging tests, via matched filtering of the received and reference waveforms. Azimuth compression was achieved via matched filtering of the radar signal's phase history with a reference phase history, as described, for example, in Soumekh [32]; however, instead of Doppler centroid estimation (which is required in a general case for phase history reconstruction of a wideband signal), we implemented a novel method of single subcarrier phase recovery, suitable for OFDM:

$$\hat{\Phi}_m(k) = \tan^{-1}\left(\frac{\mathrm{Im}\left(\hat{X}_m(k)\right)}{\mathrm{Re}\left(\hat{X}_m(k)\right)}\right), \tag{6.5}$$

where $\hat{\Phi}_m(k)$ is the phase estimate of kth OFDM subcarrier at the mth radar position. In our experiment we have arbitrarily selected the 10th sub-band (centered on $f_{10} = 38.7597$ MHz) for phase reconstruction.

Then, the cross-range profile for a kth sub-band and range bin value is found conventionally via the matched filtering of the received signal's phase history and the reference phase history [32]. Finally, the SAR image is formed via multiplication of the range profile data and the cross-range profile data. A simulated ideal SAR

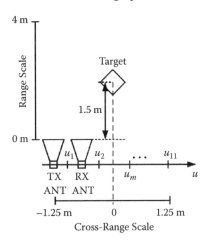

FIGURE 6.4 Short-range SAR experiment setup.

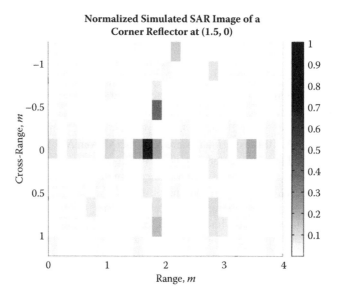

FIGURE 6.5 Normalized simulated SAR image of a 1 ft corner reflector at (1.5, 0) meter coordinates.

image for the geometry described in Figure 6.4 was generated; no noise and clutter were added and the target was modeled as a point reflector. As in experiment, the 10th sub-band of the OFDM signals was used for phase history reconstruction and its coefficient was set to a constant value for all modeled pulses, whereas the rest of the sub-bands' coefficients were generated randomly for each pulse. A representative resultant image is shown in Figure 6.5. The experimental SAR image generated using actual test data is shown in Figure 6.6. It is seen that the images are fairly similar, thus confirming the functionality of the system and the processing method.

It is also seen that the true position of the target is reconstructed fairly well, with less than 10 cm of error in both range and cross range. An interesting possibility that arises from the use of a multifrequency waveform such as OFDM is cross-range profile reconstruction using several various sub-carriers.

6.3.2 ANGLE-FREQUENCY RECONSTRUCTIONS

We further explored the benefits of the multifrequency signal transmission and reception afforded by the UWB-OFDM system architecture by reconstructing target scenes in angle and frequency domains. Matched-filtered returns were collected from a single corner reflector (described in the preceding subsection), as well as from a piece of drywall for each of the 16 subcarriers that together composed a 500 MHz baseband spectrum of the OFDM signal. The experimental setup is illustrated in Figure 6.7.

After the data were averaged over 100 trials, the angle-frequency graphs were plotted; two examples are shown in Figure 6.8. It can be seen that the signatures for two different targets can, indeed, be discerned, making it potentially possible to

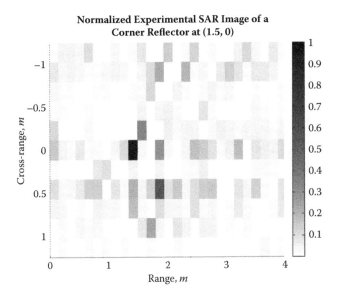

FIGURE 6.6   Normalized experimental SAR image of a 1 ft corner reflector at (1.5, 0) meter coordinates.

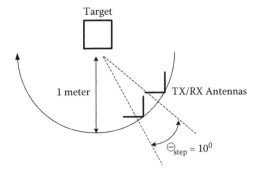

FIGURE 6.7   Data collection setup for angle-frequency target reconstructions.

perform automatic target detection and identification based on the observed angle-frequency profiles.

6.3.3   DATA COMMUNICATIONS EXPERIMENT SETUP

In this experiment we performed wireless data transfer from the transmitter to the receiver without any hardware adjustments in either of them, and also without changing the basic radar geometry: transmit and receive antennas both faced a single corner reflector. Randomly generated binary messages were coded onto the OFDM pulses using a simple on/off keying (OOK) modulation technique. For this experiment the number of sub-bands was chosen to be $N = 64$ and the total pulse length was 128 ns. Through initial tests it was found that transmission quality of several

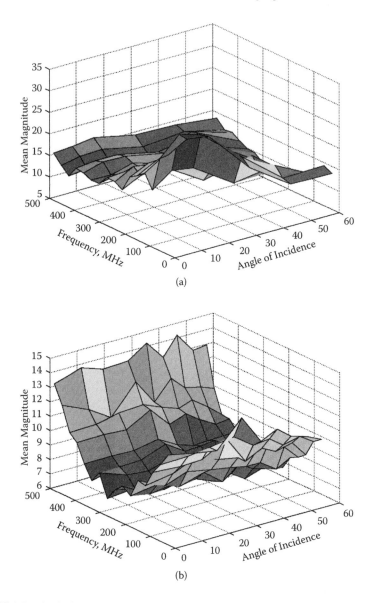

FIGURE 6.8 Angle-frequency profiles of (a) corner reflector; (b) drywall segment at 1 m range.

sub-bands was significantly degraded by systemic in-band noise caused by the DC offset and switching noise in the DAC/ADC; specifically, sub-bands in the range from DC to approximately 100 MHz and also from 470 to 500 MHz were found unusable. After usable sub-band allocation, the total number of bits per each OFDM pulse was set to 19. Additionally, the guard interval between any two OFDM message pulses was set to be four times the length of the typical delay spread of UWB signals in an indoor environment [33], or $L_{guard} = 200$ ns.

The resultant data rate achievable under this scenario is 57 Mb/s. Communication tests were run for the antenna to target distances varying from 1 to 5 m. Signals reflected from the corner reflector were collected by the receive antenna and compared to the original message after detection. The results of the simulations, using the work of Santhananthan and Tellambura [34] and Dharmawansa, Rajatheva, and Minn [35] for various numbers of OFDM subcarriers, along with the overlaid experimental bit error rate (BER) curve, are shown in Figure 6.9. Here, the deterioration factor was the intercarrier interference (ICI), expressed in terms of frequency offset resultant from the loss of subcarrier orthogonality ε.

It can be seen that the frequency offset-induced ICI is, indeed, a strong factor in BER degradation. At $\varepsilon \approx 0.25$, the dependence of BER on the number of subcarriers is quite evident as well: The more subcarriers are used, the less BER depends on the range and the faster it converges to a vicinity of 1% to 2%. This characteristic is not observed for the cases with $\varepsilon \approx 0.5$, however. It can also be seen from the plot that within the tested ranges, experimental BER stays below approximately 3% and does not exhibit strong dependence on distance, leading us to believe that the frequency offset in our system is probably close to 0.25, or approximately 2 MHz.

Thus, compared to the results reported in Saddik et al. [13], we have demonstrated true wireless communication performance of a dual-use system in an indoor laboratory environment—as opposed to a direct transmitter–receiver connection via an attenuator—while achieving the functionality in which the same signals could be used for high-resolution radar and communication purposes in the presence of multiple system platforms. This capability makes the proposed system very viable for implementation in radar sensor networks.

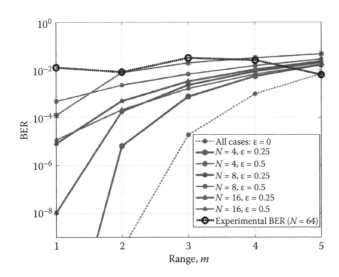

FIGURE 6.9   Simulated and experimental BER performance.

6.4 UWB-OFDM RADAR-ASSISTED NAVIGATION

The UWB-OFDM radar system was simulated to measure its performance as an integrated sensor in a navigation system. In the simulation, the radar is modeled using the parameters of the experimental system, and is mounted as a side-looking radar on board an aerial vehicle (AV). As the AV flies with continuously changing velocity, the radar collects backscatter from reflective surfaces on the ground. The collected backscatter (analogous to the phase history in SAR imaging) is then used to gain information about the relative position of the (assumed) stationary reflectors on the ground and the AV. This relative motion information can be used in a navigation filter to provide additional information about the velocity and position of the AV.

6.4.1 RAW DATA PROCESSING

The raw data are the baseband samples collected at the receiver. For navigation, it is necessary to use a two-channel in-phase and quadrature-phase (I/Q) demodulator, yielding samples from two channels at the baseband. The detection of persistent, strong, stationary reflectors on the ground is performed via thresholding the normalized MF output between the received signal and the OFDM reference pulse. Let $x_k(n) = x_{I,k}(n) + x_{Q,k}(n)$ be the two-channel raw data collected at a particular time instant k. Then the MF output is

$$m_k = IFFT\left(FFT(|x_k|)^* FFT(x)\right) \tag{6.6}$$

where x are time-domain samples of an OFDM signal and $*$ is the complex conjugate. Assuming that a single reflector is present, the detection statistic is the peak output of $m_k(n)$ over all n normalized by the noise level is

$$m_{\max,k} = \frac{P(\max m_k)}{P(m_k)} \tag{6.7}$$

where P is the average power of the samples. The detection is then performed on $m_{\max,k}$ at every time k, using a threshold for detection that sets the probability of false alarm as desired. Figure 6.10 shows a histogram of the values calculated for $m_{\max,k}$ with no target and a single target present. In general, multiple reflectors may be present. Multiple reflectors are detected by repeatedly selecting the next highest peak output of the MF and performing threshold detection as described earlier.

Simple detection as described before produces a set of detection locations at every time instant. A tracking algorithm must be used to associate detections at one time instance with those at the next, which is vital for navigation as we are interested in the position of the reflectors relative to ours over time. This problem was solved by using a global nearest neighbor (GNN) tracking algorithm combined with an M/N detector to validate detected reflector tracks as persistent. In order to be used in the navigation filter, a tracked reflector must have had an observation (MF output peak above detection threshold) associated with it by the GNN association algorithm at M

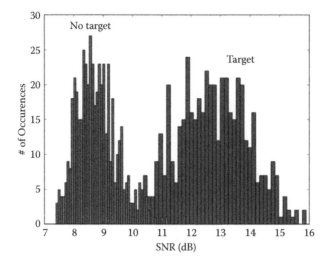

FIGURE 6.10   Histogram of calculated MF SNRs with true MF SNR of 18 dB.

of the last N data collections. When this condition is met, the track is inserted into the navigation filter as used to correct position updates, as discussed in the next section.

6.4.2 NAVIGATION FILTER DESIGN

The navigation filter combines information from an inertial navigation system (INS) and the tracked reflectors from the previous section to generate an optimal estimate of the position of the AV. Figure 6.11 illustrates the integration process. This design is called an error-state model, as the quantity being estimated by an extended Kalman filter (EKF) is actually the error in the INS position estimate. To generate the best position estimate known to the filter, the calculated INS error is subtracted from the INS position estimate. The performance of the filter is measured by the error between this estimate and the true location of the AV.

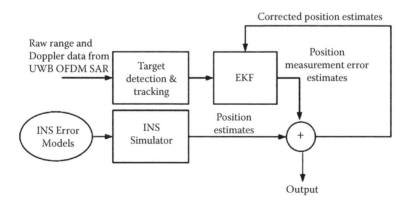

FIGURE 6.11   Block diagram of INS and radar integration.

TABLE 6.1

**Position Error in Latitude (Initial AV Direction of Travel)
after 10 Minutes of Flight**

	Aiding	No aiding
Commercial grade	4.0 m	1700 km
Tactical grade	1.2 m	600 km
Navigation grade	0.5 m	15 m

TABLE 6.2

Position Error in Longitude after 10 Minutes of Flight

	Aiding	No aiding
Commercial grade	15 m	1700 km
Tactical grade	10 m	600 km
Navigation grade	4 m	15 m

The navigation filter performance was simulated for two cases: INS-only position estimates and INS + radar aiding position estimates. Tables 6.1 and 6.2 show the resulting error between truth and estimate for varying grades of INS. We see that aiding greatly decreases the error in position, with a larger correction the worse the grade of the inertial becomes. Commercial grade INS would not ordinarily be appropriate for flight navigation; however, with an onboard UWB-OFDM radar aiding system, the commercial grade INS was simulated to perform better than a stand-alone navigation grade INS.

6.5   CONCLUSION

Modern technology allows for inexpensive and efficient implementation of software-defined systems with excellent potential for multiple functionalities. In this work we describe the UWB system, which was designed, constructed, and tested to achieve the goals of radar and communications without any hardware adjustments. The system described here was also intended as a test vehicle of UWB-OFDM implementation for radar imaging, angle-frequency target profile reconstruction, and radar-assisted navigation. The design is simple, fully software configurable, and cost effective.

Experimental results have confirmed the range resolution of approximately 30 cm and have shown the applicability of the system to SAR imaging. The latter was achieved via a relatively simple and computationally efficient method of range-Doppler reconstruction, where range profiles were obtained via matched filtering of the returns (range compression) and cross-range profiles were obtained via matched filtering of the phase history. The obtained image of a trihedral corner reflector shows good target localization with less than 10 cm error in range and cross range. We further investigated the topic by creating angle-frequency graphs for a variety of target scenes. Angle of incidence was varied in steps of 10° and, owing to the simplicity of extracting individual frequency components from an OFDM signal, frequencies in the baseband bandwidth of 500 MHz were used to form two-dimensional angle-frequency graphs. Communication capability using the same hardware configuration was also tested; with an OOK scheme, the data rate of approximately 57 Mb/s was experimentally confirmed in the laboratory setup in which communication signals reflected from a large-RCS target were detected and analyzed. The experimental BER was measured and contrasted with computationally feasible results obtained for OFDM OOK signals with fewer sub-carriers; the comparison suggested prevalence of frequency offset-induced ICI over the noise effects at the given ranges; the BER was found to vary between approximately 0.625% and 3.125%. Furthermore, the potential of the system to be used for navigation via referencing the position of the radar platform to the detected targets (and ranges to them) in real time is shown to be achievable.

The possibilities for future work using the designed system as a test vehicle include advanced indoor ranging and imaging at greater distances and involving multiple platforms. The ultimate goal is the creation of a unified framework for target scene reconstruction, data communications with friendly platforms, and platform self-positioning in real time, using the single software-defined UWB radar sensor system.

ACKNOWLEDGMENTS

The authors wish to thank graduate assistants Brian Jameson and Scott Spalding for help in radar testing. We are also grateful to Dr. Jon Sjogren of AFOSR for program management and reviews during the radar system design and validation.

REFERENCES

1. L. Brown, *A radar history of World War II: Technical and military imperatives.* London, UK: Institute of Physics Publishing, 1999.
2. W. L. Cowperthwait, Reconnaissance with a side-looking, modified, APQ-7 radar. *Proceedings of 1956 Symposium for Advanced Technology in Radar Components and Systems for Battlefield Surveillance,* Univ. of Michigan, Ann Arbor, MI, pp. 631–660, 1956.
3. M. Soumekh, Reconnaissance with ultra wideband UHF synthetic aperture radar. *IEEE Signal Processing Magazine,* vol. 12, no. 4, pp. 21–40, July 1995.
4. W. T. K. Johnson, Magellan imaging radar mission to Venus. *Proceedings of IEEE,* vol. 79, no. 6, pp. 777–790, June 1991.
5. R. A. Smith, Radar navigation. *Journal of Institute of Electrical Engineers,* vol. 93, no. 1, pp. 331–342, 1946.
6. P. Tait, *Introduction to automatic target recognition.* London, UK: Institution of Engineering and Technology, 2006.
7. M. Skolnik, *Introduction to radar systems.* New York: McGraw–Hill, 2002.
8. D. R. Wehner, *High resolution radar.* Norwood, MA: Artech House, 1987.
9. A. Fasih and T. Hartley, GPU-accelerated synthetic aperture radar back projection in CUDA. *Proceedings of 2010 IEEE Radar Conference,* Arlington, VA, pp. 1408–1413, 2010.
10. R. H. Cager, Jr., D. T. LaFlame, and L. C. Parode, Orbiter Ku-band integrated radar and communications subsystem. *IEEE Transactions Communications*, vol. COM-26, pp. 1604–1619, Nov. 1978.

11. W. Wiesbeck, SDRS: Software-defined radar sensors. *Proceedings of 2001 IEEE International Geoscience and Remote Sensing Symposium (IGARSS'01)*, Sydney, Australia, pp. 3259–3261, 2001.

12. G. C. Tavik, C. L. Hilterbrick, J. B. Evins, J. J. Alter, J. G. Crnkovich, Jr., J. W. de Graaf, W. Habicht II, et al., The advanced multifunction RF concept. *IEEE Transactions Microwave Theory Technology,* vol. 53, pp. 1009–1020, March 2005.

13. G. N. Saddik, R. S. Singh, and E. R. Brown, Ultra-wideband multifunctional communications/radar system. *IEEE Transactions Microwave Theory Technology,* vol. 55, pp. 1431–1437, July 2007.

14. C. Sturm and W. Wiesbeck, Waveform design and signal processing aspects for fusion of wireless communications and radar sensing. *Proceedings of IEEE,* vol. 99, no. 7, pp. 1236–1259, July 2011.

15. J. Park and C. Nguyen, An ultra-wideband microwave radar sensor for non-destructive evaluation of pavement subsurface. *IEEE Sensors Journal,* vol. 5, no. 5, pp. 942–949, Oct. 2005.

16. J. Han and C. Nguyen, Development of a tunable multiband UWB radar sensor and its applications to subsurface sensing. *IEEE Sensors Journal,* vol. 7, no. 1, pp. 51–58, Jan. 2007.

17. H. Zhang, L. Li, and K. Wu, Software-defined six-port radar technique for precision range measurements. *IEEE Sensors Journal,* vol. 8, no. 10, pp. 1745–1751, Oct. 2008.

18. A. Batra, J. Balakrishnan, G. R. Aiello, J. R. Foerster, and A. Dabak, Design of a multiband OFDM system for realistic UWB channel environments. *IEEE Transactions Microwave Theory Technology,* vol. 52, pp. 2123–2138, Sept. 2004.

19. H. Yin and S. Alamouti, OFDMA: A broadband wireless access technology. *Proceedings of 2006 IEEE Sarnoff Symposium,* Princeton, NJ, pp. 1–4, 2006.

20. N. Levanon, Multifrequency complementary phase-coded radar signal. *IEE Proceedings of Radar, Sonar Navigation,* vol. 147, pp. 276–284, Dec. 2000.

21. E. Mozeson and N. Levanon, Multicarrier radar signals with low peak-to-mean envelope power ratio. *IEE Proceedings of Radar, Sonar Navigation,* vol. 150, pp. 71–77, April 2003.

22. P. A. Antonik, H. Griffiths, D. D. Weiner, and M. C. Wicks, Novel diverse waveforms. Air Force Research Laboratory, Rome, NY, AFRL-SN-RS-TR-2001-52 in-house report, June 2001.

23. G. E. A. Franken, H. Nikookar, and P. van Genderen, Doppler tolerance of OFDM-coded radar signals. *Proceedings of 2006 European Radar Conference,* Manchester, UK, pp. 108–111, 2006.

24. M. Ruggiano and P. van Genderen, Wideband ambiguity function and optimized coded radar signals. *Proceedings of 2007 European Radar Conference,* Munich, Germany, pp. 142–145, 2007.

25. S. C. Thompson and J. P. Stralka, Constant-envelope OFDM for power-efficient radar and data communications. *Proceedings of 2009 International Waveform Diversity and Design Conference,* Kissimmee, FL, pp. 291–295, 2009.

26. S. Sen, M. Hurtado, and A. Nehorai, Adaptive OFDM radar for detecting a moving target in urban scenarios. *Proceedings of 2009 International Waveform Diversity and Design Conference,* Kissimmee, FL, pp. 264–268, 2009.

27. B. J. Donnet and I. D. Longstaff, Combining MIMO radar with OFDM communications. *Proceedings of 2006 European Radar Conference,* Manchester, UK, pp. 37–40, 2006.

28. D. Garmatyuk, J. Schuerger, and K. Kauffman, Multifunctional software-defined radar sensor and data communication system. *IEEE Sensors Journal,* vol. 11, no. 1, pp. 99–106, Jan. 2011.

29. D. Garmatyuk and M. Brenneman, Adaptive multicarrier OFDM SAR signal processing. *IEEE Transactions Geoscience and Remote Sensing,* vol. 49, no. 10, pp. 3780–3790, Oct. 2011.

30. B. Jameson, A. Curtis, D. Garmatyuk, Y. T. Jade Morton, P. Plummer, and K. Thompson, Detection of behind-the-wall targets with adaptive UWB OFDM radar: Experimental approach. *Proceedings of 2011 IEEE Radar Conference,* Kansas City, MO, May 2011.

31. K. Kauffman, D. Garmatyuk, and J. Morton, Efficient sparse target tracking algorithm for navigation with UWB-OFDM sensors. *Proceedings of 2009 IEEE National Aerospace & Electronics Conference (NAECON),* Dayton, OH, pp. 14–17, July 2009.

32. M. Soumekh, *Synthetic aperture radar signal processing with MATLAB algorithms.* New York: John Wiley & Sons, chapters 2 and 6, 1999.

33. T. Jamsa, V. Hovinen, A. Karjalainen, and J. Iinatti, Frequency dependency of delay spread and path loss in indoor ultra-wideband channels. *Proceedings of 2006 IET Seminar on Ultra Wideband Systems, Technology and Applications,* London, UK, pp. 254–258, 2006.

34. K. Santhananthan and C. Tellambura, Probability of error calculation of OFDM systems with frequency offset. *IEEE Transactions Communications,* vol. 49, no. 11, pp. 1884–1888, Nov. 2001.

35. P. Dharmawansa, N. Rajatheva, and H. Minn, An exact error probability analysis of OFDM systems with frequency offset. *IEEE Transactions Communications,* vol. 57, no. 1, pp. 26–31, Jan. 2009.

7 Large Signal and Small Signal Building Blocks for Cellular Infrastructure

Mustafa Acar, Jos Bergervoet,
Mark van der Heijden, Domine Leenaerts,
and Stefan Drude

CONTENTS

7.1   INTRODUCTION

Mobile wireless communications have become a mainstream technology for everybody; the number of subscribers is expected to hit five billion subscriptions in 2011. Network operators have experienced a strong increase in demand for fast mobile broadband applications, which are also the focus of more recent releases for cellular communications standards as frequency spectrum for cellular applications is limited. The desired high data rates come at a price: Ever more complex modulation schemes and fairly high signal-to-noise ratios in increasingly smaller cells are required to support high speed operation. Spatial diversity multiple input multiple output (MIMO) schemes using several antennas per radio channel will become the norm in cellular infrastructures (i.e., base station designs, which aim at offering the maximum throughput to the mobile user). Network operators are faced with an increasing need for flexible solutions; installation costs and operating expenses need to come down as well in order to be able to deal with the increase in number of cell sites needed to offer adequate coverage and capacity in a given area. This requirement calls for highly integrated solutions to address cost; radio frequency (RF) power amplifiers with a high DC power efficiency along with new base station partitioning options such as remote radio head units significantly reduce the operating expenses.

There are actually only a few key building blocks in the analog domain for cellular infrastructure. Low noise amplifiers, frequency conversion mixers, and radio signal drivers and power amplifiers are the main ones. Implementing these in silicon-based process technologies allows one to meet the cost and performance targets while offering a high level of integration at the same time. In the following sections a number of examples will be given that demonstrate the capabilities of recent integrated circuit designs in the field of analog components for cellular infrastructure and the methodologies applied to reach this achievement.

7.2 FULLY INTEGRATED, HIGH PERFORMANCE, LOW NOISE AMPLIFIER

Low noise amplifiers (LNAs) for wideband code division multiple access (W-CDMA) base stations need to meet two key requirements. Firstly, a noise figure below 1 dB at operating frequencies in the cellular bands from 700 up to 2700 MHz is required for optimal receiver sensitivity. Secondly, the interference resulting from operations in adjacent frequency channels demands high linearity and thus low third-order intermodulation distortion.

Two alternative semiconductors materials technologies have been applied in the past to meet these requirements. The classical solution is amplifiers based on GaAs pHEMT devices (e.g., references 1 and 2). They do not have problems achieving the low noise and high linearity requirements but suffer from the lack of protection against electrostatic discharge (ESD), low level of integration, and the need for costly external, high quality passive components to implement impedance matching while maintaining optimum noise performance. Historically, LNA designs based on pure Si process technology, the second alternative, were not very efficient or had difficulties meeting all requirements. Some designs have been reported in the past, but they either had difficulties achieving the target specifications or did not implement 50 Ω impedance matching [3–5].

We introduce a fully integrated LNA design based on advanced Si technology that is able to overcome the issues associated with III-V compound technologies mentioned earlier. A 0.25 μm SiGe:C BiCMOS technology has been used to implement a two-stage LNA design [6]. The process technology offers active devices with a cutoff frequency f_t of 130 GHz. The first stage, an NPN device with an emitter width of 0.4 μm, is optimized for a noise figure of 0.5 dB at 2 GHz operating frequency and a gain of more than 10 dB at the same frequency. It is operated at 1.8 V supply voltage. The second stage is operated at 3.0 V supply voltage and is designed to meet the linearity requirements. It provides 15 dB gain at a noise figure of 0.7 dB.

7.2.1 DETAILS OF THE DESIGN

The starting point for the development of the input stage was a single transistor design with 17 dB DC gain, an emitter length of 10 μm, and a DC bias current of 1.3 mA. Simulation results have shown that this design would not allow simultaneous input impedance match and noise match, and the OIP3 would be about −12 dBm for a gain G_{max} of close to 15 dB. In order to improve the linearity of the input stage and at the same time bring the input and optimum noise impedances back to acceptable values, 12 identical transistor stages have been connected in parallel at the expense of a 16 mA bias current. This arrangement lowers the impedances by the same factor of 12 and raises the OIP3 to +27 dBm while maintaining a noise figure close to the minimum noise figure of 0.5 dB.

Following a similar strategy for the second stage, which is optimized to match the high linearity requirement, its design employs 30 identical stages in parallel, each using a high voltage device with an emitter width of 20 μm. The second stage draws 54 mA DC current from a 3 V supply voltage and achieves an OIP3 of +40 dBm and a noise figure of around 0.7 dB. The circuit diagram of the complete low noise amplifier is shown in Figure 7.1 along with a microphotograph of the chip.

For the final design, which combines the two stages, special attention had to be paid to the impedance matching at input, interstage, and output levels. A tapped load inductor at the output of the first stage is chosen to lower the output impedance of the first stage to 5–10 Ω, which is closer to the 3 Ω input impedance of the second stage. The matching is further improved by adding a degeneration inductor in the emitter path of the first stage combined with a shunt capacitor at its collector output and a feedback capacitor from the collector output of the second stage back to its base to achieve a flatter frequency response. ESD protection circuitry has been added to all relevant pins, a feature not available in III-V compound technologies.

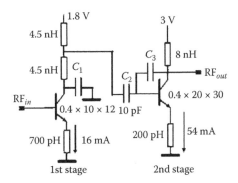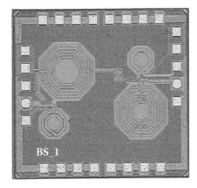

FIGURE 7.1   The low noise amplifier comprises a low noise input stage and an output stage optimized for high linearity. It is manufactured in SiGe:C BiCMOS technology, and the chip area is 1.1 × 1.3 mm². The chip microphotograph is shown on the right.

7.2.2 Measurement Results

The final design features a die size of 1.1×1.3 mm^2 and consumes a total power of 190 mW: 16 mA at 1.8 V supply voltage and 54 mA at 3 V supply. The devices were packaged in an HVSON10 package and mounted on a printed circuit board for testing. The measured noise performance indicates that an almost optimal noise match at the input has been obtained. The noise figure is 0.75 dB in the 900 MHz GSM band and 0.9 dB in the 1800 MHz band; it remains below 1 dB up to 2 GHz operating frequency (see Figure 7.2, left). At 900 MHz operating frequency a gain $|S_{21}|$ of 34 dB has been realized; at 1.8 GHz the measured gain was 24 dB. An important parameter in cellular systems design apart from the absolute gain value is its flatness within an 80 MHz wide frequency band. At frequencies of 900 MHz the gain flatness is better than 1 dB. For frequencies higher than 1.8 GHz the flatness improves to even better than 0.6 dB.

A very good input impedance match to 50 Ω has been achieved over a wide range of input frequencies; the input return loss S_{11} is better than 18 dB at the W-CDMA operating frequencies. The input return loss is better than 10 dB in the frequency range between 750 MHz and 2.4 GHz.

Two-tone linearity tests according to test standards (i.e., with tones 1.5 MHz apart) have been carried out on the devices in the 900 MHz frequency band. The devices exhibit an output IP3 of +36 dBm and a 1 dB output compression point of +19 dBm (see Figure 7.2, right). Linearity tests with two tones at 1800 MHz produced similar results. Compared to the more conventional, largely discrete solutions based on compound semiconductor technologies, the present fully integrated LNA design offers the additional benefits of including ESD protection on all its pins including the RF input/output pins and offering an input impedance matched to 50 Ω without the need for external matching components.

This Si-based two stage LNA design is an example that noise figures below 1 dB combined with high linearity performance are possible. Even though III-V compound technologies may offer better noise figures, the present design can offer more

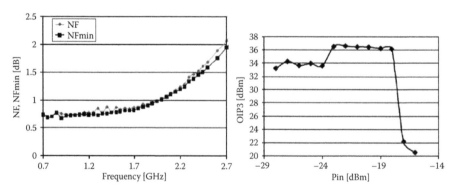

FIGURE 7.2 The low noise amplifier design meets the target specifications for application in cellular infrastructure. The noise figure stays below 1 dB up to 2 GHz. The LNA achieves an OIP3 of 36 dBm.

robustness through integrated ESD protection and a higher level of integration while meeting target specifications.

7.3 FLEXIBLE, HIGH EFFICIENCY RADIO FREQUENCY POWER AMPLIFIER DESIGNS

In modern cellular communication systems the RF power amplifier has to deal with modulated signals, which employ a high peak-to-average ratio. Complex digital modulation schemes are used to achieve communication at high data rates while using the available bandwidth efficiently. In these modulation schemes both the phase and the amplitude of the carrier signal are being modulated. As a result high requirements are imposed on the linearity of the transmitter and particularly its power amplifier. A base station transmitter design needs to address the challenge to preserve high linearity and high power efficiency over a wide operating frequency range in order to decrease overall power consumption and heat dissipation, ultimately leading to lower operating cost for the network operator.

7.3.1 Switched Mode Wideband Class-E Out-Phasing Power Amplifier

An out-phasing power amplifier is an old concept [7] applied to the challenges of current communications systems in an attempt to address the linearity–efficiency trade-off; it is based on linear amplification using nonlinear components. A switched mode power amplifier takes this concept to the extreme in that it does not even attempt to preserve any amplitude information, but rather applies hard switching of the output stages in order to achieve very high power added efficiency (PAE). In an out-phasing transmitter, the complex modulated input signal is decomposed into two signals with constant amplitude and a relative phase difference θ corresponding to the time-varying envelope of the original input signal. The two branch signals are amplified separately in highly efficient switch-mode output power stages. The two amplified signals are finally added to produce the desired complex signal with time-varying amplitude using trigonometric equalities. In the 19 W RF output power class-E power amplifier described here, the addition is implemented in a modified Chireix combiner, which offers a particularly wide operating bandwidth [8]. The output stage of the amplifier employs GaN power transistors for the nonlinear switching type amplification and a dedicated CMOS (complementary metal oxide semiconductor) driver circuit.

7.3.1.1 Details of the Design

The challenge in the design of the CMOS driver circuit is associated with the high input capacitance of the GaN power switch transistors. The gates of the GaN transistors need to be driven with pulsed signals at radio frequency with amplitudes exceeding 5 V_{pp} in order to achieve the fast switching times required for very high PAE and drain efficiency. These voltage swings are not in line with the voltage handling capabilities of regular CMOS technologies and as a result high voltage handling capabilities have to be added to the design. Instead of the device-stacking technique most commonly used to achieve high voltage performance extended drain, thin oxide MOS (EDMOS) transistors have been employed in this driver design [9]. The EDMOS transistors have been added to a 65 nm low voltage CMOS process technology without extra masks or process steps. They achieve a breakdown voltage of 12 V while offering transit frequencies f_T beyond 30 GHz. The resulting high speed driver can deliver 8 V_{pp} output swing up to 3.6 GHz [17].

The final output stage of the driver circuit consists of an EDMOS-based inverter stage. The EDMOS inverter stage can be driven directly by low voltage, high speed standard CMOS inverters, allowing for easy interfacing to the digital logic, which generates the modulated input signals, and integration on the same die. Three CMOS inverter stages in series with increasingly larger transistor geometries have been used. Throughout the design of the driver circuit a unitary P (positive) MOS to negative (N) MOS transistor size ratio has been maintained with the intention to achieve optimum layout symmetry between the two RF paths prior to the output stage. Each transistor of the four stages employed is actually made up of a large number of unit transistor parameterized cells (P-cells); they all feature minimum gate length and have gate widths of 240, 480, 1440, and 4032 μm (EDMOS device), respectively. Figure 7.3 outlines the overall block diagram of the driver circuit.

The driver design includes large on-chip AC-coupling and -decoupling capacitors. They are implemented as parallel-plate interdigitated metal-fringe capacitors. The input coupling capacitors have a capacity of 16 pF each and also implement the level shifter in combination with two DC input biasing lines. For the output pins,

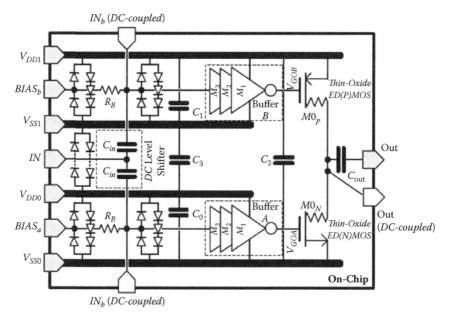

FIGURE 7.3 Circuit diagram of the CMOS driver device. Three conventional CMOS inverters amplify the input signals and drive the EDMOS output stage transistors. Level shifting and decoupling capacitors are integrated on chip.

DC-coupled and AC-coupled outputs are available, the latter using a 49 pF coupling capacitor. Besides using thick and wide power supply lines, on-chip decoupling capacitors are also integrated on chip using four capacitors for a total decoupling capacitance of about 700 pF. The total chip area of the driver device is 1.99 mm²; the EDMOS output stage and the inverter driver stages consume just 0.16 mm².

The CMOS driver has pulse-width control capabilities built in over the output square wave signal. A variable gate bias technique has been applied to achieve the required pulse-width modulation. This approach offers a way to perform fine adjustment and tuning functionality to enhance the performance of a switched mode RF power amplifier. The bias level of the first inverter of the two parallel paths shifts up/down the sinusoidal RF input signal with respect to the inverter's own switching threshold. A change on this bias voltage will vary the pulse width at the output of this inverter; the resulting pulse-modulated signals are then propagated through following two inverter stages before they are combined at the output of the two EDMOS output transistors.

For the final RF output amplifier the driver circuit is combined with two RF output power transistors (see Figure 7.6). The RF power output stage switches are realized in 28 V GaN HEMT technology; they are driven by one such CMOS driver circuit each. The CMOS drivers are capacitively coupled to the gates in order to support the negative bias voltage needed for the GaN switches. The desired amplified output signal is available after combining of the two branch signals. Ideally, a transformer would be used to merge the two signals into a combined output signal. However, a lumped element transformer is difficult to implement for high output power levels at radio frequencies. A Chireix combiner [7] attempts to implement the summation function by using transmission lines connecting the two switch outputs to the common output load; in its classical form it employs quarter wavelength transmission

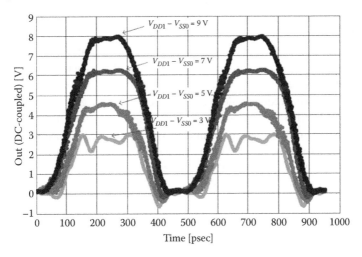

FIGURE 7.4   The CMOS driver circuit can deliver signal output swings up to 8.04 V_{pp} when operating from a 9 V supply voltage and driving a 50 Ω load. This performance can be maintained for operating frequencies reaching 3.6 GHz.

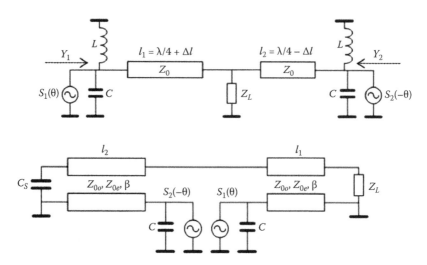

FIGURE 7.5 Schematic block diagrams of the original Chireix combiner (left) and the combiner based on coupled transmission lines (right).

lines with extra compensation elements either as lumped elements or integrated in the combiner itself. Despite its elegant design, some drawbacks in the context of switched mode out-phasing power amplifiers needed to be addressed. With the conventional Chireix combiner, the efficiency would depend not only on the out-phasing angle but also on the operating frequency because both the Chireix compensation elements and the properties of the quarter wavelength transmission lines are frequency dependent [10]. This would be highly undesirable for a class-E out-phasing amplifier, especially under the intended wide bandwidth operation.

In order to increase the operational bandwidth of the combiner, the quarter wavelength transmission lines of the conventional Chireix combiner design were replaced by coupled transmission lines meant to implement a transformer like behavior (refer to Figure 7.5). This combiner approach makes it possible to achieve improved efficiency over a wide peak-to-average ratio power range and increased RF bandwidth by removing some of the frequency dependent behavior of the original Chireix combiner.

For a more detailed analysis we start with the conventional Chireix combiner and derive some of the critical design parameters from its equivalent circuit diagram [8]. The prototype admittance conditions needed at the drain terminal of the transistor switches in order to establish the desired class-E terminations and also the Chireix compensation elements can be found to be

$$Y_{1,2} = jB_{E,\,LM} \pm jB_C + \frac{G_{E,\,LM}}{2}\left(2\sin^2\theta \pm j\sin 2\theta\right).$$

In this equation, $B_{E,LM}$ and $G_{E,LM}$ are the optimum class-E susceptance and conductance for load modulation, respectively. For an optimum design parameter $q = 1.3$, they can be approximated as

$$B_{E, LM} = -1.69 \omega C_{OUT}$$

$$G_{E, LM} = \frac{\omega C_{OUT}}{0.585}.$$

For this choice of design parameter q we observe that if the load resistance changes from its nominal to a higher value, the class-E power amplifier responds to this change by changing its turn-on voltage slope to a negative slope while keeping its turn-on voltage close to zero. By doing so, the efficiency is preserved for the varying load conditions that can occur on an out-phasing amplifier. The required Chireix compensation susceptance is then defined by the following formula:

$$B_C = \frac{G_{E, LM}}{2} sin2\Omega_C.$$

With these intermediate results, the element values for the actual Chireix combiner using coupled transmission lines can be determined by equating the prototype admittance of the original combiner to the input port admittances of the new coupled transmission line combiner. A necessary condition is that at one side of the coupled transmission line combiner is terminated with a capacitor C as given by

$$C = \frac{Y_{0o} + Y_{0e}}{2\omega} cot\beta l_2.$$

This capacitor will tune out the leakage inductance of the coupled transmission line "transformer" element. The electrical lengths βl_1 and βl_2 of the coupled transmission line combiner are unequal due to the Chireix compensation elements and can be calculated to be

$$\beta l_{1, 2} = arctan\left(\frac{2}{\left(-B_{E, LM} \pm B_C\right)\left(Z_{0e} + Z_{0o}\right)}\right).$$

The last design step for the new Chireix combiner approach is to determine the optimum even-mode and odd-mode impedances Z_{0e} and Z_{0o}. The odd-mode impedance Z_{0o} can be calculated based on the impedance transformation from the output (Z_L) to the input port impedances Z_1 and Z_2:

$$Z_{0o} = \frac{1}{G_{E, LM}} \frac{n - 2 + n^{-1}}{n + 1}.$$

In this equation, $n = Z_{0e}/Z_{0o}$ and should be large so as to maximize the magnetic coupling in the coupled transmission line "transformer" element. With these modifications we achieved an improvement in operating bandwidth by a factor of two for a coupling factor $k = 0.88$.

7.3.1.2 Measurement Results

The CMOS driver circuit was first tested in a stand-alone test setup with the die directly mounted on a printed circuit board. The tests were carried out in a 50 Ω load environment. The input of the device was driven with a sinusoidal input signal at 2.1 GHz. The output signal at the DC-coupled output pin was monitored using a high speed digital sampling oscilloscope for supply voltages between 3 and 9 V. The maximum output voltage swing observed was 8.04 V_{pp} for a 50 Ω load and a 9 V supply voltage. The output signal traces for a variety of supply voltages are shown in Figure 7.4. An on-resistance of the driver output of 4.6 Ω was measured. The ability to apply pulse width modulation to the input signal was also tested. At 2.4 GHz operating frequency and with 5 V supply voltage, a control range of the duty cycle between 30.7% and 71.5% was measured. The driver circuit achieves a much higher output voltage swing and higher operating frequency than previously reported state-of-the-art designs in CMOS [11]. Its performance is comparable to that achieved with designs in SiGe BiCMOS technology [12] with the extra benefit of pulse width control capability and better efficiency.

A demonstrator out-phasing switched mode RF power amplifier has been built using the aforementioned CMOS driver circuit in combination with GaN devices and the modified coupled transmission line Chireix combiner discussed before. A fourth-order Butterworth matching filter connected to the output of the combiner transforms the 50 Ω antenna impedance to the required class-E load and also sets the required loaded quality factor. In this demonstrator the Chireix combiner is built

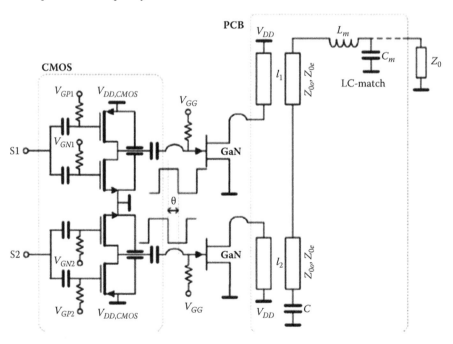

FIGURE 7.6 The out-phasing class-E RF power amplifier consists of the three stages CMOS input driver, GaN output switch transistors, and Chireix combiner using coupled transmission lines. It can achieve a high DC power efficiency over a broad operating frequency range.

with symmetrical broadside coupled transmission lines, which are implemented between the top metal layers in a dual-layer Rogers laminate with a dielectric constant $\varepsilon_r = 3.5$. The top layer is 0.1 mm thick and nearly level with the surface of the GaN switch device dies that are attached directly to the flange. The bottom layer is 1 mm thick and terminated with a metal ground plane. The layer thickness aspect ratio basically determines the maximum ratio between the even-mode and odd-mode impedance of the coupled transmission line implementation.

The CMOS driver devices are mounted close to the two GaN output transistor switches in order to allow for short bond wires when connecting the two devices via chip-to-chip bonding. This will minimize the inductance of the bond wires and helps to maximize the signal bandwidth of the interface. A similar approach is used for the wiring to the laminate input and output connections. Figure 7.7 shows the CMOS driver chips, the two GaN power switch transistors, and the bond wire connections in more detail.

The demonstrator class-E out-phasing switched mode power amplifier module has been characterized using a dedicated measurement setup capable of acquiring single-tone data and analyzing the performance of the power amplifier with complex modulated input signals including predistortion. The module has been evaluated by sweeping a single tone from 1.75 to 2.1 GHz. The drain efficiency is defined as the RF output power at the 50 Ω connector interface divided by the DC power supplied to the drains of the GaN switch transistors. The total lineup efficiency also includes the power dissipated by the CMOS driver and the RF input power delivered to the inputs of the module. The peak power and drain efficiency have been measured over the given frequency range for a number of back-off scenarios. The peak power is about 42.4 dBm and varies only 0.7 dB over the frequency range (see Figure 7.8). The drain efficiency varies with the power back-off as a result of the change in load impedance of the combiner with the phase difference applied and is better than 60% over a bandwidth larger than 250 MHz for 6 dB back-off, 150 MHz for 8 dB back-off, and 70 MHz for 10 dB back-off. At 10 dB back-off the drain efficiency is 65% and the total lineup efficiency

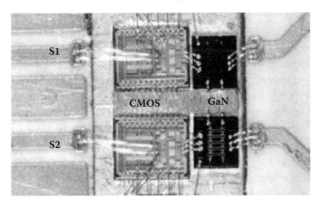

FIGURE 7.7 Chip microphotograph showing details of the assembly of the CMOS driver circuit and the GaN output switching transistors on the Rogers laminate substrate. Multiple bond wire connections are used for most of the signals so as to lower the inductance and thus improve the signal bandwidth.

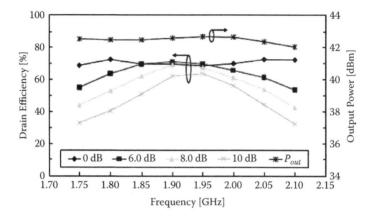

FIGURE 7.8 The out-phasing class-E RF power amplifier achieves an output power level of about 42 dBm over a wide operating frequency range. The DC power efficiency is better than 60% over a 250 MHz frequency range at 6 dB power back-off; at 10 dB back-off it can maintain this efficiency level over a range of more than 70 MHz.

is 44%. For an 8 dB back-off the results change to a drain efficiency of 70% and a total lineup efficiency of 53%. Even though the present switched mode RF power module is a two-way system, the drain efficiency at 10 dB back-off is comparable to what has been reported previously for a three-way GaN Doherty power amplifier [13]—but now with the extra benefit of better broadband capabilities and the integration of a CMOS driver stage, which enables direct digital steering of the power amplifier.

7.4 HIGHLY INTEGRATED RF POWER AMPLIFIERS IN CMOS TECHNOLOGY

For base station designs targeting very small cells such as pico and femto cells, a more fully integrated solution may be preferred by the network operators. The possibility of full integration of the RF power amplifier with the CMOS transceiver and baseband subsystems is of particular interest. The expected RF output power is usually below 30 dBm. Even at these relatively low RF output power levels, the designer needs to choose between a more linear power amplifier design with its associated poor power efficiency and a highly nonlinear switched-mode design, which is much more power efficient.

In the following sections two examples will be given for RF power amplifier designs that are implemented in bulk 65 nm CMOS technology. Besides their application in small cell base stations, they can also serve as amplifiers in mobile terminals. The implementation of CMOS RF power amplifiers in a standard CMOS technology is a challenging task due to the associated low breakdown voltage of the active devices. Lately, devices with high breakdown voltage, such as extended drain MOS (EDMOS) and thick-oxide MOS transistors, have gained increased interest [9]. Such devices can withstand a higher drain-source voltage than regular CMOS transistors and are therefore able to deliver more RF output power reliably.

7.4.1 RF Power Amplifier with Wideband Transformer Combiner

The same out-phasing technology discussed before for high RF output power levels can also be applied at lower output levels to achieve linear amplification using non-linear components. The two branch signals are being amplified in highly nonlinear amplifier stages and are later combined to produce the desired original amplified output signal. In order to avoid the bandwidth limitations of the Chireix combiner with two quarter wavelength transmission lines, a new approach using a wideband transformer-based design has been applied [10]. Such an approach becomes practical now that we are focusing on much lower RF output power levels.

7.4.1.1 Details of the Design

There are two basic circuits that can be used to combine two output signals into a single load impedance without any bandwidth limitations. We either connect the two signal sources to a floating load or we connect the signal sources in series to feed the load impedance. Despite the advantage of being very wideband because they do not employ any frequency limiting passive components, both solutions suffer from the disadvantage of requiring either a floating load or signal source.

Transformer designs have been reported that can provide large bandwidth [14], and we have designed an RF power amplifier using a transformer as a balun to avoid the need for floating components while still achieving a large operating bandwidth [10]. Two transformer configurations have been analyzed, each turning the floating component of the original configuration mentioned before into a single ended component (see Figure 7.9, top left and right).

Analysis of the first circuit, in which the transformer replaces the floating load, has shown that the conductance seen by one of the signal sources can still get

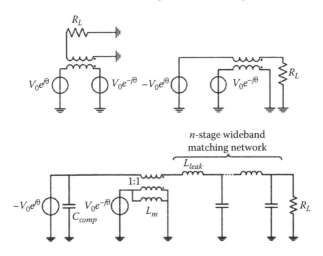

FIGURE 7.9 Two alternative options can be used to drive a single ended load from two single ended sources using a transformer (top left and right). The actual transformer-based combiner design includes the transformation and leak inductances in the analysis along with the output matching network (bottom).

negative for some operating frequencies and out-phasing angles. This is the result of the transformer's magnetizing inductance. As negative conductance is one of the issues to be avoided in switched mode power amplifier design, this topology was not considered worth pursuing further.

The alternative transformer topology turns the single ended signal source into a floating signal source such that the two signal sources combined can feed a single ended load impedance. This approach turned out to be more promising even when including also the parasitic elements of the transformer in the analysis. For the final analysis, not only have the transformer's magnetizing and leakage inductance been included but also a wideband matching network at the secondary side of the transformer (Figure 7.9, bottom). The leakage inductance in the equivalent circuit diagram is taken as the first reactive element of the matching network.

In order for the out-phasing transmitter concept to work properly, one of the sources needs an inductive compensation susceptance while the other requires a capacitive compensation component. The transformer magnetizing inductance can be used as the inductive compensation element, while the capacitive compensation element has to be added separately. It can be shown that the transformer's transformation ratio must be equal to one for correct operation. In order to achieve this ratio, the primary inductance L_p and the secondary inductance L_s must satisfy the following equation:

$$ n = \sqrt{\frac{L_s}{L_p}} = \frac{1}{k} $$

with k being the coupling coefficient of the transformer. The secondary inductance L_s can be derived from this formula; the coupling coefficient k will be typically fixed for a given implementation technology and L_p is constrained by the required compensation susceptance. The resulting leakage inductance can then be calculated as follows:

$$ L_{leak} = L_p \left(\frac{1}{k^2} - 1 \right) $$

In order to achieve broadband operation of the combiner, L_p must be small, and k should be close to one. The final transformer-based combiner design has been tested with a class-E power amplifier device built with NXP Semiconductors' 65 nm CMOS technology and EDMOS devices. The series resonators for both class-E output stages are shifted to the secondary side of the combiner in series with the transformer leakage inductance. A two-stage wideband impedance matching network was also added to offer the correct load impedance to the active devices over a wide operating bandwidth. The transformer was implemented using a three-layer printed circuit board with minimum feature sizes of 100 μm. The transformer coupling coefficient is rather low at 55%; it is the main bandwidth limiting factor.

7.4.1.2   Measurement Results

The output power and drain efficiency values of the transformer-based power amplifier design have been tested under various driving conditions; the supply voltage has been 3.6 V. The peak output power of 33.9 dBm was measured at 700 MHz; the maximum power gain observed was 20.1 dB at the same frequency. The maximum drain efficiency for 10 dB power back-off was measured at 650 MHz and was better than 50%.

7.4.2   BROADBAND CLASS-E RF POWER AMPLIFIER

Cellular network operators have a clear preference for flexible radio solutions. They would like to use as few different components in their system deployment as possible. From an RF power amplifier design point of view, this translates into a requirement for flexibility in operating frequencies and supported modulation schemes while maintaining high DC efficiency levels. A class-E RF power amplifier can serve as a building block for a variety of different cellular standards and modulation schemes. The challenge in designing a class-E power amplifier for broadband operation lies in the requirement for high efficiency and flat RF output power over a relatively large output frequency range.

In the analytical model used in the design of the amplifier, the active device is replaced by a switch with resistance values of R_{on} and R_{off} for the on and off states of the switch, respectively [15, 16]. The analysis further considers the DC feed inductance L, the loaded quality factor of the output tank circuit Q_0, and operating mode (see Figure 7.10). Suboptimum class-E variable voltage operation (E_{VV} operation) and nonzero switching voltage and zero switching slope have been chosen in order to reduce the peak voltage across the transistor terminals and to protect from overvoltage conditions. For class-E_{VV} operation, at the switching time T, the following conditions must be met for the voltage $V_C(t)$ across the switch:

$$V_C(t)\,|_{t=T} = \alpha\ V_{DD}$$

$$\frac{dV_C(t)}{dt}\,|_{t=T} = 0,$$

with $\alpha\ V_{DD}$ being the voltage across the switch at the moment in time when the switch is closed (i.e., $t = T = 1/f_o$). The peak voltage across the switch can decrease from 3.6 V_{DD} to 2.5 V_{DD} when α increases from $\alpha = 0$, which corresponds to the ideal class-E operation, to $\alpha = 2$ for certain nonideal switch, on-resistance R_{on}, as shown in Figure 7.10 (bottom). This indicates that class-E_{VV} operation will lead to lower peak voltage across the switch and as a result the power amplifier can use a higher supply voltage and thus deliver more RF output power at the cost of a slight decrease in efficiency.

The power output stage of the amplifier is designed using a single, extended drain, thick oxide NMOS transistor device implemented in 65 nm CMOS technology. It achieves an off-state breakdown voltage of 15 V. The output transistor is actually built using an array of smaller unit cell transistors; the total gate width of the transistor is

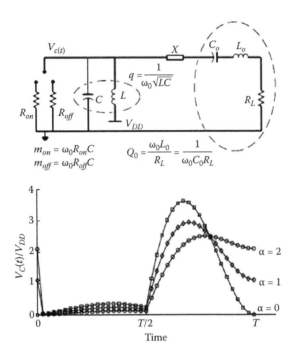

FIGURE 7.10 Schematic diagram of the analytical model used in the analysis and optimization of the class-E RF power amplifier (top). The timing diagram at the bottom shows the trace of the drain voltage of the transistor V_{DD} over time normalized to the supply voltage for various values of the parameter a. After time T, the cycle repeats again.

3.84 mm, and the channel length is 0.28 μm. This leads to an on-resistance R_{on} of 0.7 Ω, an off-resistance R_{off} of 10 kΩ, and a total off-state output capacitance of 4.14 pF. The square wave signal required to drive and operate the transistor as a switch is generated by an inverter-based driver implemented using standard thick-oxide MOS devices with a gate length of 0.28 μm. With proper biasing and a 2.4 V supply voltage, the driver stage can deliver a 2.4 V_{pp} square wave signal to the EDMOS output device.

With the desired target output power of 30 dBm at 1 GHz and 5 V supply voltage and the switch parameters as indicated previously, we can move on to determine the remaining values of the class-E circuit elements. In simulations it was found that for values α > 1.5, the peak voltage V_{DS} across the EDMOS transistor is less than its breakdown voltage. We also noticed that the highest drain efficiency is not obtained for α = 0, the ideal class-E case, due to higher switching losses under this condition. In this design, α = 1.5 was chosen as an optimum compromise between output power, drain efficiency, and peak source drain voltage V_{DS}. The corresponding optimum output load impedances for the fundamental ($Z_{1,opt}$) and second harmonic ($Z_{2,opt}$) output frequency were found to be

$$Z_{1,\,opt} = 5.43 + j14.3 \ \Omega$$

$$Z_{2,\,opt} = -j24.1 \ \Omega.$$

With these values as a starting point, the design of the two-stage LC-ladder matching network is relatively straightforward. The matching network needs to provide the desired load impedances from a 50 Ω termination within the frequency range of interest. During the optimization process more emphasis was put on a flat magnitude response for the fundamental frequency than a flat phase response.

7.4.2.1 Measurement Results

For test and measurement purposes the CMOS power amplifier chip was mounted on a printed circuit board made from Rogers substrate with $\varepsilon_r = 3.5$ and a thickness of 0.203 mm. The bond wires used to connect the chip to the substrate were kept as short as possible in order to keep their influence on the load impedances limited. The measurements were taken with a supply voltage of 5 V for the EDMOS output stage and 2.4 V for the driver stage.

In the frequency range between 550 and 1050 MHz a flat characteristic was measured for the available RF output power at a value of 30.5 dBm ± 0.5 dBm (see Figure 7.11). A power gain of about 16.5 dB was measured over the same frequency range. The drain efficiency and PAE remain above 67% and 52%, respectively, across the same bandwidth. The peak drain efficiency of 77% and peak PAE of 65% were observed at an operating frequency of 700 MHz and 31 dBm output power.

The class-E RF power amplifier can also be used as part of a polar transmitter architecture. In this case the drain efficiency with a reasonable power back-off is more critical. To enable output power control of the class-E amplifier and to improve its PAE at power back-off, supply voltage modulation has been applied. A power added efficiency of 45% and a drain efficiency of 68% were achieved with 10 dB power back-off from the reference of 30.2 dBm at 900 MHz operating frequency.

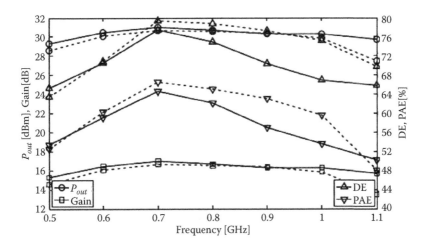

FIGURE 7.11 The class-E RF power amplifier can deliver about 30 dBm of output power over a wide operating frequency range. The drain efficiency and power added efficiency remain above 67% and 52%, respectively, over a frequency range of 550 to 1050 MHz.

7.5 SUMMARY

We have presented several silicon-based solutions for large signal and small signal building blocks aimed at cellular infrastructure applications. They replace the more conventional GaAs technology-based designs except for the GaN transistors used for efficiency reasons in RF power amplifiers. These solutions can be seen as a first step toward further integration of the radio components in a cellular base station. The demonstrated measurement results can match up with state of the art in III-V compound technology solutions, but offer lower cost, higher level of integration, and additional features such as ESD robustness, calibration, and reconfigurability over a broad range of operating frequencies. The out-phasing RF power amplifier approach allows one to achieve high levels of DC power efficiency over a broad range of RF output power levels and operating frequencies. This enables the step up toward next generation cellular technologies such as LTE and phased array antenna approaches.

REFERENCES

1. Avago Technologies, MGA-631P8. Low noise, high linearity, active bias low noise amplifier. Data sheet.
2. T. Chong et al., Design and performance of a 1.6–2.2 GHz low-noise high gain dual amplifier in GaAs e-pHEMT. *Proceedings APMC,* pp. 1–4, 2005.
3. O. Boric-Lucbeke et al., Si-MMIC BiCMOS low-noise high linearity amplifiers for base-station applications. *Proceedings Asia-Pacific Microwave Conference,* pp. 181–184, 2000.
4. V. Aparin et al., Highly linear SiGe BiCMOS LNA and mixer for cellular CDMA/AMPS applications. *Proceedings RFIC,* pp. 129–132, 2002.
5. L. Belostotski, J. W. Haslett, Sub-0.2 dB noise figure wideband room temperature CMOS LNA with non-50 ohm signal source impedance. *IEEE Solid State Circuits,* vol. 42, pp. 2492–2502, 2007.
6. D. Leenaerts, J. Bergervoet, J-W. Lobeek, M. Schmidt-Szalowski, 900 MHz/1800 MHz GSM base station LNA with sub-1 dB noise figure and +36 dBm OIP3. *IEEE RFIC Symposium,* pp. 513–516, 2010.
7. H. Chireix, High power out-phasing modulation. *Proceedings of the IRE,* pp. 1370–1392, Nov. 1935.
8. M. P. van der Heijden, M. Acar, J. S. Vromans, D. A. Calvillo-Cortes, A 19 W high-efficiency wide-band CMOS-GaN class-E Chireix RF out-phasing power amplifier. IEEE MTT-S Symposium, 2011.
9. J. Sonsky, A. Heringa, J. Perez-Gonzales, J. Benson, P. Chiang, S. Bardy, I. Volokhine, Innovative high voltage transistors for complex HV/RF SoCs in baseline CMOS. *Proceedings International Symposium on VLSI Technology, Systems and Applications,* 2008.
10. M. van Schie, M. van der Heijden, M. Acar, A. de Graauw, L. de Vreede, Analysis and design of a wideband high efficiency CMOS out-phasing amplifier. *Proceedings 2010 IEEE Radio Frequency Integrated Circuits Symposium,* p. 399 ff., 2010.
11. B. Serneels et al., A high-voltage ouput driver in a standard 2.5 V 0.25 µm CMOS technology. *Digest of Technical Papers, International Solid State Circuits Conference,* Feb. 2004.
12. S. Heck et al., A switching mode amplifier for class-S transmitters for clock frequencies up to 7.5 GHz in 0.25 µm SiGe BiCMOS. *Digest IEEE RFIC Symposium,* May 2010.

13. M. J. Pelk et al., A high-efficiency 100-W GaN three-way Doherty amplifier for base station applications. *IEEE Transactions on Microwave Theory and Technology,* vol. 56, no. 7, pp. 1528–1591, July 2008.

14. J. R. Long, Monolithic transformers for silicon RF IC design. *IEEE Journal of Solid State Circuits,* vol. 53, Sept. 2000.

15. M. Acar, A. Annema, B. Nauta, Variable-voltage class-E power amplifiers. *Digest IEEE MTT-S International Microwave Symposium,* pp. 1095–1098, June 2007.

16. R. Zhang, M. Acar, M. van der Heijden, M. Apostolidou, L. de Vreede, D. Leenaerts, A 500–1050 MHz +30 dBm class-E power amplifier in 65 nm CMOS. *IEEE RFIC Symposium,* 2011.

17. D. Calvillo-Cortes, M. Acar, M. van der Heijden, M. Apostolidou, L. de Vreede, J. Sonsky, A 65 nm CMOS pulse-width-controlled driver with 8 V_{pp} output voltage for switch-mode RF paths up to 3.6 GHz. *Proceedings IEEE International Solid States Circuits Conference,* p. 12, 2011.

8 Efficient MIMO Receiver Design for Next Generation Wireless Systems

Ji-Woong Choi, Hyukjoon Kwon,
Jungwon Lee, and Inyup Kang

CONTENTS

Recently, there has been growing interest in spatial multiplexing (SM) for multiple-input multiple-output (MIMO) systems since it can increase the data rate linearly with the number of antennas [1]. To achieve the promised gain of MIMO SM, it is crucial to handle the interference among multiple spatial streams wisely. Given the channel knowledge at the transmitter, the spatial streams can be precoded such that interference among multiple streams at the receiver is mitigated. The required channel knowledge is fed back from the receiver to the transmitter. However, it is not easy to obtain the perfect channel knowledge at the transmitter. For example, when the channel fluctuates rapidly, the amount of feedback from the receiver becomes prohibitively high. Since it is difficult to achieve the instantaneous channel knowledge perfectly at the transmitter, correspondingly there is a limit to cancelling the interference at the transmitter. Instead, it is necessary for a receiver to handle the interference among spatial streams with ease of obtaining the channel knowledge. Thus, the design of an intelligent receiver becomes critical.

This chapter introduces the advanced SM detection scheme over conventional detection schemes. Section 8.1 describes the trend on MIMO systems in recent standards. Section 8.2 explains basic principles of SM technologies for both single-user and multiuser SM, the advanced detection combined with decoding on SM, and the introduction of both hard- and soft-detection receivers. In Section 8.3, various SM detection schemes are depicted, such as equalization-based schemes and maximum likelihood (ML) schemes. In particular, we develop an ML detection algorithm named dimension reduction soft detector (DRSD) that achieves the near-ML performance with a significantly lower complexity than existing receivers. Finally, an iterative detection and decoding (IDD) scheme, which recently has attracted more attention due to higher performance at the cost of additional complexity, is discussed.

8.1   RECENT TRENDS ON MIMO SYSTEMS

MIMO technologies have been adopted rapidly in wireless communications since they can significantly increase data throughput or reliability without paying the cost of additional bandwidth or additional power at the transmitter. Such a potential benefit is achieved by distributing the same total transmit power over multiple antennas. These MIMO technologies lead to an array gain or diversity gain that improves the spectral efficiency or link reliability, respectively. SM techniques increase data rate performance by simultaneously sending multiple streams of information. If these signals arrive at the receiver with sufficiently different spatial signatures, the receiver can separate these streams into parallel channels. On the other hand, spatial diversity (SD) techniques enable the streams to be reliable and the coverage to be enhanced on the received signal by sending redundant streams of information. The received signal coming from multiple paths causes it to be less likely that all of the paths are simultaneously degraded [2]. As a result, MIMO techniques along with SM and SD increase the minimum cell spectral efficiency, the cell edge user spectral efficiency, and the peak spectral efficiency, all of which are considered as important quality of service parameters [3]. Because of these benefits, multiple antenna techniques are widely used in various modern wireless communication standards such as IEEE 802.11n, third generation partnership project (3GPP), long term evolution (LTE), worldwide interoperability for microwave access (WiMAX), high speed packet access plus (HSPA+), and so forth.

8.1.1   CELLULAR SYSTEMS

The IEEE 802.16e (i.e., WiMAX Profile 1.0) and 3GPP evolved universal terrestrial radio access (E-UTRA) LTE (releases 8 and 9) standards have been developed so as to be a part of the IMT-2000 third generation (3G) technologies [4]. IEEE 802.16m (WiMAX Profile 2.0) and 3GPP E-UTRA LTE-advanced (LTE-A) (release 10 or beyond) are two evolving standards targeting 4G wireless systems to meet or exceed the requirements of ITU 4G technologies [5,6], where MIMO technologies play an essential role in meeting 4G requirements. They not only enhance the performance of the conventional point-to-point link, but also effectively support multiuser channels such as broadcast channels and multiple access channels through multiuser MIMO. A large family of MIMO techniques has been developed for various links and modes, including space frequency block coding (SFBC), frequency-switched transmit diversity (FSTD), cyclic delay diversity (CDD), and open-loop and closed-loop beam-forming and precoding [7]. For example, 802.16m and LTE-A MIMO schemes have various configurations of two, four, or eight transmit antennas and a minimum of two receive antennas in the downlink, and one, two, or four transmit antennas with the minimum of two receive antennas in the uplink, where up to eight streams can be transmitted for a single-user downlink SM [2,11].

8.1.2 WIRELESS LAN AND PAN

One of the most prominent features in IEEE 802.11n, which was ratified in 2009, is the MIMO capability. A flexible MIMO concept allows the array of up to four antennas. These multiple antennas enable use of SM or beam-forming technologies along with wider bandwidths and a number of MAC layer algorithms that achieve a peak theoretical throughput in the order of 600 Mbps [8]. Since the ratification of 802.11n, a new task group, 802.11ac, has worked on the next standard for even higher throughput within 6 GHz spectra. New features in this standard include the bandwidths of 80 and 160 MHz, while 802.11n offered only 40 MHz; 256 quadrature amplitude modulation (QAM); up to eight antennas; and multiuser MIMO.

Another task group, 802.11ad, is looking at very high throughput (e.g., 7 Gbps) for spectrum at 6 GHz or higher frequency such as the 60 GHz ISM band. Besides 802.11ad, several standards have been introduced in the 60 GHz band, such as IEEE 802.15.3c, WirelessHD, Wireless Gigabit Alliance (WiGig), and ECMA-387 [9]. With much shorter wavelength at 60 GHz, MIMO techniques are extended to overcome the range issue through beam-forming algorithms with tens of or more antennas. For instance, a 32×32 antenna array can be implemented on a CMOS (complementary metal oxide semiconductor) process. Similarly, a large number of receive and transmit chains required for such a system are pushing the standards committee to consider MIMO techniques other than beam-forming techniques such as space-time block coding (STBC) [10].

8.2 BASIC MIMO SM TECHNOLOGIES

This section mainly discusses transmitter and receiver technologies for SM MIMO. The SM transmission can be classified into single-user and multiuser SM where spatial streams are allocated to a single user or multiple users, respectively. At the receiver, the detection of the received signal is first applied and optionally followed by decoding the detected output bits when channel coding is employed. Whether the detection delivers a hard symbol, one of the candidate symbols determined on the constellation, or a soft symbol, the likelihood for each message bit calculated instead of being determined to the decoder, the detection is categorized into hard and soft detection, respectively. These basic concepts regarding the SM detection and decoding are to be explained in detail in the following.

8.2.1 SINGLE-USER AND MULTIUSER SM

In single-user SM, all the spatial streams are allocated to a single-user. They are used in conjunction with beam-forming techniques to maximize the throughput. As higher peak throughput is required, the larger number of streams has been exploited to meet the requirement. The advanced receiver technology enables achieving such a requirement in practice. For example, it is known that the higher order modulation substantially increases the peak data rate. The recently implemented receiver starts to support such a higher order modulation with a reliable error rate. On the other hand, SM techniques in a multiuser scenario deliver data streams in a spatially multiplexed fashion to different users. Under these techniques, all the degrees of freedom of a MIMO system are known to be achieved. These SM techniques operate with the idea that the applied multiuser beam-forming or precoding schemes cause the radiation pattern of the base station to be adapted to each user for obtaining the highest possible gain in the direction of the user. Akyildiz, Gutierrez-Estevez, and Reyes [11] and Sugiura, Chen, and Hanzo [12] show that multiuser SM offers better trade-off between complexity and performance than single-user SM.

Originally, the single-user and multiuser SM were designed to be applied to a single-cell configuration. However, they can be used for a multicell condition in a straightforward manner in order to boost the cell-edge user throughput with the coordination in transmission and reception of signals among different base stations.

8.2.2 SM DETECTION AND DECODING

SM technologies have been extended to use multiple dimensions together. For example, the multicarrier system divides a single carrier into multiple subcarriers so as to transmit multiple symbols simultaneously. The orthogonal frequency division multiplexing (OFDM) scheme [13] is a representative technique to operate on multicarrier systems. On the other hand, the space dimension can be also used with the time dimension to achieve the throughput improvement over the benefit of SD gain. Alamouti [14] converts a channel matrix to be orthogonal among its channel vectors, and Foschini [15] forms an architecture to take advantage of spatial multiplexing over multiple antennas, named Bell Lab layered space-time (BLAST). In Wolniansky et al. [16], it is serially extended to eliminate interference between interstreams and successively detects each stream. This detecting algorithm is called the vertical BLAST (V-BLAST) and becomes a pioneering algorithm to use successive interference cancelation (SIC).

Another line of research seeking to achieve the channel capacity has been on the track in order to improve the channel coding gain. The coding theorem has provided redundancy to the coded bit sequence so that the sequence becomes robust to the channel variance with fading effects. Moreover, the introduction of both an interleaver and a deinterleaver to the coded bit sequence enables each bit to have strong independence of neighbor bits. As a result, it is less likely that an entire code word falls into the unsound status of channels. The bit interleaved coded modulation (BICM) scheme [17] is designed on the base of two principles—that is, the bit information is redundantly coded and interleaved so as to achieve the coding gain.

Then, space-time detecting algorithms are combined with channel coding schemes to achieve the channel capacity. The simplest method to combine both space-time algorithms and channel coding schemes is to connect them serially so that the output of the space-time detector is directly passed to the channel decoder. The bit information delivered to the decoder from the detector is measured as the ratio of bit probabilities called the maximum a posteriori (MAP) or a posteriori probability (APP). In detail, this is expressed as the log likelihood ratio (LLR) called L-values [18]. The large absolute value of LLRs indicates that the detected or decoded bit information is reliable. On the other hand, the zero value of LLRs implies that the bit still remains random.

This one-way connection, however, has a limit that a priori information of each bit cannot be used at the detector while it is less complicated to implement. In order to overcome the issue of the one-way connection from the detector to the decoder, the combined scheme evolves to the iterative detection and decoding (IDD) scheme where the feedback path from the decoder to the detector is added [19], which will be described in more detail in Section 8.4.

8.2.3 System Model

This section describes an SM MIMO system model that will be used as a baseline afterward.

Although this section assumes to use a single-user SM configuration, it can be easily extended to the multiuser SM case. In the considered SM MIMO system, a transmitter with N_T antennas sends N_S spatial streams to a receiver with N_R antennas over a flat fading channel, where $N_S \leq \min\{N_R, N_T\}$.

Figure 8.1 shows a MIMO transmitter model. At the transmitter, data bits are encoded using a coding scheme such as convolutional codes, turbo codes, and low density parity check (LDPC) codes. The encoder block can include any necessary interleaving operation. Depending on the encoder type, this block can consist of multiple parallel subencoders. The coded bits are grouped into N bits and are mapped to a modulation symbol for M-ary modulation with $M = 2^N$. For example, binary phase shift keying (BPSK) modulates one bit, i.e., $M = 2$ and $N = 1$. Then, a set of N_S

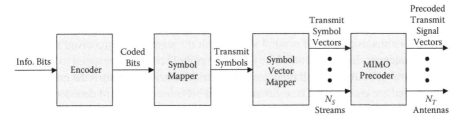

FIGURE 8.1 MIMO SM transmitter.

modulation symbols form a transmit symbol vector. This vector is preprocessed to generate a precoded transmit signal vector consisting of N_T elements. Although the precoding is optional with $N_S = N_T$, even the case of no precoding can be represented as a trivial linear precoding with an identity matrix. This precoded transmit signal vector is then transmitted over N_T transmit antennas, which goes through a wireless channel. While it is assumed for simplicity that the same modulation order M is used for all spatial streams, the generalization to the case of different modulation orders for different streams is straightforward.

At the receiver side, the received signal at a given time can be represented as

$$\mathbf{y} = \mathbf{Hx} + \mathbf{z}, \tag{8.1}$$

where $\mathbf{y} = [y_1 \cdots y_{N_R}]^T$ is an $(N_R \times 1)$ receive signal vector, $\mathbf{x} = [x_1 \cdots x_{N_S}]^T$ is an $(N_S \times 1)$ transmit symbol vector, z is an $(N_R \times 1)$ noise vector, and $\mathbf{H} = [\mathbf{h}_1 \ \mathbf{h}_2 \cdots \mathbf{h}_{N_S}]$ is an $(N_R \times N_R)$ effective channel matrix with $\mathbf{h}_m$ representing an $(N_R \times 1)$ channel gain vector from the mth stream to all receive antennas. The effective channel matrix combines the effect of the transmitter precoding and the wireless channel. The time index is omitted to simplify the notation.

In the preceding receive signal model, it was assumed that the channel is flat fading. When the channel is frequency selective, orthogonal frequency division multiplexing (OFDM) can be employed. Then, the preceding channel model still applies for each subcarrier of the OFDM system. The noise vector is an independent, identically distributed (i.i.d.) circularly symmetric complex Gaussian random vector. Without loss of generality, the variance of each element of the noise vector is set to $\sigma_z^2 = 1$. Then the probability density function (pdf) of the noise vector z is given as

$$f(\mathbf{z}) = \frac{1}{\pi^{N_R}} \exp\left(-\|\mathbf{z}\|^2\right). \tag{8.2}$$

This noise variable can be modeled to follow a Gaussian distribution with any covariance matrix after whitening and normalization. Thus, the effective channel includes the whitened noise with the transmitter precoding over wireless channels.

8.2.4   HARD-DETECTION AND SOFT-DETECTION RECEIVERS

In this subsection, hard and soft SM detection is explained with the maximum likelihood (ML) criterion and one-way connection between the detector and decoder. Figure 8.2(a) shows a MIMO receiver model with hard decoding. The MIMO hard detector estimates a transmit symbol vector from the signal vector received at the N_R receive antennas. Each element of the transmit symbol vector is demapped to the candidate symbol on the constellation used at the transmitter, and in turn the coded bits transmitted are estimated. The estimated coded bit is used by a hard decoder to generate the data bit estimates. An uncoded system can be viewed as a special case of this hard decoding, where there exists no encoder at the transmitter and no decoder at the receiver. Thus, the same MIMO hard detection can be applied to the uncoded system.

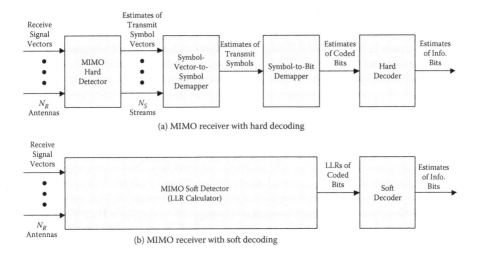

(a) MIMO receiver with hard decoding

(b) MIMO receiver with soft decoding

FIGURE 8.2   MIMO SM receiver.

One of the most complex parts for designing a MIMO receiver with hard decoding is a MIMO hard detector, which estimates a transmit symbol vector from a receive signal vector y. The ML estimate of the transmit symbol vector is given by

$$\hat{\mathbf{x}} = \underset{\mathbf{x} \in X}{\mathrm{argmax}} f(\mathbf{y}|\mathbf{x}), \tag{8.3}$$

where X is a set of all possible M^{N_S} transmit symbol vectors, and

$$f(\mathbf{y}|\mathbf{x}) = \frac{1}{\pi^{N_R}} \exp\left(-\|\mathbf{y} - \mathbf{H}\mathbf{x}\|^2\right) \tag{8.4}$$

is the conditional pdf of a receive signal vector y given a transmit symbol vector x with a known channel matrix $\mathbf{H}$. From (8.3) and (8.4), the ML estimate of the transmit symbol vector can be represented as

$$\hat{\mathbf{x}} = \underset{\mathbf{x} \in X}{\mathrm{argmin}} \|\mathbf{y} - \mathbf{H}\mathbf{x}\|^2, \tag{8.5}$$

which shows that the ML estimator is equivalent to the minimum distance estimator.

The straightforward implementation of the ML hard detector (8.5) is calculating the Euclidean distance (ED) $\|\mathbf{y} - \mathbf{H}\mathbf{x}\|^2$ for all $\mathbf{x} \in X$ and then finding the minimum ED and the corresponding $\hat{\mathbf{x}}$. However, there exist more efficient algorithms that do not require the calculation of EDs for all possible transmit symbol vectors, such as a sphere decoder with an infinite initial sphere size [20,21]. Furthermore, near-ML hard detectors also exist that further reduce the number of ED calculations [19,22,23].

Figure 8.2(b) shows a MIMO receiver model with soft decoding. In contrast to a receiver with hard decoding, the LLRs of coded bits are directly calculated from the received signal vectors. The LLR becomes an input to a soft decoder to generate the data bit estimates. The LLR for the nth bit of the sth stream, $b_{s,n}$, is

$$L(b_{s,n}) = \log\left(\frac{P\{\mathbf{y} \mid b_{s,n} = 1\}}{P\{\mathbf{y} \mid b_{s,n} = 0\}}\right). \tag{8.6}$$

Using the conditional pdf in (8.4), the LLR can be represented as

$$L(b_{s,n}) = \log\left(\sum_{\mathbf{x} \in X_{s,n}^{(1)}} \exp\left(-\|\mathbf{y} - \mathbf{Hx}\|^2\right)\right) - \log\left(\sum_{\mathbf{x} \in X_{s,n}^{(0)}} \exp\left(-\|\mathbf{y} - \mathbf{Hx}\|^2\right)\right), \tag{8.7}$$

where $X_{s,n}^{(b)}$ is the set of transmit symbol vectors with $b_{s,n} = b$. The sets $X_{s,n}^{(0)}$ and $X_{s,n}^{(1)}$ equally partition the set X of all possible M^{N_S}.

The LLR can be calculated in a straightforward way by literally evaluating (8.7) using EDs for all possible transmit vectors. However, the complexity is too high to implement using the state-of-art technology. Because of the complexity, a so-called max-log approximation [19,22–26], where the logarithm of the sum of multiple exponentials is approximated to the maximum among the arguments of the exponentials, is commonly employed in practice. With the max-log approximation, the following approximate LLR can be calculated:

$$\tilde{L}(b_{s,n}) \triangleq \min_{\mathbf{x} \in X_{s,n}^{(0)}} \|\mathbf{y} - \mathbf{Hx}\|^2 - \min_{\mathbf{x} \in X_{s,n}^{(1)}} \|\mathbf{y} - \mathbf{Hx}\|^2. \tag{8.8}$$

However, the complexity of this approximate LLR calculation is still quite high, and even further approximation is often taken in the literature [19,23–26].

8.3   DETECTION TECHNIQUES FOR MIMO SM

As mentioned before, the gain on the data rate over SM may not be as large in practice as promised by theory unless an intelligent receiver is employed. One of the biggest hurdles for increasing the data rate is the interference among multiple streams. Thus, in order to reap the benefits of SM fully, a receiver needs to use sophisticated techniques to mitigate the interstream interference.

A traditional approach for the interstream interference mitigation is to use MIMO equalization. Various equalizers have been developed for MIMO systems such as linear equalizer (LE) or decision feedback equalizer (DFE) with the criterion of zero-forcing (ZF) or minimum mean square error (MMSE) [16,17]. Although these equalizers are simple to implement, their performance is inferior to a maximum likelihood (ML) detector that directly estimates the transmit signal rather than tries to equalize. In spite of its superior performance, the ML detector is quite complex for implementation, particularly for soft detection. There have been some efforts for the development of ML SM soft demodulators. In Hochwald and ten Brink [19], a soft version sphere decoder has been developed by modifying the sphere decoder for hard detection. There also exist soft versions of various M- and K-best algorithms [22]. Moreover, in Barbero and Thompson [23], a list fixed-complexity sphere decoder has been proposed with the goal of regulating the implementation complexity.

In this section, various SM detection schemes will be introduced under the soft detection. Equalization-based schemes will be briefly explained first, followed by more detailed description on ML soft detection schemes that are of main interest.

8.3.1 EQUALIZATION-BASED DETECTORS

Although ML and near-ML detectors in the previous subsection provide optimal or suboptimal performance, implementation is challenging in practice, particularly for a large number of spatial streams (e.g., $N_S \geq 3$) or constellation points (e.g., 64 quadrature amplitude modulation [QAM]). Thus, much simpler detectors, such as linear equalizers or ordered successive interference canceller (SIC), have been usually employed in practice.

The linear equalizers determine the transmitted vector **x** by compensating channel distortions as

$$\hat{\mathbf{x}}_{LEQ} = [\hat{x}_1 \ \hat{x}_2 \cdots \hat{x}_{N_S}]^T \tag{8.9}$$

$$= [dec(\tilde{x}_1) \ dec(\tilde{x}_2) \cdots dec(\tilde{x}_{N_S})]^T$$

where $[\tilde{x}_1 \ \tilde{x}_2 \cdots \tilde{x}_{N_S}]^T = \mathbf{G}\mathbf{y}$ and $dec(\tilde{x}_s) = (\arg\min_{x \in \Omega} |x - \tilde{x}_s|)$ map equalized output $\tilde{x}_s$ into the closest constellation point (i.e., slicing). Here, Ω is the set of constellation points and G is an equalizing matrix of zero-forcing (ZF), minimum mean square error (MMSE), and maximal ratio combining (MRC) linear detectors as

$$\mathbf{G} = \begin{cases} (\mathbf{H}^H\mathbf{H})^{-1}\mathbf{H}^H, \text{ZF} \\ (\mathbf{H}^H\mathbf{H} + \sigma_z^2\mathbf{I})^{-1}\mathbf{H}^H, \text{MMSE} \\ \left(diag(\mathbf{H}^H\mathbf{H})\right)^{-1}\mathbf{H}^H, \text{MRC} \end{cases} \tag{8.10}$$

where $diag(A)$ is matrix A with zero nondiagonals and H denotes complex conjugate.

For better performance than linear equalizers with some additional complexity, conventionally ordered (CO)-SIC can also be employed by repeatedly performing nulling and cancelling where the signals are detected according to the increasing order of the noise amplification [16]. In addition, different ordering can be used for the SIC detector considering error propagation, as in Kim and Kim [30]. After the hard decision is made, it can be used for LLR calculation as in Ketonen, Juntti, and Cavallaro [29].

8.3.2   ML Type Detection

The previously mentioned linear detectors are based on the philosophy of spatial filtering, where each of the multiplexed signal streams is separated into unique spatial dimensions at the receiver. On the other hand, the ML type detector has the capability of simultaneously identifying the spatially multiplexed signals by carrying out an exhaustive search over the legitimate signal space, providing better performance [26]. Here, more detailed explanations on ML type tree-search-based soft SM detection will be given for conventional schemes and a proposed scheme named dimension reduction soft detector (DRSD).

8.3.2.1   Conventional Near-ML Detectors

In this section, the basic principle of the conventional near-ML detectors is briefly explained. Let the MIMO detector have spatial streams of x_s in a decreasing order (i.e., from $s = N_S$ to $s = 1$), without loss of generality, where a stream ordering is determined in an appropriate way [9]. For example, list fixed-complexity sphere decoder (LFSD), which provides very small and fixed implementation complexity with competitive performance, selects a stream with the smallest noise power amplification among the remaining streams to achieve the best SNR unless all the M points are chosen as candidates [31]. LFSD searches through a tree with N_s levels where k_s child nodes with the smallest d_s are retained out of M candidates from each parent node at the sth level (stream). Thus, the total number of paths in the tree from the root $s = N_S$ to the leaves $s = 1$ is

$$K_S(= \prod_{s=1}^{N_S} k_s).$$

Similar operations are performed in M-algorithm for candidate selection except that K_M candidates are retained at every tree level [21]. Other methods can be explained in a similar way using the tree search traversal with different symbol candidate selection [22,23,31].

Once all the K candidates are obtained (e.g., $K = K_S$ for LFSD or $K = K_M$ for M-algorithm), LLR can be computed based on the approximation form of (8.8) as

$$L(b_{s,n}) \approx \frac{1}{\sigma_z^2} \left[\min_{\mathbf{x} \in \bar{X}_{s,n}^{(0)}} \|\mathbf{y} - \mathbf{Hx}\|^2 - \min_{\mathbf{x} \in \bar{X}_{s,n}^{(1)}} \|\mathbf{y} - \mathbf{Hx}\|^2 \right] \qquad (8.11)$$

$$= \frac{1}{\sigma_z^2} \left(\left\| y' - R\bar{x}_{s,n}^{(0)} \right\|^2 - \left\| y' - R\bar{x}_{s,n}^{(1)} \right\|^2 \right)$$

where $\bar{X}_{s,n}^{(b)}$ is the set of transmit symbol vectors remaining after the tree search with $b_{s,n} = b$, and $\bar{X}_{s,n}^{(0)} \subset X_{s,n}^{(0)}, \bar{X}_{s,n}^{(1)} \subset X_{s,n}^{(1)}$ and $\left| \bar{X}_{s,n}^{(0)} \cup \bar{X}_{s,n}^{(1)} \right| = K$ for $s = 1, ..., N_S, n = 1, ..., N$, and

$$\overline{\mathbf{x}}_{s,n}^{(b)} = \underset{x \in \overline{X}_{s,n}^{(b)}}{\arg\min} \left\| \mathbf{y} - \mathbf{H}\mathbf{x}_{s,n}^{(b)} \right\|^2.$$

When there is no candidate in any stream, bit index, or bit value (0 or 1) (i.e., $\left|\overline{X}_{s,n}^{(b)}\right| = 0$ for any s, n, or b), a default value L_f may be assigned for this bit-wise element deficiency case to the LLR for the corresponding bit (i.e., $L(b_{s,n}) = \pm L_f$), as in Wei, Rasmussen, and Wyrwas [20] and Waters and Barry [21], resulting in performance degradation. Note that the receiver performance improves as K increases since $\overline{\mathbf{x}}_{s,n}^{(b)}$ is more likely to exist in the final candidate set $\overline{X}_{s,n}^{(b)}$ and thus the use of the default value can be avoided. However, large K increases the computation burden since more candidates have to be compared to find $\overline{\mathbf{x}}_{s,n}^{(b)}$.

8.3.2.2 Dimension Reduction Soft Detector

In this section, a recently proposed SM soft demodulator that utilizes existing hard detectors directly is introduced [28]. It can be noted that the conventional near-ML approach was to consider all transmit symbol values for all streams at the same time exhaustively. On the other hand, the equalizer-based approach that was described before does not consider any stream for exhaustive ED calculation and put all the computational burden to hard detection. A good mix of these two approaches is to consider partial numbers of streams for exhaustive ED calculation and rely on hard detectors to find the best transmit symbol subvector for the remaining streams.

First, the problem of calculating the LLRs for part of N_S streams is investigated in detail. The problem of calculating the LLRs for all N_S streams is addressed later in this section. Let N_S^{so} be the number of streams on which soft demodulation will be performed, and let N_S^{ha} be the number of remaining streams (i.e., $N_S^{ha} = N_S - N_S^{so}$). Let $\mathbf{x}^{so}$ be the transmit symbol subvector corresponding to soft demodulation streams, and $\mathbf{x}^{ha}$ be the transmit symbol subvector corresponding to the remaining streams. Finally, a rearranged transmit symbol vector can be formed by stacking $\mathbf{x}^{ha}$ on top of $\mathbf{x}^{so}$:

$$\tilde{\mathbf{x}} = \begin{bmatrix} \mathbf{x}^{ha} \\ \mathbf{x}^{so} \end{bmatrix} \tag{8.12}$$

For example, consider the case of demodulating the first stream and the third stream softly out of $N_S = 4$ streams. Then, $N_S^{so} = 2$, $\mathbf{x}^{so} = [x_1 \ x_3]^T$, $\mathbf{x}^{ha} = [x_2 \ x_4]^T$ and $\tilde{\mathbf{x}} = [x_2 \ x_4 \ x_1 \ x_3]$.

The rearranged transmit symbol vector $\tilde{\mathbf{x}}$ can be represented using some permutation matrix P as follows:

$$\tilde{\mathbf{x}} \triangleq \mathbf{Px}, \tag{8.13}$$

where permutation matrix P is, by definition, a square matrix that has exactly one entry with a value of 1 in each row and each column and all the other entries with a value of 0. Similarly, the columns of the channel matrix are rearranged using the permutation matrix P:

$$\tilde{\mathbf{H}} \triangleq \mathbf{H}\mathbf{P}^T = [\mathbf{H}^{ha} \ \mathbf{H}^{so}], \tag{8.14}$$

where $\mathbf{H}^{so} \in C^{N_R \times N_S^{so}}$ and $\mathbf{H}^{ha} \in C^{N_R \times N_S^{ha}}$ represent the channel submatrices for the soft-demodulation streams $\mathbf{x}^{so}$ and the remaining streams $\mathbf{x}^{ha}$.

With the preceding permutation on the transmit symbol vector and the channel matrix, the received signal can be represented as

$$\mathbf{y} = \mathbf{H}\mathbf{x} + \mathbf{z} \tag{8.15}$$

$$= \tilde{\mathbf{H}}\tilde{\mathbf{x}} + \mathbf{z}$$

$$= \mathbf{H}^{ha}\mathbf{x}^{ha} + \mathbf{H}^{so}\mathbf{x}^{so} + \mathbf{z}.$$

The second equality holds because $\tilde{\mathbf{H}}\tilde{\mathbf{x}} = (\mathbf{H}\mathbf{P}^T)(\mathbf{P}\mathbf{x}) = \mathbf{H}(\mathbf{P}^T\mathbf{P})\mathbf{x} = \mathbf{H}\mathbf{x}$, which uses a basic permutation matrix property that the inverse of a permutation matrix is the transpose of the permutation matrix.

Then the LLR for the nth bit of the sth stream is

$$\tilde{L}(b_{s,n}) = \min_{\mathbf{x}^{so} \in X_{s,n}^{so,(0)}, \mathbf{x}^{ha} \in X^{ha}} \left\| \mathbf{y} - \mathbf{H}^{ha}\mathbf{x}^{ha} - \mathbf{H}^{so}\mathbf{x}^{so} \right\|^2 \tag{8.16}$$

$$- \min_{\mathbf{x}^{so} \in X_{s,n}^{so,(1)}, \mathbf{x}^{ha} \in X^{ha}} \left\| \mathbf{y} - \mathbf{H}^{ha}\mathbf{x}^{ha} - \mathbf{H}^{so}\mathbf{x}^{so} \right\|^2$$

where $X_{s,n}^{so,(b)}$ is the set of soft-demodulation transmit symbol subvectors $\mathbf{x}^{so}$ with $b_{s,n} = b$, and X^{ha} is the set of all hard-detection transmit symbol subvectors $\mathbf{x}^{ha}$. The implicit assumption in the preceding LLR expression is that the sth stream belongs to the set of soft-demodulation streams. Calculating the minimum ED over the sets $X_{s,n}^{so,(b)}$ and X^{ha} can be performed in two steps: Calculate the minimum ED over the set X^{ha} for each $\mathbf{x}^{so}$ and then take the minimum ED among all soft-demodulation transmit symbol subvectors in $X_{s,n}^{so,(b)}$. Thus, the LLR can be represented as

$$\tilde{L}(b_{s,n}) = \min_{\mathbf{x}^{so} \in X_{s,n}^{so,(0)}} \left(\min_{\mathbf{x}^{ha} \in X^{ha}} \left\| \mathbf{y}^{ha}(\mathbf{x}^{so}) - \mathbf{H}^{ha}\mathbf{x}^{ha} \right\|^2 \right) \tag{8.17}$$

$$- \min_{\mathbf{x}^{so} \in X_{s,n}^{so,(1)}} \left(\min_{\mathbf{x}^{ha} \in X^{ha}} \left\| \mathbf{y}^{ha}(\mathbf{x}^{so}) - \mathbf{H}^{ha}\mathbf{x}^{ha} \right\|^2 \right)$$

where

$$\mathbf{y}^{ha}(\mathbf{x}^{so}) \triangleq \mathbf{y} - \mathbf{H}^{so}\mathbf{x}^{so}, \tag{8.18}$$

which is formed by subtracting the influence of the transmit symbol subvector $\mathbf{x}^{so}$ from the received signal.

A MIMO ML hard detector efficiently finds the transmit symbol vector $\hat{\mathbf{x}}^{ha}(\mathbf{x}^{so})$ that minimizes ED:

$$\hat{\mathbf{x}}^{ha}(\mathbf{x}^{so}) = \underset{\mathbf{x}^{ha} \in X^{ha}}{\arg\min} \| \mathbf{y}^{ha}(\mathbf{x}^{so}) - \mathbf{H}^{ha}\mathbf{x}^{ha} \|^2. \tag{8.19}$$

The corresponding minimum ED is calculated as follows:

$$D(\mathbf{x}^{so}) = \| \mathbf{y}^{ha}(\mathbf{x}^{so}) - \mathbf{H}^{ha}\hat{\mathbf{x}}^{ha}(\mathbf{x}^{so}) \|^2. \tag{8.20}$$

Then, the LLR can be calculated using the following equation:

$$\tilde{L}(b_{s,n}) = \min_{\mathbf{x}^{so} \in X_{s,n}^{so,(0)}} D(\mathbf{x}^{so}) - \min_{\mathbf{x}^{so} \in X_{s,n}^{so,(1)}} D(\mathbf{x}^{so}). \tag{8.21}$$

When LLR needs to be calculated for all streams, the preceding LLR calculation can be performed multiple times with different permutation matrices such that each stream belongs to soft-demodulation streams at least once. Partitioning the streams into soft-demodulation streams can be performed in many different ways. For example, when the total number of streams is four, then the soft-demodulation dimension can be chosen as one such that the partial LLR calculation is performed four times. Another option is to choose the soft-demodulation dimension as two such that the partial LLR calculation is performed twice.

Figure 8.3 shows a block diagram of the partial demodulator with the proposed dimension reduction approach. The first block is a candidate subset generation block for soft-demodulation transmit symbol subvectors. The second block forms a new received signal vector that removes the effect of the soft-demodulation transmit symbol subvectors from the original received signal vector. Then, a hard-detection problem is solved with the dimension of N_S^{ha} in the third block. In the following block, based on the estimate of the hard-detection transmit symbol subvector, ED for each soft-demodulation transmit symbol subvector is calculated. Lastly, the LLR is calculated based on the EDs. More details on the DRSD including the comparison of the performance and complexity with other conventional schemes are available in Lee, Choi, and Lou [28].

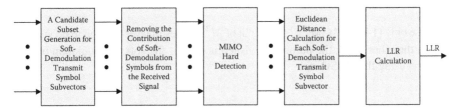

FIGURE 8.3 Partial MIMO soft demodulator with dimension reduction approach.

8.3.2.2.1 Performance and Complexity of DRSD

In this section, the complexity of the proposed DRSD is evaluated first. As a complexity measure, the number of visited nodes is chosen when QR decomposition is used and the demodulation problem is formulated as a tree search problem. With the use of QR decomposition, ED can be computed recursively from $s = N_S$ to $s = 1$ based on

$$\|\mathbf{w} - \mathbf{Rx}\|^2 = \sum_{j=1}^{N_S} \left(w_j - \sum_{i=j}^{N_S} r_{ji} x_i \right)^2 \tag{8.22}$$

where $\mathbf{H} = \mathbf{QR}$, $\mathbf{y}' = \mathbf{Q}^H \mathbf{y}$, Q is an $(N_R \times N_R)$ unitary matrix, and R is an $(N_R \times N_S)$ upper triangular matrix; r_{ji} denotes the entry in the jth row and the ith column of R. This ED calculation can be represented as a tree structure where layer 1 has M nodes representing M different values of x_{N_S} and each node in layer i has M children nodes in layer $i + 1$ with M different values of x_{N_S-i} for $1 \le i \le N_S - 1$. With this complexity measure, it is inherently assumed that the complexity associated with each node in a tree is the same regardless of the node location. Although the actual number of operations per node can vary depending on the node location, choosing the number of operations as the complexity measure makes the analysis too complicated. Thus, in this section, the number of visited nodes has been chosen as the complexity measure (see Jalden and Ottersten [31]).

With the conventional exhaustive search method, the number of visited nodes is

$$C_{exh}(N_S) = \sum_{i=1}^{N_S} M^i = \frac{M(M^{N_S} - 1)}{M - 1}, \tag{8.23}$$

since the exhaustive search visits all the nodes of the tree that has M^i nodes in layer i for $i = 1, 2, \cdots, N_S$. With the near-exhaustive search method in Lee et al. [32] that reduces the computation at the final stream, the number of visited nodes is

$$C_{near-exh}(N_S) = \sum_{i=1}^{N_S-1} M^i + (1 + \log_2 M) M^{N_S-1} \tag{8.24}$$

$$= \frac{M(M^{N_S-1} - 1)}{M - 1} + (1 + \log_2 M) M^{N_S-1}$$

because only $(1 + \log_2 M) M^{N_S-1}$ nodes out of M^{N_S} nodes in layer N_S need to be visited.

On the other hand, the number of visited nodes for the soft demodulation of N_S^{so} streams using the DRSD is

$$C_{DRSD,part}(N_S, N_S^{so}) = \frac{M(M^{N_S^{so}} - 1)}{M - 1} + M^{N_S^{so}} C_{hard}(N_S - N_S^{so}) \tag{8.25}$$

where $C_{hard}(N)$ is the number of visited nodes of the considered hard detector with N streams. This number of visited nodes can be derived by noting that all the nodes in layer 1 to layer N_S^{so} are visited and the number of visited nodes in layers $s > N_S^{so}$ depends on the hard detector.

When N_S^{so} streams are demodulated softly at one time and the same partial demodulator is used multiple times with reordering of streams, the partial demodulator should be used $\lceil N_S / N_S^{so} \rceil$ times, where $\lceil x \rceil$ represents the ceiling operation (i.e., it is the smallest integer that is not less than x). For example, when three streams are demodulated at the same time out of a total of five streams, the partial demodulator needs to be used twice, such as $\mathbf{x}^{so} = [x_1 \ x_2 \ x_3]^T$ for the first time and $\mathbf{x}^{so} = [x_4 \ x_5 \ x_1]^T$ for the second time. In this case, the LLR for x_1 is calculated twice. With the repeated use of the same partial DRSD $\lceil N_S / N_S^{so} \rceil$ times, the total number of visited nodes for the demodulation of all N_S streams is

$$C_{DRSD,tot}(N_S, N_S^{so}) = \lceil N_S / N_S^{so} \rceil C_{DRSD,part}(N_S, N_S^{so}). \tag{8.26}$$

When a different partial demodulator is allowed to be used each time, the total number of ED can be expressed as

$$C_{DRSD,tot} = \sum_{p=1}^{P} M^{N_{S,p}^{so}} C_{DRSD,part}(N_S, N_S - N_{S,p}^{so}). \tag{8.27}$$

The optimality of the proposed DRSD can be maintained by the use of the optimal hard detector. However, by sacrificing the optimality, further reduction in complexity can be achieved—that is, lower $C_{hard}(N)$. One easy way of reducing complexity is using a suboptimal hard detector such as numerous existing near-ML hard detectors, linear equalizers, and decision feedback equalizers. The use of suboptimal hard detectors may cause significant performance loss. In this case, as a compromise of complexity and performance, multiple suboptimal hard detectors can be used instead of one suboptimal hard detector [35]. Another way of reducing complexity is considering less than $M^{N_S^{so}}$ candidates for soft-demodulation transmit symbol subvectors [36].

To see the performance of the DRSD scheme, simulation results of the optimal DRSD and low complexity DRSDs are given in the setting of WiMAX radio conformance tests [33] (RCT) for IEEE 802.16-based systems [34]. For the low complexity DRSDs, suboptimal hard detectors are considered. Orthogonal frequency division multiple access (OFDMA) modes in IEEE 802.16 are chosen for the simulation. With the OFDMA mode, a convolutional turbo code (CTC) is used for channel coding along with the bit-interleaved coded modulation (BICM), and MIMO soft demodulation is performed subcarrier by subcarrier. Among the various channel models defined in WiMAX RCT, a vehicular-A 60 km/h channel is chosen for the determination of the power delay profile and the Doppler spread of the multipath fading channel. For antenna correlation, both the low and high correlation models of WiMAX RCT are considered.

Figure 8.4(a) shows packet error rate (PER) curves for various soft demodulators for 4 QAM and coding rate of 1/2 with four transmit and four receive antennas having the high spatial correlation. Here, four streams are transmitted, and the number of soft demodulation streams is chosen as one for the DRSD. As can be seen in the figure, the suboptimal DRSD that uses multiple SIC hard detectors with all possible detection orders [35] (represented as "mixed SIC" in the figure) as a suboptimum hard detector shows little performance loss compared to the optimal soft demodulator. The suboptimal DRSD based on ordered SIC [22] shows some performance degradation of approximately 0.7 dB at 10% PER but still outperforms

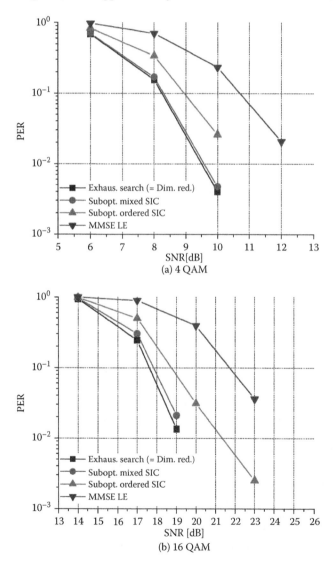

FIGURE 8.4   PER curves with four transmit and four receive antennas.

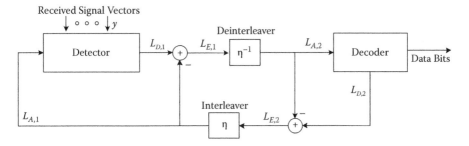

FIGURE 8.5   A block diagram of iterative detecting and decoding receiver.

MMSE LE by approximately 1.7 dB. In Figure 8.4(b), PER curves for 16 QAM and coding rate of 1/2 are shown when the other simulation parameters are the same as in Figure 8.4(a). Figure 8.4(b) exhibits similar trends as Figure 8.4(a) but with more pronounced differences among the soft demodulators at 10% PER. Similar tendencies can be observed for spatially low correlated channels. More evaluation results for the performance and complexity can be obtained in Lee et al. [28].

8.4   ITERATIVE DETECTION AND DECODING

The iterative detection and decoding (IDD) scheme exploits the feedback path from the decoder to the detector [19]. The IDD scheme iteratively exchanges soft information between the detector and the decoder in a two-way algorithm as shown in Figure 8.5. Even though the IDD scheme is not strictly proved as an optimal algorithm, it is very effective to decode the symbol and to achieve the near-optimal result. Under the IDD scheme, the kind of LLR is divided into three types: a priori LLR, extrinsic LLR, and a posteriori LLR. The output of a detector or a decoder denotes a posteriori LLR that is generated by using the input, a priori LLR. Therefore, only the extrinsic LLR obtained by subtracting a priori LLR from a posteriori LLR is exchanged between the detector and the decoder. While iterating to detect and decode the receive signal, a posteriori LLR becomes accurate and correspondingly more accurate a priori LLR can be used to detect and decode. The iterating process continues until the stopping criterion is satisfied or the improvement over the IDD scheme is negligible.

8.4.1   SYSTEM MODEL

As introduced in Hochwald and ten Brink [19], the IDD architecture is designed with both the detector and the decoder, and the set of an interleaver and a deinterleaver on the path where the soft information is exchanged. As these authors explain [19], the IDD scheme can be introduced from the received signal $\mathbf{y}$ as

$$\mathbf{y} = \mathbf{Hx} + \mathbf{z} \tag{8.28}$$

where $\mathbf{H}$ is a complex channel matrix whose dimension is the number of receiver antennas, N_R, times the number of spatial streams N_S. In general, N_R and N_S are not restricted to one such that the detector can be extended to the MIMO detector explained in the previous section. The noise $\mathbf{z}$ is a vector whose element follows an independent complex Gaussian distribution with mean zero. The signal vector $\mathbf{x}$ has its elements as discrete modulated symbols with energy constraint as

$$\mathbf{x} = \begin{bmatrix} x_1 \\ \vdots \\ x_{N_S} \end{bmatrix}, \tag{8.29}$$

where x_i is the ith element of $\mathbf{x}$. It is noted that a symbol consists of N bits and, correspondingly, a signal vector consists of $N_S N$ bits. In particular, the nth bit of the sth stream is expressed as $b_{s,n}$. Thus, using any of the detection algorithms such as MAP detecting, the signal vector could be estimated and the estimated signal symbols associated with a single codeword could be collected accordingly. The estimated symbols are equivalently transformed into the soft bit information and passed to the decoder as extrinsic LLRs.

8.4.2 SOFT INFORMATION

In detail, the soft information obtained from the detector can be explained [19] as follows. Given the received signal $\mathbf{y}$, the a posteriori LLR of the bit $b_{s,n}$ is

$$L_D(b_{s,n} \mid \mathbf{y}) = \ln \frac{P(b_{s,n} = +1 \mid \mathbf{y})}{P(b_{s,n} = -1 \mid \mathbf{y})}, \tag{8.30}$$

where the subscript D stands for a posteriori LLR. Under the assumption that the interleaver provides fully independent output bits after scrambling input bits, each bit $b_{s,n}$ is considered statistically independent of neighbor bits. Then, using Bayes's theorem, the a posteriori LLR can be divided into the sum of a priori LLR, L_A, and the extrinsic LLR, L_E, as

$$L_D(b_{s,n} \mid \mathbf{y}) = L_A(b_{s,n}) + \ln \frac{\displaystyle\sum_{\mathbf{b} \in \beta_{s,n,+1}} P(\mathbf{y} \mid \mathbf{b}) \exp\left(\displaystyle\sum_{(i,j) \in \vartheta_{s,n,+1}} L_A(b_{i,j})\right)}{\displaystyle\sum_{\mathbf{b} \in \beta_{s,n,-1}} P(\mathbf{y} \mid \mathbf{b}) \exp\left(\displaystyle\sum_{(i,j) \in \vartheta_{s,n,-1}} L_A(b_{i,j})\right)} \tag{8.31}$$

$$= L_A(b_{s,n}) + L_E(b_{s,n} \mid \mathbf{y})$$

where the set $\beta_{s,n,+1}$ is defined as

$$\beta_{s,n,+1} = \{\mathbf{b} \mid b_{s,n} = +1\} \tag{8.32}$$

$$\beta_{s,n,-1} = \{\mathbf{b} | b_{s,n} = -1\}. \tag{8.33}$$

The bit vector b indicates the concatenated bits of x in a vector form. Also, the set $\vartheta_{s,n,\mathbf{b}}$ is given by

$$\vartheta_{s,n,k} = \{(i,j) | i = 1,\dots,N_S, j = 1,\dots,N,(i,j) \ne (s,n), b_{(i,j)} = k\}, k = +1 \text{ or} -1. \tag{8.34}$$

The a priori LLR is defined as the ratio of the bit probability and is expressed as

$$L_A(b_k) = \ln \frac{P(b_{s,n} = +1)}{P(b_{s,n} = -1)}. \tag{8.35}$$

The conditional probability $P(\mathbf{y} | \mathbf{b})$ depends on the detector. For the case of the MAP detector, the conditional probability is a function of exponential terms. As explained in Hochwald and ten Brink [19], the extrinsic LLR is converted by multiplying the constant

$$\sum_s \sum_n L_A(b_{s,n})$$

into the following:

$$L_E(b_{s,n} | \mathbf{y}) = \ln \frac{\displaystyle\sum_{\mathbf{b} \in \beta_{s,n,+1}} P(\mathbf{y} | \mathbf{b}) \exp\left(\frac{1}{2} \mathbf{b}_{[s,n]}^{\dagger} \mathbf{L}_{A[s,n]}\right)}{\displaystyle\sum_{\mathbf{b} \in \beta_{s,n,-1}} P(\mathbf{y} | \mathbf{b}) \exp\left(\frac{1}{2} \mathbf{b}_{[s,n]}^{\dagger} \mathbf{L}_{A[s,n]}\right)}, \tag{8.36}$$

where the subscript $[s,n]$ denotes that the nth bit of the s-stream is excluded in the vector associated with the subscript. The extrinsic LLR is interpreted as the soft information of each bit so as to be passed to the decoder. Figure 8.5 depicts a block diagram of the IDD receiver in detail. The subscript 1 attached in the L-values implies that the L-values work for the detector. On the other hand, the subscript 2 denotes that the L-values are with the decoder. As shown in Figure 8.5, the decoder output is treated in the same way as the detector output; that is, a priori LLR of the decoder is subtracted from the decoder output, a posteriori LLR, and the resultant extrinsic LLR is fed back to the detector after being descrambled at the deinterleaver. This process is iteratively repeated until the decoding performance or the stopping criterion is satisfied.

Figure 8.6(a) and (b) shows PER of the IDD receiver using a MAP detector and a turbo decoder in terms of the number of iterations between the detector and the decoder. The 4×4 MIMO channel is considered with two code words such that one codeword is assigned over two layers. The turbo code is generated with the polynomial $(7,5)$ and two code rates, 0.33 and 0.83. It is observed that the PER

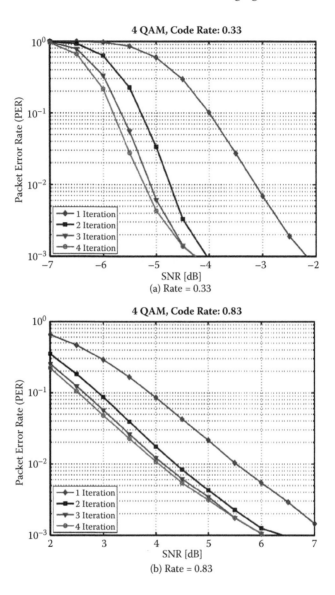

FIGURE 8.6 The PER of a 4×4 IDD receiver using a MAP detector and a turbo decoder with 4 QAM.

curves quickly drop as the number of iterations between the detector and the decoder increases. Since the bit sequence having a low code rate is encoded with more redundancy, the PER slope of the corresponding bit sequence is shown to be much more rapid. Also, when the number of iterations is large enough, the updated LLRs from the IDD receiver become accurate, so the PER is also saturated. In this case, no more gain is achieved.

8.5   CONCLUSIONS AND FUTURE DIRECTIONS

In this chapter, we examined various MIMO SM receiver schemes. We provided a comprehensive review of conventional detectors as well as advanced detectors for both hard and soft decoding. In particular, we presented a dimension reduction soft detector for soft decoding that achieves near-ML performance at a low complexity. We also described an iterative detection and decoding architecture as an advanced receive technique to achieve the performance promised by MIMO SM. In real-world communication systems, many antennas have been adopted to meet the peak throughput requirement. Therefore, it is vital to continue to develop a receiver that obtains the best trade-off between performance and complexity.

REFERENCES

1. I. E. Telatar, Capacity of multi-antenna Gaussian channels. *European Transactions on Telecommunications,* vol. 10, pp. 585–595, 1999.
2. Q. Li, G. Li, W. Lee, M. Lee, D. Mazzarese, B. Clerckx, and Z. Li, MIMO techniques in WiMAX and LTE: A feature overview. *IEEE Communications Magazine,* vol. 48, no. 5, pp. 86–92, 2010.
3. ITU-R Rep. M.2134, Requirements related to technical system performance for IMT-Advanced radio interface(s), Nov. 2008.
4. ITU-R Rec. M.1457-8, Detailed specifications of the radio interfaces of international mobile telecommunications-2000 (IMT-2000), May 2009.
5. ITU-R SG WP 5D, Acknowledgment of candidate submission from IEEE under step 3 of the IMT-advanced process (IEEE Technology), Doc. IMT-ADV/4-E, Oct. 2009.
6. ITU-R SG WP 5D, Acknowledgment of candidate submission from 3GPP proponent (3GPP Organization Partners of ARIB, ATIS, CCSA, ETSI, TTA AND TTC) under step 3 of the IMT-advanced process (3GPP Technology), Doc. IMT-ADV/8-E, Oct. 2009.
7. 3GPP TS 36.101, Evolved universal terrestrial radio access (E-UTRA); user equipment (UE) radio transmission and reception (Rel. 8), March 2010.
8. G. Heirtz, D. Denteneer, L. Stibor, Y. Zhang, X. P. Costa, and B. Walke, The IEEE 802.11 universe. *IEEE Communications Magazine,* vol. 48, pp. 62–70, Jan. 2010.
9. S. Yong, P. Xia, and A. Valdes-Garcia. *60 GHz Technology for Gbps WLAN and WPAN: From theory to practice.* New York: Wiley, 2011.
10. X. Zhu, A. Doufexi, and T. Kocak, Throughput and coverage performance for IEEE 802.11ad millimeter-wave WPANs. *Proceedings of IEEE Vehicular Technology Conference (VTC),* pp. 1–5, May 2011.
11. F. Akyildiz, D. M. Gutierrez-Estevez, and E. C. Reyes, The evolution to 4G cellular systems: LTE-Advanced. *Physical Communication,* vol. 3, no. 4, pp 217–244, Dec. 2010.
12. S. Sugiura, S. Chen, and L. Hanzo, A universal space-time architecture for multiple-antenna-aided systems. *IEEE Communications Surveys & Tutorials,* vol. 14, pp. 401–420, 2012.
13. L. J. Cimini, Jr., and N. R. Sollenberger, OFDM with diversity and coding for high-bit-rate mobile data applications. *Mobile Multimedia Communication,* vol. 1, pp. 247–254, 1997.
14. S. M. Alamouti, A simple transmit diversity technique for wireless communications. *IEEE Journal on Selected Areas in Communications,* vol. 16, pp. 1451–1458, Oct. 1998.
15. G. J. Foschini, Space-time block codes from orthogonal designs. *Bell Labs Technical Journal,* vol. 2, pp. 41–59, 1996.

16. P. Wolniansky, G. Foschini, G. Golden, and R. Valenzuela, V-BLAST: An architecture for realizing very high data rates over the rich-scattering wireless channel. *Proceedings of URSI International Symposium on Signals, Systems, and Electronics (ISSSE),* pp. 295–300, Sept. 1998.

17. G. Caire, G. Taricco, and E. Biglieri, Bit-interleaved coded modulation. *IEEE Transactions on Information Theory,* vol. 44, no. 3, pp. 927–946, May 1998.

18. J. Hagenauer, E. Offer, and L. Papke, Iterative decoding of binary block and convolutional codes. *IEEE Transactions on Information Theory,* vol. 42, no. 2, pp. 429–445, March 1996.

19. B. M. Hochwald and S. ten Brink, Achieving near-capacity on a multiple-antenna channel. *IEEE Transactions on Communications,* vol. 51, pp. 389–399, March 2003.

20. L. Wei, L. K. Rasmussen, and R. Wyrwas, Near optimum tree-search detection schemes for bit-synchronous multiuser CDMA systems over Gaussian and two-path Rayleigh-fading channels. *IEEE Transactions on Communications,* vol. 45, pp. 691–700, June 1997.

21. D. W. Waters and J. R. Barry, The chase family of detection algorithms for multiple-input multiple-output channels. *IEEE Transactions on Signal Processing,* vol. 56, pp. 739–747, Feb. 2008.

22. Y. L. C. de Jong, and T. J. Willink, Iterative tree search detection for MIMO wireless systems. *IEEE Transactions on Communications,* vol. 53, pp. 930–935, June 2005.

23. L. G. Barbero and J. S. Thompson, Extending a fixed-complexity sphere decoder to obtain likelihood information for turbo-MIMO systems. *IEEE Transactions on Vehicular Technology,* vol. 57, pp. 2804–2814, Sept. 2008.

24. C. Studer, A. Burg, and H. Bölcskei, Soft-output sphere decoding: algorithms and VLSI implementation. *IEEE Journal on Selected Areas in Communications,* vol. 26, no. 2, pp. 290–300, Feb. 2008.

25. L. Milliner, E. Zimmermann, J. R. Barry, and G. Fettweis, A fixed-complexity smart candidate adding algorithm for soft-output MIMO detection. *IEEE Journal on Selected Topics in Signal Processing,* vol. 3, pp. 1016–1025, Dec. 2009.

26. J.-S. Kim, S.-H. Moon, and I. Lee, A new reduced complexity ML detection scheme for MIMO systems. *IEEE Transactions on Communications,* vol. 58, pp. 1302–1310, April 2010.

27. J. M. Cioffi and G. D. Forney, Generalized decision-feedback equalization for packet transmission with ISI and Gaussian noise. In *Communication, computation, control and signal processing,* ed. A. Paulraj, V. Roychowdhury, and C. Schaper. Boston: Kluwer, pp. 79–127, 1997.

28. J. Lee, J.-W. Choi, and H.-L. Lou, MIMO maximum likelihood soft demodulation based on dimension reduction. *Proceedings IEEE Global Telecommunications Conference (GLOBECOM),* pp. 1–5, Dec. 2010.

29. J. Ketonen, M. Juntti, and J. R. Cavallaro, Performance-complexity comparison of receivers for a LTE MIMO–OFDM system. *IEEE Transactions on Signal Processing,* vol. 58, no. 6, pp. 3360–3372, June 2010.

30. S. W. Kim and K. P. Kim, Log-likelihood-ratio-based detection ordering in V-BLAST. *IEEE Transactions on Communications,* vol. 54, pp. 302–307, Feb. 2006.

31. J. Jalden, and B. Ottersten, On the complexity of sphere decoding in digital communications. *IEEE Transactions on Signal Processing,* vol. 53, pp. 1474–1484, April 2005.

32. J. Lee, J.-W. Choi, H. Lou, and J. Park, Soft MIMO ML demodulation based on bitwise constellation partitioning. *IEEE Communications Letters,* vol. 13, pp. 736–738, Oct. 2009.

33. WiMAX forum mobile radio conformance tests (MRCT) release 1.0. (http://www.wimaxforum.org).

34. IEEE Std 802.16-2009, IEEE standard for local and metropolitan area networks, part 16: Air interface for broadband wireless access systems. IEEE, May 2009.
35. J.-W. Choi, J. Lee, H.-L. Lou, and J. Park, Improved MIMO SIC detection exploiting ML criterion. *Proceedings IEEE Vehicular Technology Conference (VTC)*, pp. 1–5, Sept. 2011.
36. J.-W. Choi, J. Lee, J. P. Choi, and H.-L. Lou, MIMO soft near-ML demodulation with fixed low-complexity candidate selection. To be published, *IEICE Transactions on Communications,* vol. E-95B. pp. 2884–2891. Sept. 2012.

9 A Low Noise, Low Distortion Radio Design

Ahmet Tekin and Hassan Elwan

CONTENTS

9.1   NOISE–LINEARITY TRADE-OFF IN RADIO DESIGN

Various wireless standards have emerged in recent years as a result of strong consumer demand for wireless applications [1,2]. The introduction of digital data communication with digital signal processing has fueled the development of numerous wireless standards and applications ranging from cellular, cordless phones and mobile TV to short range home RF, wireless LAN, and Bluetooth technologies.

While abundance of these wireless applications drives the technology to its limits to catch up with the demand, the crowding of the spectrum poses additional challenges for the designers. These emerging wireless standards have to be compatible with the existing standards. Hence, very strong interferers can coexist in the nearby channels, whereas the desired signal in the channel of interest might be very weak. As a result, the classical noise–linearity–power area trade-off becomes an even more pronounced challenge in wireless receiver design. Moreover, because of the trend to integrate multiple applications into a single wireless device, the battery life becomes an even more significant concern. Hence, any design solution for portable devices should offer low power operation. Since device size is also of greater concern in such portable devices, the designs should be reconfigurable for different frequency bands and applications to minimize the component count in the final design.

One of such emerging technologies, mobile TV, involves bringing TV services to mobile phones. It combines the services of a mobile phone with television content and represents a logical step for consumers, operators, and content providers. Mobile TV over cellular networks allows viewers to enjoy personalized, interactive TV with content specifically adapted to the mobile medium. The services and viewing experience of mobile TV over cellular networks differ in a variety of ways from traditional TV viewing. In addition to mobility, mobile TV delivers a variety of services, including video on demand.

Mobile TV is one of the most challenging wireless applications in terms of design requirements and specifications for the actual hardware designers. Once more, spread of numerous incompatible mobile TV standards across the globe could not be prevented, as was also the case with 2G and 3G mobile systems. These include digital video broadcasting-handheld (DVB-H), digital multimedia broadcasting (DMB), TDtv (based on TD-CDMA technology), 1seg (based on Japan's integrated services digital broadcasting-terrestrial (ISDB-T)), digital audio broadcasting (DAB), and MediaFLO. None is ideal as all have drawbacks of one kind or another: spectral frequencies used or needed, signal strength required, new antennas and towers, network capacity. The main challenge now is the design of a universal multistandard receiver that is low cost and low power and can work over all the standards mentioned before. It is not only the variety of standards with different signal bandwidths and carrier frequencies that makes the design difficult to achieve but also the frequency band that every individual standard needs to cover. The frequency allocation for DVB-T/H system, for example, is shown in Figure 9.1. The DVB-T/H channels in the upper UHF band suffer from very strong nearby transmit-path GSM interferers residing at 880–915 MHz band. Due to strong antenna-to-antenna coupling in the devices accommodating both standards, the linearity requirement on the mobile TV receiver is very demanding because of the intermodulation (IM) products of the GSM interferers. Moreover, the DVB-T/H channels reside throughout most of the UHF band and part of the VHF band. Other analog and digital standards will continue to broadcast in these bands; hence, after the down-conversion, these adjacent blocker signals have to be filtered out with a sharp filter. The effect of intermodulation products is illustrated in Figure 9.2. The presence of strong adjacent channel blockers along with

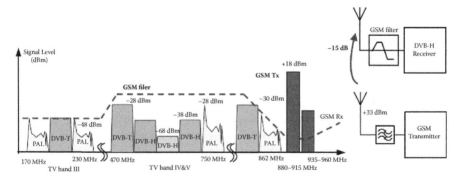

FIGURE 9.1   Frequency allocation plot for DVB-T/H.

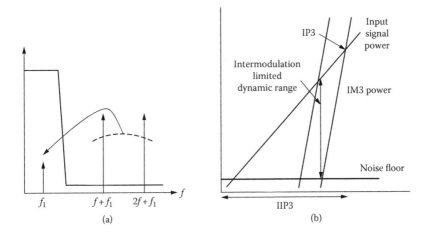

FIGURE 9.2 Intermodulation; (a) IM mixing due to nonlinearity, (b) implications of inter-modulation on dynamic range.

the desired signal at the baseband requires the design of a filter with high linearity and dynamic range. The filter must be able to process large signals with little inter-modulation distortion. Harmonics of the signal will remain in the filter stopband, where they are automatically attenuated. However, it is very possible that third-order intermodulation between particular combinations of two tones in the stopband generates significant products in the passband as shown in Figure 9.2(a).

A mobile TV receiver design for any of the mentioned standards must take the interferers into account. As a concrete example, ISDB-T, channel 7 in VHF band, overlaps with the National Television System Committee (NTSC) channel 8. Figure 9.3 illustrates this particular case. For the 3-Seg standard, the adjacent NTSC

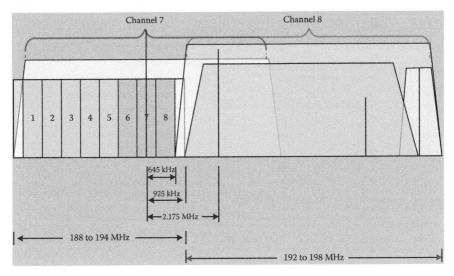

FIGURE 9.3 Channel allocation in 3-Seg (ISDB-T).

analog blocker channel can start as close as 300 kHz from the desired ISDB-T channel. Once more, the adjacent channel-filtering requirement is stringent. As a result, the receiver system design for mobile TV applications must consider not only power consumption and device size, but also multiple interferers that do exist in the same UHF and VHF bands. Moreover, mobility-dictated impairments, such as Doppler and multipath interference, set additional constraints for mobile TV receiver designers. Hence, new circuit architectures must be investigated to find receiver solutions for mobile TV technology that are low cost, low power, and high performance.

9.2   FRONT-END LNA NOISE

9.2.1   LNA NOISE CONSIDERATION

A typical direct conversion receiver (DCR) block diagram with corresponding signal and noise profiles is shown in Figure 9.4. The most critical block in the design is the first stage low noise amplifier (LNA), which generally sets the performance of the radio. In most cases, the gain of this first stage is set high, reducing the contribution of the following blocks to the overall system noise figure. In the case of nearby large blockers, though, linearity limitation in this block may dictate reduction in gain

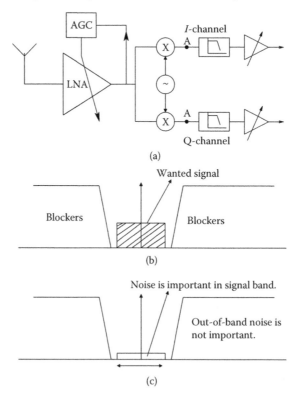

FIGURE 9.4   Typical DCR receiver front end; (a) block diagram, (b) signal profile around the desired channel, (c) Noise profile around the desired channel.

through a front-end AGC loop. Hence, the noise of the following stages will start to come into the picture. Thus, one should as well be careful in designing subsequent baseband blocks such as filters, variable-gain amplifiers (VGAs), analog-to-digital converters (ADCs), etc. As a result, one of the most important performance metrics for a radio is its sensitivity in environments with strong interferers.

The point immediately after RF down-conversion ("A" in the figure) is as well a critical point and determines the required circuitry down the chain depending on the signal and blocker profiles in the band of interest. A topology with a gain stage followed by filtering can again yield better sensitivity, as long as the blocker levels in the adjacent channels are limited and noise is the main factor to consider (Figure 9.5a). In the case of strong blockers, however, such architecture may not be optimum, due to the resulting stringent linearity requirement of the amplifier.

In the case of a gain-filtering interleaved architecture as shown in Figure 9.5(c), the design of the first amplifier stage can still be demanding. Thus, a filtering block might be required ahead of the gain stage as shown in Figure 9.5(b). The filter noise should be minimal for such a topology so as not to degrade the sensitivity due to noise. This can be satisfied with an additional cost of area and power in the filter if classical filter circuit topologies are to be employed [3–11].

9.2.2 NOISE-CANCELING LNA

The noise-canceling wideband CG-CS LNA topology of Figure 9.6 is utilized in the proposed work [12–14]. Features like noise-canceling, single-ended to differential conversion and relatively wideband input matching make this topology suitable for many UHF applications.

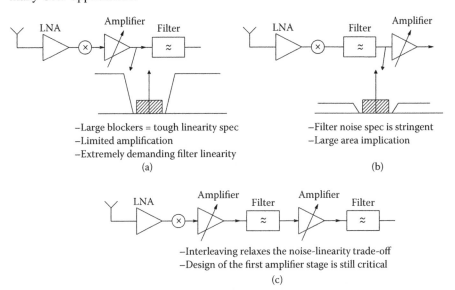

–Large blockers = tough linearity spec
–Limited amplification
–Extremely demanding filter linearity
(a)

–Filter noise spec is stringent
–Large area implication
(b)

–Interleaving relaxes the noise-linearity trade-off
–Design of the first amplifier stage is still critical
(c)

FIGURE 9.5 Noise–linearity trade-off in different architectures; (a) gain-filtering topology, (b) filtering-gain topology, (c) interleaved gain-filtering topology.

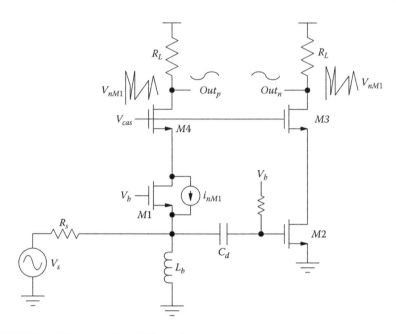

FIGURE 9.6   Noise-canceling LNA topology.

In this configuration, the noise generated in CG amplification device M_1 is mirrored to the other CS amplification path and amplified. By simply matching the voltage gains through both paths, the noise of the device M_1 can effectively be canceled. Hence, the main noise contributor remains M_2, of which the noise can independently be optimized. The cascode devices M_3 and M_4 provide Miller isolation and gain boost in case the device output impedances are a limiting factor in the overall gain of the stage. L_b serves as an RF choke whose value sets the low frequency end of the gain characteristics.

9.3   BASEBAND NOISE

As it has been pointed out in the previous sections, not only LNA noise but also the noise of mixers and other subsequent baseband blocks, in particular that of filters, may become a significant noise contributor in hostile spectrum conditions with strong blockers. This section is devoted to discussion of some noise-shaping analog baseband filtering techniques that can help to handle such interferers early along the chain and hence maintain a good sensitivity across a wide range of blocker profiles. The challenge in this, though, is to achieve this target without a significant impact on cost and power consumption.

9.3.1   Noise-Shaping, Blocker-Filtering Technique for Low Noise Integrated Receivers

The FDNR (frequency dependent negative resistance)-based filtering technique that is described in detail in this section offers unique advantages in terms of noise with its noise-shaping characteristics [15–18]. Moreover, the circuit provides this high-order filtering at the mixer output, protecting this node against blockers—a feature that traditional filter topologies cannot offer. Filtering at this critical node relaxes the mixer linearity spec and allows the mixer to have a higher gain. The input noise spec of postmixer blocks can thus be further relaxed. Hence, employing the technique in the receive chain of a radio can enhance the overall performance of the radio, an improvement that cannot be achieved using classical gain-filtering techniques unless area is sacrificed. The proposed technique, however, needs to be analyzed and the trade-offs should be clarified in terms of noise, linearity, power, area, and, most importantly, stability. Coexistence of multiple interdependent feedback paths dictates a careful analysis of the stability of the proposed circuit. To prove the clear benefits of the concept, detailed noise, linearity, and stability analysis are carried on in Section 9.3.1.1, Section 9.3.1.2, and Section 9.3.1.3, respectively. The details of the class-AB op-amp and the bandwidth calibration circuit are presented in Section 9.3.1.4. Section 9.3.1.5 presents the measurement results of a 65 nm CMOS (complementary metal oxide semiconductor) proof-of-concept test chip.

9.3.1.1 FDNR-Based Third-Order Elliptic Response Circuit

In the early 70s, following their invention, FDNRs were used extensively to realize high-order filter functions [19–25]. However, some known drawbacks encountered in the filter implementations have limited their use as a filter section. In most of the target low pass implementations, for example, there is a DC response associated with the series capacitor in the signal path [23]. The FDNR circuit satisfies the negative resistance function only in reference to the circuit ground. Moreover, the number of op-amps employed in an FDNR-based filter is greater than that of an integrator-based implementation of the same transfer function. Considering these disadvantages, these FDNR-based topologies have long been abandoned. The circuit of the proposed third-order section shown in Figure 9.7, however, offers some unique features that can be utilized to design a very low noise, blocker-immune radio baseband.

The main advantage of the proposed third-order configuration of Figure 9.7 is that it uses only one resistor in the signal path, the load resistor of the preceding stage R_f, to realize the desired filter transfer function. Hence, the load resistance of a mixer or of a gm stage can be reutilized as a part of the filter transfer function, which is not the case for classical filter topologies. The noise contribution of this particular resistor is already accounted for in the mixer noise budget and the noise of the FDNR resistors R_1, R_2, and R_3 is shaped. Since the op-amps are not in the signal path, their flicker noise contributions are also shaped and hence contribute less to the overall filter noise.

Moreover, as opposed to classic filter topologies, the op-amps of the proposed third-order section are not in the signal path and hence do not contribute any IQ mismatch or DC offset, which is a much desired property in a receiver chain. The signal transfer function of this circuit from input to output can be written as follows:

$$\frac{V_{out}(s)}{I_{in}(s)} = \frac{R_f(s^2 DR_z + 1)}{s^3 DR_z R_f C_f + s^2(DR_z + DR_f) + s(R_f C_f) + 1} \tag{9.1}$$

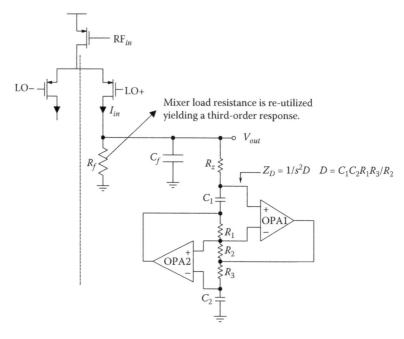

FIGURE 9.7 Simplified single-ended schematic of the proposed third-order elliptic circuit at the mixer output.

$$D = \frac{C_1 C_2 R_1 R_3}{R_2} \qquad (9.2)$$

This signal transfer function provides a notch at a frequency,

$$\omega_{notch} = \frac{1}{\sqrt{DR_z}},$$

and the notch frequency depends on the value of D and R_z. As a result, a large variety of component values can be used to realize the desired cutoff and notch frequencies. Linearity, stability, and area trade-offs associated with the proposed circuit will be investigated to narrow down to an optimum component value combination for the target application.

9.3.1.2 Noise Analysis

The schematic including the noise sources in an FDNR is shown in Figure 9.8. In the proposed circuit, the noise of all passive and active components in the FDNR section—namely, of R_z, R_1, R_2, R_3, OPA$_1$, and OPA$_2$—is shaped. Hence, the only substantial noise contributor is R_f, whose noise contribution is accounted for in the amplifier or mixer noise budget.

In order to clarify the noise-shaping concept of the proposed circuit, the noise transfer functions from each of these contributors have been calculated and presented as follows:

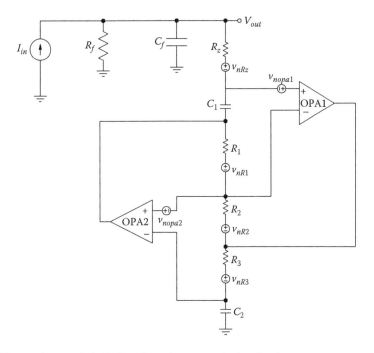

FIGURE 9.8   Schematic including the noise sources in the circuit.

The OPA$_1$ noise transfer function is

$$\frac{V_{out}(s)}{Vn_{OPA1}(s)} = \frac{sC_1R_f(1-s\,R_1R_3C_2/R_2)}{s^3DR_zR_fC_f + s^2(DR_z+DR_f)+s(R_fC_f)+1} \tag{9.3}$$

The OPA$_2$ noise transfer function is

$$\frac{V_{out}(s)}{Vn_{OPA2}(s)} = \frac{sC_1R_fR_1(sR_3C_2+1)}{R_2\left(s^3DR_zR_fC_f + s^2(DR_z+DR_f)+s(R_fC_f)+1\right)} \tag{9.4}$$

The R$_z$ noise transfer function is

$$\frac{V_{out}(s)}{Vn_{Rz}(s)} = \frac{s^2DR_f}{s^3DR_zR_fC_f + s^2(DR_z+DR_f)+s(R_fC_f)+1} \tag{9.5}$$

The R$_1$ noise transfer function is

$$\frac{V_{out}(s)}{Vn_{R1}(s)} = \frac{sR_fC_1}{s^3DR_zR_fC_f + s^2(DR_z+DR_f)+s(R_fC_f)+1} \tag{9.6}$$

The R_2 noise transfer function is

$$\frac{V_{out}(s)}{Vn_{R2}(s)} = \frac{sR_1R_fC_1}{R_2\left(s^3DR_zR_fC_f + s^2(DR_z + DR_f) + s(R_fC_f) + 1\right)} \qquad (9.7)$$

The R_3 noise transfer function is

$$\frac{V_{out}(s)}{Vn_{R3}(s)} = \frac{sR_1R_fC_1}{R_2\left(s^3DR_zR_fC_f + s^2(DR_z + DR_f) + s(R_fC_f) + 1\right)} \qquad (9.8)$$

Figure 9.9 shows the plots for the magnitude of these noise transfer functions as well as the signal transfer function. Since the noise generated by the FDNR resistors is shaped, the designer can use larger resistors (noisier) and hence can reduce the capacitor size. This results in a significant area saving. As a means of comparison, the total amount of capacitor required versus the desired noise level in a bandwidth of 1 MHz is plotted in Figure 9.10 for various third-order filter topologies (Figure 9.11) as well as the FDNR-based filter topology described in this work. It should be noted that the single op-amp topologies (Sallen-Key and multiple feedback) cannot achieve an elliptic response; hence, their overall figure of merit may not be as high for applications requiring large attenuation in the nearby channels.

For the sake of simplicity, the plots shown in Figure 9.10 reflect the noise due only to the resistors in the circuits and do not include the noise contributions of

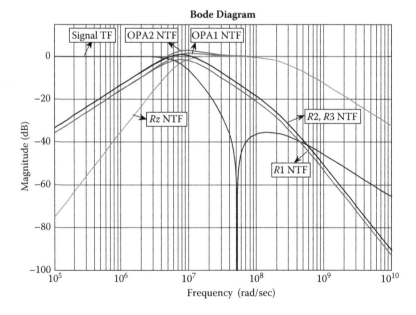

FIGURE 9.9 Signal and noise transfer functions of the proposed circuit.

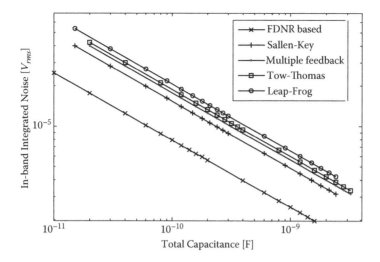

FIGURE 9.10 Total integrated in-band noise versus required capacitance for the various third-order filter topologies shown in Figure 9.11.

the op-amps used in these circuits. Since the noise of the op-amps is also shaped in an FDNR-based topology, the advantage of this topology becomes even more pronounced once the op-amp noise contributions are also included in the analysis.

9.3.1.3 Linearity Analysis

There are two important cases that need to be addressed regarding the linearity of the circuit. The first is the response of the circuit to two strong tones at the blocker frequencies. The IM3 product of these tones is minimized by providing a very large gain in the op-amps at these blocker frequencies. The other case is the nonlinearity experienced by the in-band signal. Depending on the ratio of notch frequency to cut-off frequency, the overall filter response can display some peaking around the cutoff frequency. The peaking in the signal transfer function might result in larger peaking at the internal FDNR nodes, particularly at the operational amplifier (op-amp) outputs of the proposed circuit. Thus, a full swing in-band signal might cause the FDNR op-amps outputs to have even a larger swing degrading the overall linearity of the circuit. Hence, the signal transfer function should be optimized, taking the peaking at the internal FDNR nodes into account.

The transfer function from input to op-amp2 output can be written as follows:

$$\frac{V_{opa2}(s)}{V_{in}(s)} = \frac{(s\,D/C_1+1)}{s^3 DR_z R_f C_f + s^2(DR_z + DR_f) + s(R_f C_f) + 1} \tag{9.9}$$

The magnitudes of this transfer function as well as the signal transfer function are shown in Figure 9.12(b) and Figure 9.12(a), respectively, for range filter characteristics corresponding to the same cutoff and notch frequencies. As can be observed from the plots, more rapidly decreasing filter response with more stopband attenuation

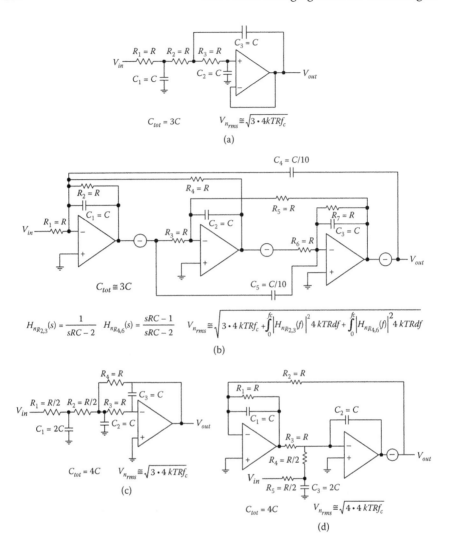

FIGURE 9.11 Integrated in-band noise for various third-order filter topologies; (a) Sallen-Key, (b) leap-frog (ladder), (c) multiple-feedback, (d) Tow-Thomas.

results in more peaking at the internal nodes. We can approximate the magnitude of the transfer function given in (9.9) around the cutoff frequency as follows:

$$\left| H_{opa2}(s) \right|_{f=f_c} \cong \sqrt{1 + \frac{\omega_c^2 D^2}{C_1^2}} \Big/ \sqrt{2} \tag{9.10}$$

Equation (9.10) suggests that the only way to obtain higher attenuation for fixed cutoff and notch frequencies without introducing extra peaking is to increase the value of the capacitor, $C1$. The amount of peaking that is allowed at the op-amp output sets the minimum value for this capacitor, as shown in Figure 9.13.

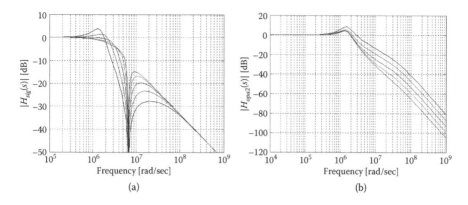

FIGURE 9.12   (a) The filter response for various peaking levels; (b) corresponding peaking levels at the FDNR internal node.

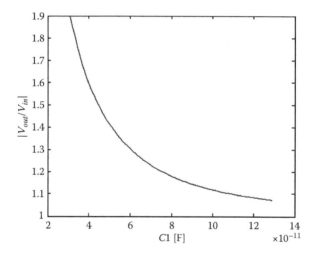

FIGURE 9.13   Minimum C_1 capacitance values corresponding to various peaking levels at the op-amp outputs.

9.3.1.4   Stability Analysis

The stability of the system should also be taken into account while optimizing the design for the desired filter response. There are multiple feedback networks that require attention. A simplified stability analysis of the system is useful since it has implications on unity gain bandwidth, DC gain, and compensation scheme of the op-amps used in the design. In order to simplify the analysis, one of the op-amps is considered to be ideal while the other one is analyzed with the network around it. If each of these independent cases provides the required margin for stability, the initial assumption for each individual case is not violated, and hence the system can be expected to be stable.

Figure 9.14(a) shows the loop around OPA1. In order to simplify the analysis, the same capacitor and resistor values are used for all of the components in the circuit.

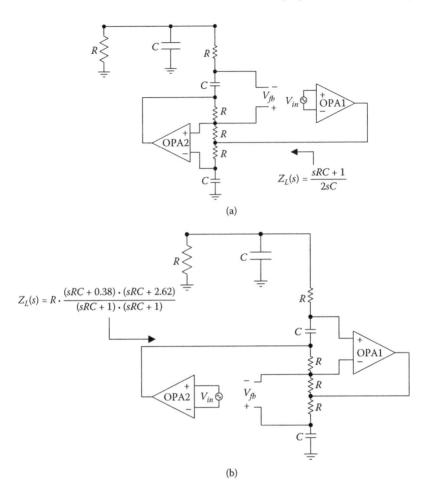

FIGURE 9.14 (a) Stability case for OPA1, where OPA2 is assumed to be ideal; (b) stability case for OPA2, where OPA1 is assumed to be ideal.

Switching to the actual design parameters would only move the poles and zeros around the ones corresponding to this simplified configuration. The loop transfer function around this op-amp can be written as follows:

$$L(s) = A(s) \cdot \frac{(sRC+1)(sRC+0.12+j0.8)(sRC+0.12-j0.8)}{(sRC+1)(sRC+0.38)(sRC+2.62)}$$

(9.11)

where A(s) is the op-amp transfer function when loaded with an impedance

$$Z_L = \frac{(1+sRC)}{2sC}.$$

This loading corresponds to a pole and a zero in the op-amp transfer function including the output impedance of the driver devices in the op-amp. All poles and zeros are in the vicinity of the filter cutoff and notch frequency and all cancel out, including the ones resulting from the op-amp loading. However, two of the zeros in the loop transfer function are complex conjugate and cause a notch in both the amplitude and phase response. If the gain of the amplifier around this notch frequency is not large enough to absorb the phase and amplitude notch with margin, system stability can be threatened. The loop transfer function is the same for the OPA2 case shown in Figure 9.14(b). However, this time the load seen by the op-amp is

$$Z_L = R \cdot \frac{(sRC + 0.38)(sRC + 2.62)}{(sRC + 1)(sRC + 1)}$$

and has two poles and two zeros. Again, the second op-amp, OPA2, needs to be able to provide sufficient gain around the cutoff frequency.

In conclusion, the op-amps in the proposed design not only should target high unity gain bandwidth, but also should provide a large gain around the desired notch frequency. This is generally the case for most of the op-amps used in filter applications but, for this particular FDNR application, this requirement has an impact on the stability of the system.

9.3.1.5   Op-amp and Bandwidth Calibration Circuit

The schematic of the op-amp used in the design is shown in Figure 9.15(a). This folded-cascode op-amp achieves a unity gain bandwidth greater than 200 MHz, consuming 380 µA from a 1.2 V supply with a phase margin of 61°. The input devices M_1 and M_2 are native (zero-Vt) devices to allow a rail-to-rail input swing. The PMOS and NMOS cascode devices are biased with a Vt-multiplier bias stage shown in Figure 9.15(a). Hence, the cascode bias points track the process variations, adjusting the headroom in these devices accordingly. The output is a class-AB stage with cascode devices biasing the floating current sources. This way, high impedance is maintained at the first stage output. Higher impedance at this node results in a higher unity gain bandwidth since the total amount of compensation capacitor needed drops with higher impedance at this node.

The bandwidth calibration circuit used in the design is shown in Figure 9.16(b). During the initial calibration a 7-bit bandwidth calibration code sets the resistive DAC targeting the desired filter bandwidth for a given crystal clock frequency. The trim range of this resistor is large enough to cover the whole filter bandwidth as well as any crystal frequency in the range of 10 to 40 MHz that is used as a reference. Following this, successive approximation logic computes a calibration code with the help of a comparator, as illustrated in Figure 9.16(b). Note that, since the resistor in this configuration is used as a trim element determining the bandwidth and crystal frequency, the capacitor is used as the calibration element with 6-bit resolution. In the filter side, in addition to a corresponding 6-bit resolution capacitor bank, there is a resistor trim to switch between the high end and the low end of the band. Tuning resolution at the low band mode is around 20 kHz.

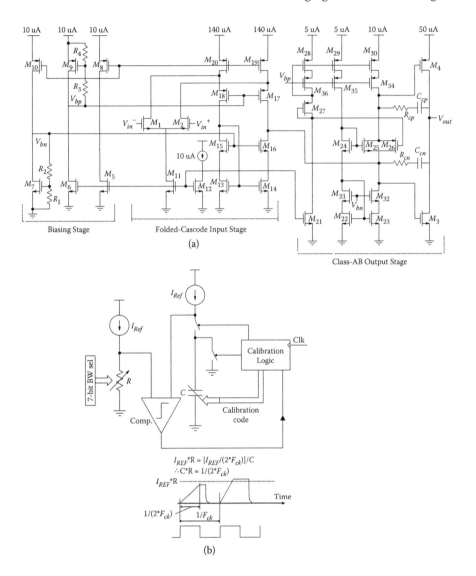

FIGURE 9.15 (a) The native input, folded-cascade, class-AB op-amp schematic; (b) bandwidth calibration circuit.

9.3.1.6 Measurements

A design covering a tuning range from 700 kHz to 5.2 MHz is fabricated in a 65 nm CMOS process and tested. The cutoff, notch, and stop-band characteristics can independently be set to satisfy the blocker and noise requirements of various applications across the band with a 7-bit control word corresponding to a tuning resolution of 20 kHz. Figure 9.16(a) shows various filter transfer function curves corresponding to different trim codes across the band. The testing and characterization are done for two distinct applications, integrated services digital broadcasting-terrestrial

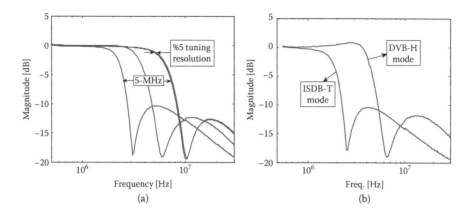

FIGURE 9.16   (a) Measured filter response for various bandwidth settings across the band; (b) filter responses corresponding to target ISDB-T and DVB-H applications.

(ISDB-T) and digital video broadcasting-handheld (DVB-H), with signal bandwidths of 750 kHz and 3.8 MHz, respectively. The filter can be tuned to provide optimum response for in-band and blocker signals corresponding to these two example applications (Figure 9.16b). In ISDB-T mode, the signal cutoff is around 750 kHz whereas the N + 1 blocker starts at 2.5 MHz. In DVB-H mode, the signal bandwidth is around 3.8 MHz, whereas the N + 1 blocker power can occur in the next 8 MHz bandwidth. The optimum case for the DVB-H has a slight peaking in the filter response, since the notch frequency is set to be relatively close to filter cutoff.

The noise characteristics of the filter for these target bands are shown in Figure 9.17. Figure 9.17(a) shows the noise in ISDB-T mode, whereas Figure 9.17(b) shows the noise in DVB-H mode. The expected noise shaping can clearly be observed for both cases. The noise density in the signal band is around 7.5 nV/sqrtHz, which is mainly due to the 2 kΩ output resistance (R_f) of the driving stage. The 20 dB/decade in-band noise roll-off stops once the noise level of this resistor is reached. This resistor is the

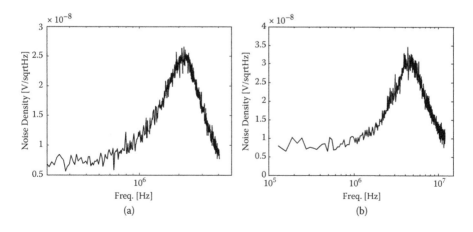

FIGURE 9.17   (a) Filter noise in ISDB-T mode; (b) filter noise in DVB-H mode.

load resistance of the mixer or the driving stage and its noise contribution is already accounted for. It is important to note that this load resistor cannot be utilized in other active filter topologies and hence additional resistors are required in the filter to achieve the desired filter characteristic. This is a unique advantage of the proposed FDNR-based topology. With regard to linearity, HD tests and two-tone IM3 tests were conducted for both target modes. Figure 9.18(a) shows that a 250 KHz, 420 mVpp differential tone in ISDB-T mode results in –57 dB HD2. A differential tone of 1.2 MHz, 420 mVpp in DVB-H mode yields –59 dB HD2 (Figure 9.18b).

In ISDB-T mode, 2.4 and 4.4 MHz blockers (420 mVpp differential each) result in –80.4 dBc IM3, which corresponds to an out-of-band IIP3 of 36.5 dBm (Figure 9.19a). In-band IIP3 for this case is 22.5 dBm. In DVB-H mode, 5 and 8 MHz blockers (420 mVpp differential each) result in –69.8 dBc IM3, which corresponds to an out-of-band IIP3 of 31.5 dBm (Figure 9.19b). In-band IIP3 for this case is 20 dBm. For both cases, the tones are swept in-band and out-of-band recording the IIP3 for each of these cases. The plots showing IIP3 with varying average two-tone frequency are shown in Figure 9.20. The design occupies a die area of 0.16 mm^2 in the mentioned 65 nm CMOS process. Total power consumption is in the range of 2 to 3.2 mA from a 1.2 V supply, depending on the received signal strength.

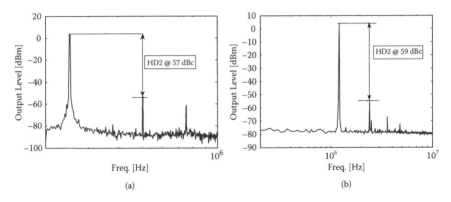

FIGURE 9.18   (a) In-band HD in ISDB-T mode; (b) in-band HD in DVB-H mode.

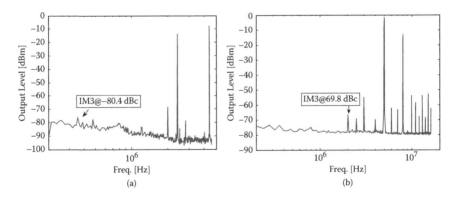

FIGURE 9.19   (a) Two-tone test in ISDB-T mode; (b) two-tone test in DVB-H mode.

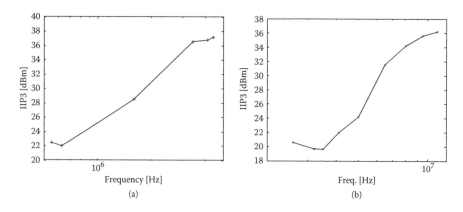

FIGURE 9.20 (a) IIP3 versus freq. sweep in ISDB-T mode; (b) IIP3 versus freq. sweep in DVB-H mode.

The technique enhances the sensitivity of a receiver chain significantly, particularly in an environment where strong blockers can coexist. Performance metrics of a 65 nm CMOS test chip are summarized in Table 9.1. Comparison of the literature in terms of noise and required capacitance per pole is shown in Figure 9.21. It can be seen that the approach proposed in this work results in lower noise for a given amount of capacitance relative to published work to date. The die photo of the design is shown in Figure 9.22.

9.3.2 Low Noise Gain Stage with Noise-Shaping Blocker Suppression

In this section, a blocker-aware gain stage is presented. The asymmetric floating frequency dependent negative resistance (AFFDNR) circuit in the feedback of an amplifier is introduced as a unique solution for providing gain in the signal band while simultaneously implementing a third-order, low pass elliptic response for the blocker signals. Due to its noise-shaping characteristics, the noise contribution due to the filtering action is insignificant.

The technique proposed in this section offers a solution with simultaneous gain and noise-shaping filtering—namely, amplification only in the band of interest. Low noise amplification of only the desired signal is proposed against the classical approach of amplifying everything, including the blockers, and then trying to filter out the blockers in the subsequent filtering stage. Avoiding the amplification of the blockers not only relaxes the linearity spec of the amplifier, but also relieves the filtering requirements of the following filtering stage.

9.3.2.1 AFFDNR-Based Gain Stage

A new circuit topology based on the previously mentioned FDNR structure, the AFFDNR used in the feedback of a programmable gain amplifier (PGA) is shown in Figure 9.23. Figure 9.23(a) shows an instrumentation topology with high input impedance, while Figure 9.23(b) shows a fully differential implementation with finite input impedance. Two single-ended op-amps are used in instrumentation topology,

TABLE 9.1

Performance Summary of the Proposed Baseband Prefilter

Technology	65 nm CMOS
Die area	0.24 mm^2
Power supply	1.2 V
Current consumption	
Max	3.2 mA
Min	2 mA
DC-gain	0 dB
f_c	
ISDB-T mode	750 kHz
DVB-H mode	3.8 MHz
Out-of-band IIP2	
ISDB-T mode	57 dBm
DVB-H mode	50.5 dBm
Tuning range	700 kHz–5.2 MHz
Tuning resolution	3%
Noise density	7.5 nV/sqrtHz
SFDR	
ISDB-T mode	84 dB
DVB-H mode	76.6 dB
DR (at HD3 = 40 dB)	
ISDB-T mode	92.7 dB
DVB-H mode	85.5 dB
In-band IIP3	
ISDB-T	22.5 dBm
f_1 = 500 kHz, f_2 = 600 kHz	
DVB-H	20 dBm
f_1 = 2.4 MHz, f_2 = 2.8 MHz	
Out-of-band IIP3	
ISDB-T	36.5 dBm
f_1 = 2.4 MHz, f_2 = 4.4 MHz	
DVB-H	31.5 dBm
f_1 = 5 MHz, f_2 = 8 MHz	
In-band HD2	
(420 mVpp diff.)	
ISDB-T(f = 500 kHz)	−57 dB
DVB-H(f = 1.2 MHz)	−59 dB

whereas, one fully differential op-amp with common mode feedback is used in the fully differential case.

The PGA op-amp input stage should be able to handle signal swing in the instrumentation topology, while in the fully differential case, the op-amp inputs are at virtual ground and do not experience any voltage swing. The circuits are composed of the main PGA op-amps that realize the gain with the use of the feedback resistance R_f and input resistance R_{in}. Although linearity of such a system is relaxed,

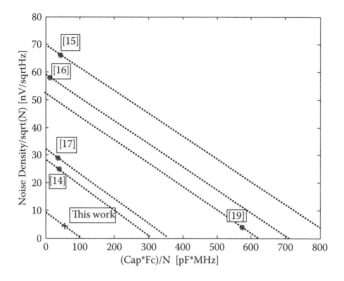

FIGURE 9.21   Literature comparison in terms of noise and required capacitance per pole.

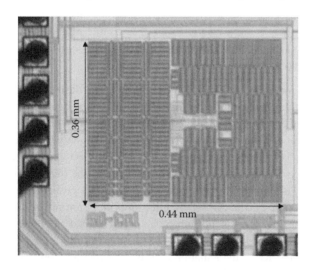

FIGURE 9.22   Die microphotograph of the design.

allowing larger gain in the amplifier, certain applications still require gain variability depending on the received signal strength. Input resistor R_{in} can be trimmed for PGA operation if gain variation is desired. The filtering stage is placed in the feedback path in parallel with R_f. R_f is incorporated into the filter transfer function and hence serves a dual function: providing gain and adding filtering. When the signal is applied to the input terminal, the feedback path with AFFDNR presents an impedance of R_f for the in-band signal, whereas the blocker sees a short to the output. The blocker signals do not experience any gain in the signal path. Thus, the linearity spec of the amplifier is relaxed since the output cannot see a large

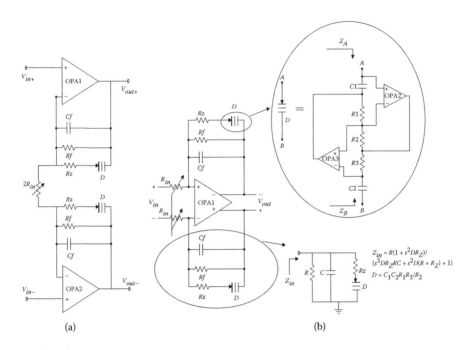

FIGURE 9.23 Amplifiers with AFFDNR feedback; (a) instrumentation topology, (b) fully differential implementation.

blocker voltage swing. More precisely, the signal in the desired channel of interest is amplified with a third-order elliptic filter characteristic due to the proposed frequency selective feedback. It should be noted that the proposed AFFDNR is not a reciprocal circuit and the mentioned filtering action can only be obtained provided that the polarity is as shown in Figure 9.23(b). Namely, the impedance looking into the node A, Z_A is desired negative resistance, whereas the impedance looking into the node B, Z_B is inductive when the opposing port is grounded for each of the cases. In this topology, the AFFDNR filter amplifiers are not in the signal path, so no additional DC offset or IQ imbalance is introduced as opposed to common amplifier-filter topologies.

Using KVL and KCL, the equation defining the relation between the input and output of this AFFDNR feedback circuit can be written as

$$\frac{V_{in}(s) \cdot R_f}{R_{in}} = \frac{\left(V_{out}(s) - V_{in}(s)\right) \cdot \left(s^3 DR_z RC + s^2 (DR_z + DR) + sRC + 1\right)}{s^2 DR_z + 1} \quad (9.12)$$

$$= \left(V_{out}(s) - V_{in}(s)\right) \cdot F(s)$$

where F(s) is the transmission function and

$$D = \frac{C_1 C_2 R_1 R_3}{R_2}. \quad (9.13)$$

Looking into the relation between input and output provided in (9.12), for the low frequency in-band signals, F(s) is unity and hence an in-band signal experiences expected noninverting gain of

$$A_V = 1 + \frac{R_f}{R_{in}}.$$

The input–output relation around the notch frequency, however, yields an interesting case. One would expect a minimum gain of unity from this amplifier topology—worst case being the zero feedback impedance. Looking into the response of the amplifier around the notch frequency, however, we can even observe some attenuation (Figure 9.26). The reason for this becomes clear once the F(s) in (9.12) is analyzed. Figure 9.24 shows the bode plot of the function F(s). The phase response reaches 180° very quickly, while the amplitude response is in the ramp-up and still has a relatively finite value. Such a finite magnitude with a negative sign corresponds to attenuation in the overall signal transfer function given in (9.12).

9.3.2.2 Noise Analysis
The detailed schematic of the simplified single-ended design including the noise sources is shown in Figure 9.25. The noise of the amplifier elements, the input resistance R_{in} and the feedback gain resistor R_f, remains the same and sets the noise floor for this topology. The noise of the AFFDNR filtering section elements (R_z, R_1, R_2, R_3, OPA1, and OPA2) is, however, shaped and does not contribute to the overall noise figure significantly. Thus, one can obtain a relative high-order filtering without extra noise penalty. To demonstrate the noise-shaping concept of the proposed circuit explicitly, the noise transfer functions from each of these contributors to the output have been calculated and are as follows.

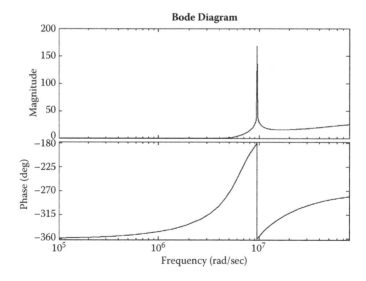

FIGURE 9.24 Bode plot for the function F(s) given in Equation (9.12).

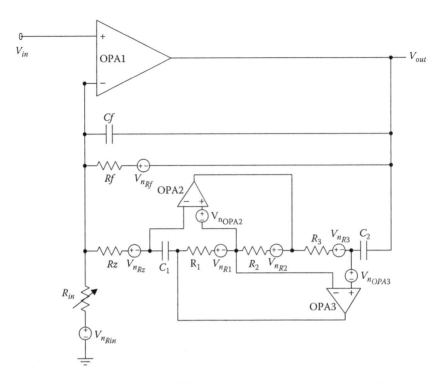

FIGURE 9.25 Schematic of a simplified single-ended circuit including the noise sources.

The OPA2 noise transfer function is

$$\frac{V_{out}(s)}{Vn_{OPA2}(s)} = \frac{sC_1R_f(1 - s\,R_1R_3C_2/R_2)}{s^3DR_zR_fC_f + s^2(DR_z + DR_f) + s(R_fC_f) + 1} \tag{9.14}$$

The OPA3 noise transfer function is

$$\frac{V_{out}(s)}{Vn_{OPA3}(s)} = \frac{sC_1R_fR_1(sR_3C_2 + 1)}{R_2\left(s^3DR_zR_fC_f + s^2(DR_z + DR_f) + s(R_fC_f) + 1\right)} \tag{9.15}$$

The R_z noise transfer function is

$$\frac{V_{out}(s)}{Vn_{Rz}(s)} = \frac{s^2DR_f}{s^3DR_zR_fC_f + s^2(DR_z + DR_f) + s(R_fC_f) + 1} \tag{9.16}$$

The R_1 noise transfer function is

$$\frac{V_{out}(s)}{Vn_{R1}(s)} = \frac{sR_fC_1}{s^3DR_zR_fC_f + s^2(DR_z + DR_f) + s(R_fC_f) + 1} \tag{9.17}$$

The R_2 noise transfer function is

$$\frac{V_{out}(s)}{Vn_{R2}(s)} = \frac{sR_1R_fC_1}{R_2\left(s^3DR_zR_fC_f + s^2(DR_z + DR_f) + s(R_fC_f) + 1\right)} \tag{9.18}$$

The R_3 noise transfer function is

$$\frac{V_{out}(s)}{Vn_{R3}(s)} = \frac{sR_1R_fC_1}{R_2\left(s^3DR_zR_fC_f + s^2(DR_z + DR_f) + s(R_fC_f) + 1\right)} \tag{9.19}$$

The signal transfer function from input to output can be written as follows:

$$\frac{V_{out}(s)}{V_{in}(s)} = 1 + \frac{R_f(s^2DR_z + 1)}{R_{in}\left(s^3DR_zR_fC_f + s^2(DR_z + DR_f) + s(R_fC_f) + 1\right)} \tag{9.20}$$

Figure 9.26 shows the plots for the magnitude of these noise transfer functions as well as the signal transfer function.

It is also possible to use the noise-shaping characteristic of the proposed technique to reduce overall chip area. Since the noise generated by the AFFDNR resistors is shaped, this enables the designer to use larger resistors (noisier) and hence reduce the capacitor size. This approach can result in significant area saving for a particular noise level. As a means of comparison, the total amount of capacitor required versus the desired in-band input referred noise level is plotted in Figure 9.27 for various third-order filter topologies as well as the circuit topology described in this work.

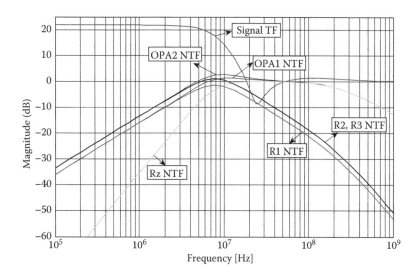

FIGURE 9.26 Signal and noise transfer functions showing the noise shaping.

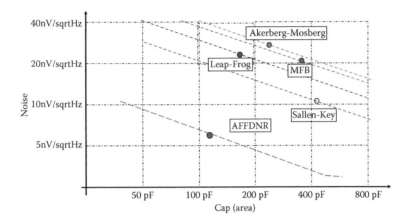

FIGURE 9.27 Desired noise level versus required capacitor value for a range of filter topologies targeting the same cutoff frequency of around 3.5 MHz.

9.3.2.3 Measurement Results

A differential instrumentation stage covering a frequency range from 700 kHz to 4.2 MHz is fabricated in a 65 nm CMOS process and tested (Figure 9.28). The cutoff, notch, and stop-band characteristics can independently be set to satisfy the blocker and noise requirements of various applications across the band with a 7-bit control word

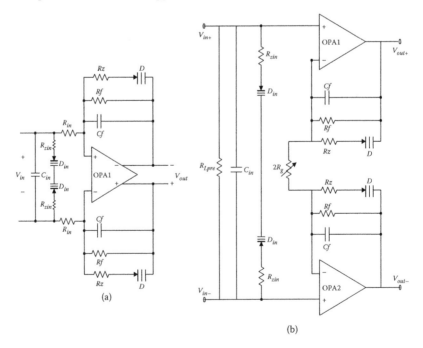

FIGURE 9.28 (a) Fully differential cascade topology; (b) cascade instrumentation topology.

corresponding to a tuning resolution of 20 kHz. The design consumes a die area of 0.12 mm² and the total power consumption is in the range of 3 to 4.8 mA from a 1.2 V supply, depending on the received signal strength. Figure 9.29(a) shows various amplifier transfer function curves corresponding to different trim codes across the band.

The testing and characterization is done for ISDB-T and DVB-H, with signal bandwidths of 750 kHz and 3.8 MHz, respectively. The radio chain employing the circuit requires that IM3 levels resulting from –3.5 dBm N + 1 blockers be below –70 and

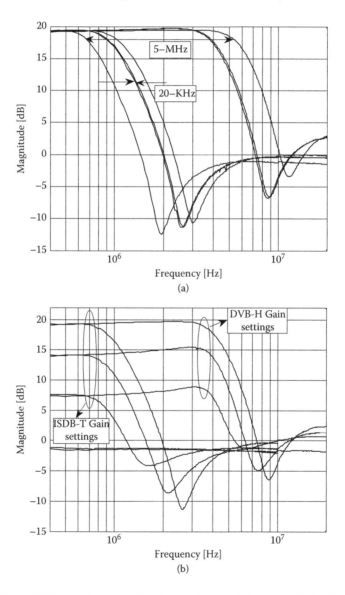

FIGURE 9.29   (a) Measured amplifier bandwidth characteristics across the band; (b) amplifier response for various gain settings in ISDB-T and DVB-H modes.

−60 dBc for ISDB-T and DVB-H modes, respectively. The input referred noise requirements are 12 and 9 nV/sqrtHz for ISDB-T and DVB-H modes, respectively. The transfer characteristics with various gain steps corresponding to these target applications
are shown in Figure 9.29(b). In ISDB-T mode, the signal cutoff is around 750 kHz,
whereas the N + 1 blocker is centered around 2.5 MHz. In DVB-H mode, the signal
bandwidth is around 3.8 MHz, whereas the N + 1 blocker power can occur in the next
8 MHz bandwidth. The optimum case for the DVB-H has a slight peaking in the filter
response, since the notch frequency is set to be relatively close to filter cutoff.

Regarding the noise budget allocation, setting the noise contribution of the
AFFDNR and of the gain amplifier is not the optimum choice since the power is
more of a concern in the amplifier design. An optimum design strategy for the target
design was to allocate most of the in-band noise budget to the amplifier to reduce its
power consumption and limit the FDNR contributions. The noise characteristics of
the filter for these target bands are shown in Figure 9.30. Figure 9.30(a) shows the
noise in ISDB-T mode, whereas Figure 9.30(b) shows the noise in DVB-H mode. It
can be observed from these plots that the shaped AFFDNR contribution becomes
dominant only at the edge of the band. The measured noise density in the signal band
is around 9.5 nV/sqrtHz in DVB-H mode and 14 nV/sqrtHz in the ISDB-T mode. Total
capacitance required in the design corresponding to these noise levels was 120 pF.

With regard to linearity, HD tests and two-tone IM3 tests were conducted for
both target modes. Figure 9.31(a) shows that a 250 kHz, 750 mVpp differential tone
at the output in ISDB-T mode results in −57.4 dB HD2 and −58.9 dB HD3. In DVB-H
mode a 1.2 MHz, 750 mVpp differential tone at the output yields −60 dB HD2 and
−57 dB HD3 (Figure 9.31b).

In ISDB-T mode, 2.25 and 3.75 MHz blockers (420 mVpp differential each) result
in −78 dBm IM3 at the output, which corresponds to an out-of-band IIP3 of 37 dBm
(Figure 9.32a). In DVB-H mode, 7 and 11 MHz blockers (420 mVpp differential
each) result in −65.6 dBm IM3 at the output, which corresponds to an out-of-band
IIP3 of 29.7 dBm (Figure 9.32b). In Figure 9.33, measured IIP3 interpolation plots

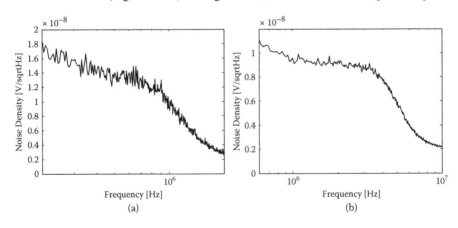

FIGURE 9.30 Total input referred noise measured; (a) in ISDB-T mode, (b) in DVB-H
mode.

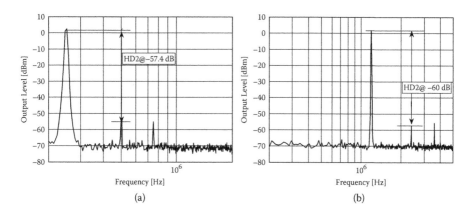

FIGURE 9.31   In-band distortion levels measured; (a) in ISDB-T mode, (b) in DVB-H mode.

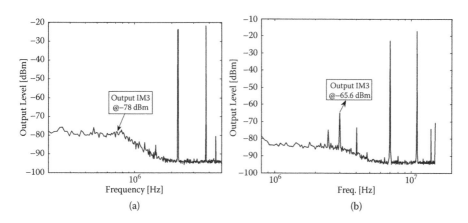

FIGURE 9.32   Two-tone tests; (a) output spectrum for 240 mVpp diff. each, 2.25 MHz and 3.75 MHz input blockers in ISDB-T mode; (b) output spectrum for 240 mVpp diff. each, 7 MHz and 11 MHz input blockers in DVB-H mode.

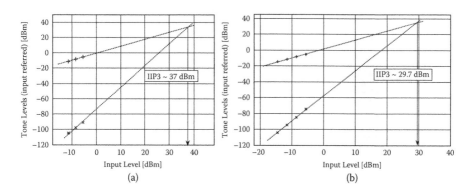

FIGURE 9.33   Measured IIP3; (a) in ISDB-T mode, (b) in DVB-H mode.

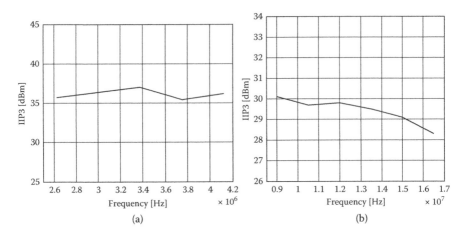

FIGURE 9.34   IIP3 across the blocker band; (a) in ISDB-T mode, (b) in DVB-H mode.

are presented for both target bands. Moreover, the tones are swept across the blocker band recording the IIP3 for each of these cases. The plots showing IIP3 with varying average two-tone frequency are shown in Figure 9.34.

Performance metrics of this test circuit are summarized in Table 9.2. The die picture is shown in Figure 9.35.

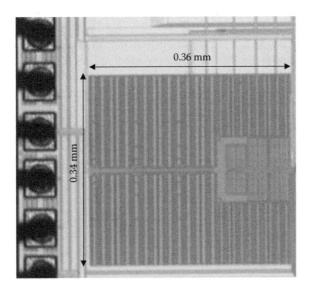

FIGURE 9.35   Die microphotograph of the PMA.

TABLE 9.2

Performance Summary of the Proposed Baseband PMA

Technology	65 nm CMOS
Die area	0.13 mm^2
Power supply	1.2 V
Current consumption	
Max	4.8 mA
Min	3 mA
Gain settings	20-15-8-0 dB
f$_c$	
ISDB-T mode	750 kHz
DVB-H mode	3.8 MHz
Out-of-band IIP2	
ISDB-T mode	75 dB
DVB-H mode	65.5 dB
Tuning range	700 kHz–4.2 MHz
Tuning resolution	3%
Noise density	
ISDB-T mode	14 nV/sqrtHz
DVB-H mode	9.5 nV/sqrtHz
DR (at HD3 = 40 dB)	
ISDB-T mode	87.6 dB
DVB-H mode	83.3 dB
Out-of-band IIP3	
ISDB-T	37 dBm
f$_1$ = 2.25 MHz, f$_2$ = 3.75 MHz	
DVB-H	29.7 dBm
f$_1$ = 7 MHz, f$_2$ = 11 MHz	
HD2 (750 mVpp diff.)	
ISDB-T(f = 500 kHz)	−57.4 dB
DVB-H(f = 1.2 MHz)	−60.5 dB
Total cap size	120 pF

9.3.3 CURRENT-MODE NOISE-SHAPED FILTERING

Current-mode noise-shaping filtering has as well been proven to another low noise, low distortion filtering technique that is suitable for wireless communication applications. The circuit converts a simple gm-C type of cascade into a more effective second-order filter with complex poles by uniquely cross coupling the intermediate nodes intp and intn (Figure 9.36) [26]. Intuitively, device noise is pushed back at low frequencies while it finds a low impedance path into the capacitors at high frequencies and hence gains up at the load. Assuming the same gm for all of the active MOS devices, the filter response can be written as follows:

$$\frac{I_{out}(s)}{I_{in}(s)} = \frac{(gm^2/C_{f1}C_{f2})}{s^2 + s(gm/C_{f1}) + (gm^2/C_{f1}C_{f2})} \tag{9.21}$$

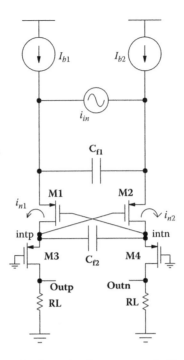

FIGURE 9.36 Detailed circuit schematic of a current-mode filter.

Again, the same outstanding feature in this noise-shaping filter as in the case of FDNR-based ones is that the blocker currents are not allowed to reach high impedance nodes without filtering. Figure 9.37 shows the filter transfer function as well as the output noise profile.

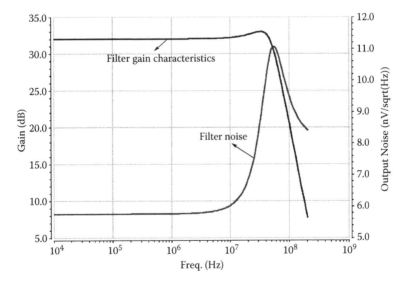

FIGURE 9.37 Filter gain response and noise characteristics across the channel.

9.4 BLIXELTER: SINGLE-STAGE BALUN–LNA–MIXER–FILTER COMBO TOPOLOGY

The circuit technique that is introduced in this section provides balun, LNA, mixer, and noise-shaped filter functions all in one folded circuit stage as shown in Figure 9.38. In this architecture, using minimal numbers of transistors and achieving a high gain and high-order filtering stage results in a low noise, high-sensitivity receiver in the existence of strong adjacent in-band blockers as well as other out-of-band interferers. In this topology, which is referred to as Blixelter, the RF input drives the noise-canceling LNA device pair M_{10} and M_{11}. This common-gate, common-source amplifier pair as well provides balun functionality, converting the single-ended signal to differential. LO clock source mixes the RF signal down through mixer devices M_{12} through M_{15}. A larger portion of the LNA device currents are provided through I_{LNA} current sources and hence do not flow through the mixers, reducing the mixer noise contribution. Current source I_{LNA} thermal noise can be limited and its flicker noise contribution is upconverted and hence will not fall in-band. The current sources I_{b1} and I_{b2} provide DC bias current for the front-end LNA–mixer pair and some remaining current flows through M_6 and M_7, biasing the folded filter section. This section implements a second-order pipe filter through C_{f1} and C_{f2} and additional third-order notch filtering at the load through FDNR and C_x. Since both of these techniques provide noise-shaped filtering, the strong interferers at the adjacent channels are attenuated without significant addition of filter noise in-band. The current sources I_{b1} and I_{b2} remain significant flicker noise contributors and hence they occupy a relatively large die area not to degrade noise at the lower end of the channel

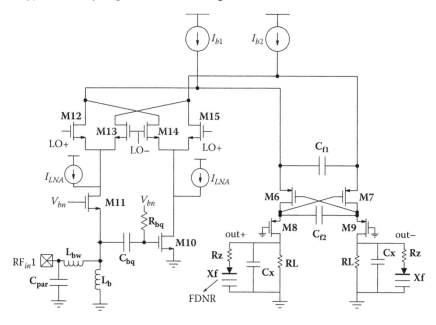

FIGURE 9.38 Detailed circuit schematic of Blixelter.

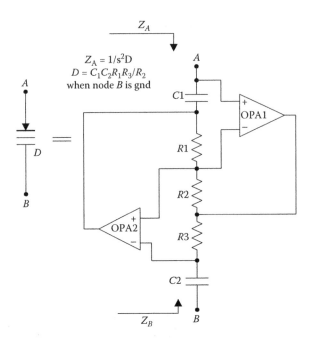

FIGURE 9.39 FDNR detailed circuit schematic.

band. The receiver front ends implementing such topology can allow higher gain without suffering from adjacent interferers and hence can achieve better sensitivity for a given current and area budget. The detailed schematic of the FDNR section is shown in Figure 9.39. The overall filter transfer function for the combiner section can be written as

$$\frac{V_{out}(s)}{I_{in}(s)} = \frac{(gm^2/C_{f1}C_{f2})R_f(s^2DR_z+1)}{\left(s^3DR_zR_LC_x+s^2(DR_z+DR_L)+s(R_LC_x)+1\right)\left(s^2+s(gm/C_{f1})+(gm^2/C_{f1}C_{f2})\right)} \quad (9.22)$$

where

$$D = \frac{C_1C_2R_1R_3}{R_2}$$

of FDNR and g_m is the transconductance for the devices M_6 through M_9. The filter capacitors C_x and C_2, which are shown to be single ended conceptually here in the schematic, are implemented fully differential in the actual design.

The simulated frequency response and noise characteristics of the folded filtering section in a 28 nm CMOS process are shown in Figure 9.40 for a 5 MHz DVBH channel. An average attenuation of 30 dB is achieved immediately at the adjacent channel with the notch corresponding to the center of the adjacent channel at around 8 MHz. This amount of filtering is achieved with less than 10 nV/sqrtHz average noise density in-band corresponding to a total of 150 pF capacitor area.

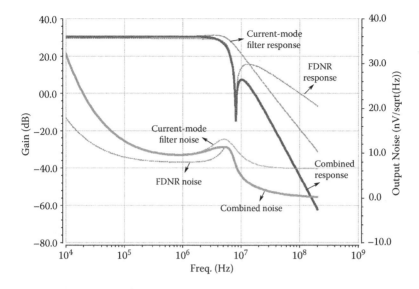

FIGURE 9.40 Noise and frequency response of the folded filtering section.

At the input of the proposed front end, a 1 uH RF choke is used, providing a wideband matching down to VHF frequencies. S_{11} across the whole UHF band is better than −25 dB in simulations assuming 0.8 nH bondwire inductance and 0.5 pF ESD and package parasitics (Figure 9.41).

The simulated conversion gain of the Blixelter front end is 26.8 dB for a UHF DVBH channel at 470 MHz with adjacent blocker IIP3 of −11.48 dBm, which corresponds to an effective out-of-band OIP3 of 15.32 dBm (Figure 9.42 and Figure 9.43). The tones for this analysis were chosen at the center of the immediate adjacent channel, at 2.5 and 3.5 MHz away from the 5 MHz channel edge. The notch response around the desired channel is visible at around the carrier frequency of 470 MHz in Figure 9.42.

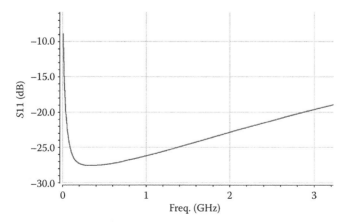

FIGURE 9.41 Simulated S_{11} of the Blixelter front end.

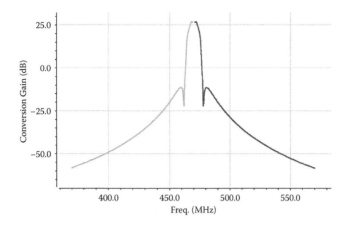

FIGURE 9.42   Simulated conversion gain around 470 MHz UHF channel.

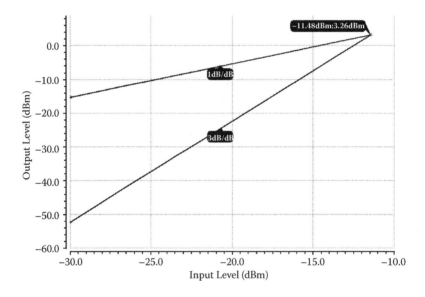

FIGURE 9.43   Adjacent blocker channel linearity.

Finally, the noise figure plot of the proposed Blixelter front end is shown in Figure 9.44. The NF across the channel is around 10.8 dB. The value goes up due to noise-shaping, reaching 11 dB at the high end of the band at 5 MHz. The value deteriorates to 14 dB at the flicker-noise dominant frequencies below 200 kHz. The total current consumption of the proposed circuit is 10.4 mA from a 1.8 V supply.

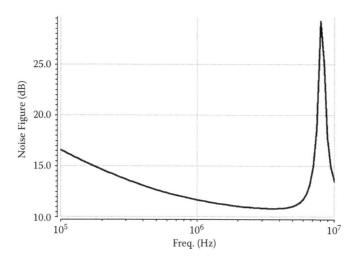

FIGURE 9.44   Blixelter noise figure.

9.5   CONCLUSION

In conclusion, with an increasing demand for wireless/mobile applications, deep, submicron CMOS has become a very attractive platform to design SOC radios for the emerging standards. Although it has indisputable dominance for such applications due to the amount of DSP that can be integrated, CMOS poses significant noise and linearity challenges to the front-end designers, for it has low supply voltage requirements. Due to rapid spectrum crowding and demand for increased communication range, this historic noise–linearity–power trade-off is pronounced even more. This chapter has presented some noise-shaping, noise-canceling techniques at front end and baseband, focusing particularly on FDNR-based noise-shaping filtering techniques in a 65 nm CMOS process. Finally, all of the techniques presented are combined, forming what is called "Blixelter," which includes balun, LNA, mixer, and high-order noise-shaped filter functionalities in one single folded circuit stage. Implementing fifth-order noise-shaped filtering, all in current domain, helps to attenuate nearby blockers before they can reach the LNA–mixer load. Thus, the topology not only eliminates the need for a complex front-end AGC loop, but also relaxes the noise–linearity specification of the subsequent baseband blocks.

REFERENCES

1. P. Chaudhury et al., The 3GPP proposal for IMT-2000. *IEEE Communications Magazine,* vol. 37, pp. 71–81, Dec. 1999.
2. M. Zeng et al., Recent advance in wireless communications. *IEEE Communications Magazine,* vol. 37, pp. 128–138, Sept. 1999.
3. J. F. Fernandez-Bootello et al., A 0.18 μm CMOS low-noise elliptic low-pass continuous-time filter. *Proceedings of IEEE International Symposium on Circuits and Systems,* pp. 800–803; vol. 1, 23–26 May 2005.

4. A. Yoshizawa et al., A channel-select filter with agile blocker detection and adaptive power dissipation. *IEEE Journal of Solid-State Circuits,* vol. 42, no. 5, pp. 1090–1099, May 2007.

5. C. H. J. Mensink et al., A CMOS soft-switched transconductor and its application in gain control and filters. *IEEE Journal of Solid-State Circuits,* vol. 32, no. 7, pp. 989–998, July 1997.

6. C. Yoo et al., A ±1.5 V, 4 MHz CMOS continuous-time filter with single-integrator based tuning. *IEEE Journal of Solid-State Circuits,* vol. 33, no. 1, pp. 18–27, Jan 1998.

7. F. Krummanecher et al., A 4 MHz CMOS continuous-time filter with on-chip automatic tuning. *IEEE Journal of Solid-State Circuits,* vol. 32, no. 7, pp. 989–998, July 1997.

8. W. Schelmbauer et al., An analog baseband chain for a UMTS zero-IF receiver in a 75 GHz BiCMOS technology. *MTT-S Digest of Technical Papers,* vol. 1, pp. 13–16, June 2002.

9. S. D'Amico et al., A 4.1 mW, 10 MHz fourth-order source-follower based continuous-time filter with 79-dB DR. *IEEE Journal of Solid-State Circuits,* vol. 41, no. 12, pp. 2713–2719, Dec. 2006.

10. H. A. Alzaher et al., A CMOS highly linear channel-select filter for 3G multistandard integrated wireless receivers. *IEEE Journal of Solid-State Circuits,* vol. 37, no. 1, pp. 27–37, Jan. 2002.

11. S. Kousai et al., A 19.7 MHz, fifth-order active-RC Chebyshev LPF for draft IEEE 802.11n with automatic quality-factor tuning scheme. *IEEE Journal of Solid-State Circuits,* vol. 42, no. 11, pp. 2326–2337, Nov. 2007.

12. F. Bruccoleri et al., Wideband CMOS low-noise amplifier exploiting thermal noise canceling. *IEEE Journal of Solid State Circuits,* vol. 39, pp. 275–282, Feb. 2004.

13. S. C. Blaakmeer et al., Wideband balun-LNA with simultaneous output balancing, noise cancelling and distortion canceling. *IEEE Journal of Solid State Circuits,* vol. 43, pp. 1341–1350, June 2008.

14. S. C. Blaakmeer et al., The BLIXER, a wideband balun–LNA–IQ–mixer topology. *IEEE Journal of Solid State Circuits,* vol. 43, no. 12, pp. 2706–2715, Dec. 2008.

15. A. Ismail, J. E. Vasa, B. Ramachandran, Low noise filter for a wireless receiver. US Patent no. 10/949,534.

16. A. Ismail, E. Youssoufian, H. Elwan, F. Carr, System and method for performing RF filtering. US Patent no. 11/377,721.

17. A. Ismail, E. Youssoufian, H. Elwan, F. Carr, System and method for performing RF filtering. US Patent no. 11/378,558.

18. A. Tekin et al., Noise-shaping gain-filtering techniques for integrated receivers. *IEEE Journal of Solid-State Circuits,* vol. 44, no. 10, pp. 2689–2701, Oct. 2009.

19. A. Antoniou, Realization of gyrators using operational amplifiers, and their use in RC-active network synthesis. *Proceedings of Institute of Electrical Engineers,* vol. 116, pp. 1838–1850, Nov. 1969.

20. A. Antoniou, Bandpass transformation and realization using frequency-dependent negative resistance elements. *IEEE Transactions on Circuit Theory,* vol. CT-18, pp. 297–299, March 1971.

21. L. T. Bruton, Network transfer functions using the concept of frequency dependent negative resistance. *IEEE Transactions on Circuit Theory,* vol. CT-16, pp. 406–408, Aug. 1969.

22. K. Martin et al., Optimum design of active filters using the generalized immitance converter. *IEEE Transactions on Circuits Systems,* vol. CAS-24, pp. 495–502, Sept. 1977.

23. J. Hutchison et al., Some notes on practical FDNR filters. *IEEE Transactions on Circuits Systems,* vol. CAS-28, pp. 242–245, March 1981.

24. N. C. Bui et al., On Antoniou's method for bandpass filters with FDNR and FDNC elements. *IEEE Transactions on Circuits Systems,* vol. CAS-25, pp. 169–172, March 1978.

25. L. T. Bruton et al., Electrical noise in low-pass FDNR filters. *IEEE Transactions on Circuit Theory,* vol. CT-20, pp. 154–158, March 1973.
26. A. Pirola et al., Current-mode, WCDMA channel filter with in-band noise shaping. *IEEE Journal of Solid-State Circuits,* vol. 45, no. 9, pp. 1770–1780, Sept. 2010.

10 Ultralow Power Radio Design for Emerging Healthcare Applications

Maja Vidojkovic, Li Huang, Julien Penders,
Guido Dolmans, and Harmke de Groot

CONTENTS

10.1 INTRODUCTION

The rapid growth in physiological sensors, low power integrated circuits, and wireless communication has enabled a new generation of wireless sensor networks for healthcare. The wireless body area network (WBAN) is a new wireless network that could allow inexpensive continuous health monitoring and early detection of medical patient conditions via the Internet. Intelligent physiological sensors can be placed on, in, or around the human body to collect various physiological changes in order to monitor the patient's health status no matter where the patient is [1,2]. The information will be transmitted wirelessly to an external base station, mostly in the form of a smart phone. The base station will instantly transmit all information in real time to the healthcare providers. If an emergency is detected, the patient and his or her caregivers will be immediately informed to take the required actions. The development of telehealth solutions is pushed by large industry alliances such as Continua [3]. And in the last few years, telemonitoring products have been introduced in the area of cardiac health management by companies like Cardionet [4], Corventis [5], AliveCor [6], and Airstrip [7]. In parallel, several pilot trials have been reported for the use of WBAN technologies in various applications (e.g., for sleep [8] or epilepsy monitoring [9]). The WBAN concept allows patients to stay in their home environment and hence have a better quality of life with lower costs involved. The reduced cost of this concept is a necessity as the cost of healthcare in First World countries is increasing dramatically as a result of advances in medicine, a population that is becoming older, and an increasingly unhealthy lifestyle.

The benefits from these technological advances highlight the need for rapid development of WBAN technology. This chapter presents a research overview in the field of wireless body area networks. More specifically, in the chapter an ultralow power radio design for these emerging healthcare applications is presented. To provide proper context of the research we will start by discussing some prototypes of current health monitoring systems and discuss the requirements of wireless systems for future WBAN healthcare applications. Second, we will present link budget analysis that is carried out in order to determine the expected performance of such systems. Third, we will discuss the design and the implementation of the ultralow power BAN radio. Also, we will show the evaluation of the BAN radio system level performances in the BAN environment. The chapter will end with a summary and conclusions.

10.1.1   WBAN and Healthcare Applications

The technology breakthroughs in WBAN can lead to new application opportunities in the areas of health, wellness, and lifestyle. Examples of applications include

- Vital signals monitoring (EEG, ECG, EMG, temperature, respiratory, heart rate, pulse oximeter, blood pressure, oxygen, pH value, glucose, cardiac arrhythmia)
- Wireless capsule endoscopes (gastrointestinal)
- Wireless capsules for drug delivery
- Deep brain stimulators
- Cortical stimulators (visual neurostimulators, audio neurostimulators, Parkinson's disease)
- Remote control of medical devices (pacemaker, actuators, insulin pumps)
- Hearing aids
- Retina implants
- Disability assistance (muscle tension sensing and stimulation, wearable weighing scale, fall detection, aiding sport training).

In this section, we will focus on three prototypes of the current health monitoring systems and discuss the requirements of wireless systems for future WBAN healthcare applications.

The first prototype is a wireless electrocardiography (ECG) system for cardiovascular disease monitoring. Cardiovascular disease is the number one cause of death and disability in the United States and most European countries. Heart diseases, among which are cardiac arrhythmias, are estimated to account for 30% of all death in the United States. Nowadays, diagnosis of cardiac arrhythmias is performed by point-of-care ECG monitoring or using Holter devices. Very recently, a new wave of portable patient monitors has been introduced, targeting outpatient monitoring with embedded detection of arrhythmia [4]. It is recognized that treatment and prevention of many cardiovascular diseases would benefit from a wireless ECG system for long-term continuous monitoring, but technology barriers have so far prevented the widespread use and acceptance of such a continuous ECG monitor on a daily basis. Such a device should be a very small wireless device, with no required battery replacement, that analyzes the data online and takes appropriate actions in case of emergency. The device should not affect the wearer in his daily life activity while constantly looking over his health.

A wireless ECG patch prototype [10] is illustrated in Figure 10.1. This prototype combines technologies for ultralow power biopotential readout, optimized power management, and advanced electronic integration on flexible substrate. Depending on the application, the ECG patch streams the one-channel ECG data to a receiver within a 10 m range, or performs local analysis on the data to extract R-peak and other fiducial points. Power consumption of the system has been reported to be 1.17 mW in data streaming mode, and 1.74 mW if the data are processed locally and sent at every beat [10]. The latter value is due to a heavy usage of the microcontroller resources, pointing to an important limitation of today's off-the-shelf microcontrollers. Targeted at very low processing duty cycles, they become quickly limiting for applications requiring quite advanced embedded digital signal processing (DSP) capabilities. This reinforces the need for ultralow power DSP technology in the future.

The second prototype is a wireless device for sleep monitoring. Sleep disorders are known to affect a significant part of the population: up to 10% of the American population and 4% of the European population. Typical diagnosis of sleep disorders is performed using polysomnography tests at the point of care. Ambulatory sleep monitoring devices have been introduced for home monitoring and prescreening. However, they suffer from important burdens such as their weight (mainly due to the battery) and the high density of wires going from the head to the data acquisition box (often located around the belt). Centers for sleep disorders would benefit from a miniaturized, wire-free, sleep-staging system, targeting the monitoring of the patient's hypnogram—that is the sequence of sleep stages overnight. The development of a prototype WBAN for wireless sleep staging was reported in Romero Legarreta et al. [11]. It relies on the ultralow power single-channel biopotential readout chip

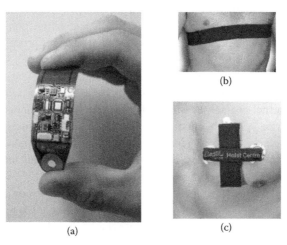

(a)   (c)

FIGURE 10.1   (a) Wireless ECG patch integrated on flexible substrate; (b) chest belt package; (c) textile pocket package.

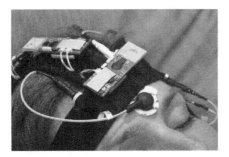

FIGURE 10.2 Prototype wireless body area network for sleep staging.

described in Yazicioglu et al. [12]. Its low power consumption (60 µW) allows dramatic reduction of the size of the battery, hence of the entire system, while maintaining an autonomy suitable for sleep analysis (>12 hours).

The system, illustrated in Figure 10.2, consists of a body sensor network composed of three wireless sensor nodes, collecting data from two-channel electroencephalography (EEG), two-channel electro-oculography (EOG), and one-channel electromyography (EMG), and sending it wirelessly to a receiver located in the patient's room. These particular signals were selected according to the Rechtschaffen and Kales standards for sleep staging [13]. The data can then be analyzed, on- or off-line, by the clinical staff. Each node achieves a power consumption of 15 mW, for a sampling rate of 200 Hz. Thanks to their small size and light weight, the three sensor nodes can easily be integrated in a headband, hence increasing patient comfort and acceptance.

The third prototype is the system for emotion monitoring. Monitoring the emotional state of human beings is becoming increasingly popular, with applications in psychophysiology (e.g., stress management), gaming, and human–computer interaction. The autonomic nervous system (ANS) is part of the body control system responsible for maintaining stability in the body. It affects the regulation of body parameters such as heart rate, respiration rate, salivation, and perspiration in response to external conditions and events. This unconscious process reveals valuable information to interpret the emotional state of a human being, such as arousal. The concept of WBAN is applicable to monitoring physiological changes in the autonomic nervous system that carry much of the information related to arousal responses to external or internal stimuli. Brown et al. [14] developed a customized emotion monitoring system by monitoring four physiological signals: ECG, respiration, skin conductance, and skin temperature. Each of these modalities is known to be regulated by the autonomic nervous system and thus represents interesting candidates to capture emotional responses to external stimuli [14,15].

The system, illustrated in Figure 10.3, consists of two low power miniaturized body sensor nodes that communicate with a receiver connected to a PC or to a data logger. The first node is integrated in a wireless chest belt and monitors ECG (lead-I) and respiration. The second node is integrated in a wireless wrist sensor and monitors skin conductance and skin temperature. Each node is composed of a generic wireless node, to which are connected the corresponding sensor front-end boards.

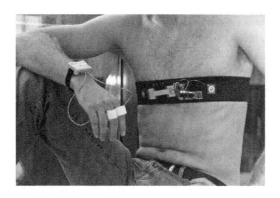

FIGURE 10.3 Integrated body area network for ambulatory monitoring of physiological responses from the autonomic nervous system.

From this, it is clear that WBAN is an interdisciplinary area that could allow inexpensive and continuous health monitoring with real-time updates of medical records. The number of applications where WBAN can be applied is very large. Therefore, a universal WBAN system should allow scalable data rates ranging from 10s of kbps to 1 Mbps such that most of the WBAN applications could be supported. Furthermore, the prototypes presented earlier can only operate from a few days to a week at full functionality. This is not enough, especially for implantable communications where months or years of autonomy are expected. Early demonstrators of autonomous wireless health monitors have shown that harvesting energy from the environment during the operation of the system will allow the system to run eternally with a battery or a supercapacitor acting only as a temporary energy buffer. But as the overall size and weight of sensor nodes are also an important design issue, it is impossible for the energy harvester to generate infinitely large energy. For example, a 1 cm² area of the current energy harvesters can only generate power in the range of 10 μW to 1 mW [2]. Thus, the ultralow power wireless system consuming less than 1 mW is expected to enable autonomy and miniaturization of WBAN.

10.1.2 Link Budget

Successful design of wireless systems for healthcare applications involves many factors. However, a top-level link budget analysis is the first step to be carried out in order to determine the expected performance of such systems in a given channel scenario. It is also an excellent means to the understanding of various factors that must be traded off in terms of complexity and performance.

At higher data rates of up to 1 Mbps, the state-of-the-art super-regenerative front ends can achieve sensitivity from −60 dBm [16,17] up to −70 dBm [18], while keeping the power consumption sufficiently low. Taking this into account in the link budget analysis, the target specification of the transceiver for health applications can be finalized. Normally the receiver sensitivity could be computed as follows:

$$P_r = P_t - PL + G_t + G_r \tag{10.1}$$

where P_t, PL, G_t, and G_r represent the transmit power, path loss, transmit antenna gain, and receiver antenna gain, respectively. From (10.1) it is clear that the path loss is a very important factor determining the receiver power. Please note that path loss is rapidly changed in close spatial proximity to a particular location caused by the multipath. This variability is called the small-scale fading and is often characterized by a channel model. Thus, to take the variation of the path loss into account for link budget calculation, we should choose practical channel models.

The channel model considered here is the one for the body surface to external communication at 2.4 GHz proposed from the IEEE 802.15.6 working group [19]. This channel model assumes that the transmitter is placed on the chest and the receiver is away from the body with various distances. It considers both the standing and walking scenarios. Thus, the effects of the body movement on system performance could also be investigated. Note that different distributions (e.g., gamma distribution, lognormal distribution) are used to characterize the small-scale fading for different scenarios.

For determining the target receiver sensitivity and transmit power, we define that the system is reliable if the channel path loss is sufficiently small such that the received signal strength exceeds the receiver sensitivity [20]. The probability of this reliability is dependent on the channel small-scale fading. To decouple the antenna effect, we assume that the antenna gain of 0 dBi is here. From Table 10.1 we find that 0 dBm transmit power and –75 dBm receiver sensitivity are sufficient since the probability of reliability is always larger than 95%, which is a typical requirement in WBAN [21].

However, when the receiver sensitivity is –65 dBm, the system is not always reliable, especially in the walking scenario at 4 m transmission distance, as shown in Table 10.2. Therefore, we choose the 0 dBm transmit power with a receiver sensitivity of –75 dBm as our target specification in wireless system design for WBAN.

TABLE 10.1
Probability of Reliability When Transmit Power is 0 dBm and Receiver Sensitivity is –75 dBm

Distance (m)	1	2	3	4
Standing	~100%	~100%	~100%	~100%
Walking	99.72%	99.61%	~100%	99.99%

TABLE 10.2
Probability of Reliability When Transmit Power is 0 dBm and Receiver Sensitivity is –65 dBm

Distance (m)	1	2	3	4
Standing	~100%	~100%	~100%	93.97%
Walking	98.13%	95.87%	98.39%	79.57%

10.1.3 BAN RADIO

It is known that the radio often consumes between 50% and 85% of total sensor power consumption for the kind of BAN applications mentioned earlier [2,22]. As an example, an ECG necklace has been integrated (see Figure 10.4a). This two-lead ECG necklace transmits an ECG signal without any motion artifact compensation or other algorithms done on the necklace. The ECG necklace was created using off-the-shelf but best-in-class minimal power consumption components. The sensor readout was specifically developed [12] because it was impossible to find sensor readout with sufficient specifications and low enough power consumption. When optimally using all available power modes, the average power consumption of 0.903 mW* is achieved. In Table 10.3 the ECG application parameters are summarized. Table 10.4 shows the list of the components used in the ECG necklace, while Figure 10.4(b)

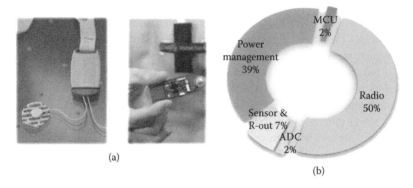

(a)

(b)

FIGURE 10.4 Sensor node: (a) wireless ECG necklace; (b) power budget with average power optimally using all available modes and duty cycling.

TABLE 10.3
ECG Application Parameters

Application parameter at the electronics level	Value
Listening interval (ms)	100
Reception time slot (ms)	2
Transmission interval (ms)	35
Number of bits per sample (transmission) (bit)	12
Sampling interval (ms)	2.5
Transmission/reception range (m)	10

---

* This power consumption is simulated taking real measurement values of the single components running at 2.2 V. Compared to the complete ECG necklace implementation, this is excluding the power consumption of several components. Most noticeably in terms of power consumption are voltage regulators and the three-dimensional accelerometers that are also present. In addition, the internal ADC reference voltage of the MSP430 consumes an extra 500 mA. Exclusion of these components is done to make replacing components and comparing their impact later in this chapter as fair as possible. The measured power consumption of the original ECG necklace including all components is 1.05 mA at 3 V.

TABLE 10.4

Components Used in ECG Necklace Made with Off-the-Shelf Components and Imec Biopotential

Component
μC: MSP430
Radio: Nordic nRF24L01
ADC: MSP430
Power manager: TP780
Biopotential: Imec

shows the breakdown chart for the power consumption. Please note that the radio transmission rate for this example is extremely low and that, for higher radio transmission rates, the power consumption of the radio further increases.

The analysis of the sensor power budget shows that the radio is the most power hungry block. Therefore, a new radio can reduce the overall power consumption of the system significantly. The design of an ultralow power radio is a challenging task. To reduce power as much as possible, a low complexity, nonlinear architecture seems a good choice. A combination of on/off keying (OOK) modulation, a direct modulation transmitter, and super-regenerative receiver principle [23] enables ultralow power consumption, thus providing a suitable architecture for autonomous body area network (BAN) sensor nodes.

10.1.3.1 System Architecture

A block diagram of the BAN transceiver that we have implemented is shown in Figure 10.5 [24,25]. Next to the direct modulation transmitter (Tx) and super-regenerative receiver (Rx) front end, this work integrates analog and digital baseband and PLL functionality. The transceiver supports additional programmability for flexible data rates and achieves ultralow power consumption for the overall system. For lower power consumption and better performance, the Tx and Rx matching networks are placed externally. The serial peripheral interface (SPI) host interface links our designed chip with the MSP 430 microprocessor for data and control information communication.

10.1.3.2 Transmitter Front End

A block diagram of the implemented direct modulated Tx is presented in Figure 10.6 [26]. The direct modulation transmitter architecture simplifies the stage preceding the power amplifier (PA). This reduces the power overhead from the pre-PA stages [27] and increases the Tx power efficiency. In the Tx, the 2.4 GHz carrier is generated by the integrated LC voltage-controlled oscillator (VCO), and amplitude modulation is applied on the PA. Between the PA and the VCO, a low power buffer is inserted to avoid the dynamic loading effect from the PA. This improves the frequency stability of the VCO. To avoid spectrum artifacts during the VCO start-up period, only the PA is switched on and off in our architecture. This also increases the data rate since the start-up time of an LC VCO with reasonable Q-factor is much longer than the

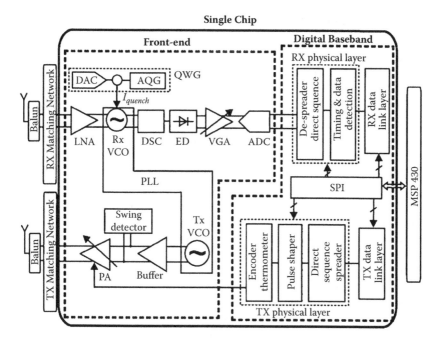

FIGURE 10.5　Block diagram of the implemented ULP OOK transceiver.

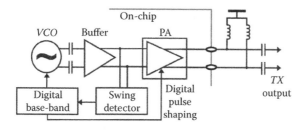

FIGURE 10.6　Block diagram of the implemented direct modulated transmitter.

time needed to switch the PA on or off. To ensure optimal power efficiency despite PVT variations, we have designed a duty cycled swing detector that will monitor and adjust the driving level of the PA. The complete transmitter also implements digital pulse shaping for improved spectral efficiency. The digital baseband oversamples the baseband bit stream by 6× and shapes the OOK pulse with a predetermined filter (e.g., the raised cosine (finite impulse response) FIR filter).

The Tx front end is optimized to transmit 0 dBm output power as it is required by the link budget analysis. It consumes 2.53 mW at 1 V supply voltage for 50% OOK. The overall efficiency of the Tx front end is 24% when transmitting continuous ones, while the efficiency for both zeros and ones reaches 40%. More details regarding the Tx front-end implementation and the achieved results are presented in references 24–26.

10.1.3.3   Receiver Front End

Due to its simplicity and low power expenditure, super-regeneration is a commonly employed principle in the Rx of sensor networks [16,17,28]. A block diagram of the implemented ULP super-regenerative Rx front end is shown in Figure 10.7 [24,25]. The LNA provides input matching for the antenna, amplifies the RF signal, and improves the isolation between the antenna and the VCO. The VCO is the main part of the super-regenerative receiver. It is biased with a time-varying quench current I_{quench} [23] from the quench waveform generator (QWG). In the presence of an input signal, the VCO oscillations build up faster. The signal detection can be based on a difference in the VCO start-up time for the logarithmic mode or on a difference in the VCO amplitude for the linear mode [25].

In our implementation, the Rx is optimized for the linear mode. The linear mode will provide a higher data rate because the VCO does not need additional time to reach saturation, and the VCO discharging is shorter. Further, the quench current I_{quench} is generated in a combined analog and digital fashion to achieve both low power consumption and higher data rate [29]. The differential-to-single-ended converter (DSC) in the Rx provides the differential to single-ended conversion of the signal and suppresses the common-mode level from the VCO. The envelope detector (ED) performs the down-conversion of the wanted signal to DC. The variable gain amplifier (VGA) adjusts the signal to fit into the dynamic range of the analog-to-digital converter (ADC). The VGA is partially open loop to reduce the power consumption [25]. The 8-bit successive approximation (SAR) ADC [30] uses asynchronous dynamic logic and custom-built 0.5 fF capacitors to achieve low power at high speed.

The Rx front end operates up to 5 Mbps, for which it achieves −75 dBm sensitivity for bit error rate (BER) of 10^{-3} while it consumes 534 µW. At 0.5 Mbps it achieves −78 dBm sensitivity. More details regarding the Rx front-end implementations and achieved results are summarized in references 24, 25, and 29.

10.1.3.4   PLL

The phase-locked loop (PLL) sets the desired channel in the radio. In the proposed transceiver architecture, the VCO is part of the signal path in both the Tx and Rx. Because the Tx and the Rx have different constraints for the VCO, they use a separate VCO, while sharing the same PLL. Figure 10.8 shows the architecture of the integer-N PLL implemented in this prototype [24,25]. The phase/frequency detector (PFD), the charge pump (CP), the loop filter (LPF), and the divider-by-M are shared

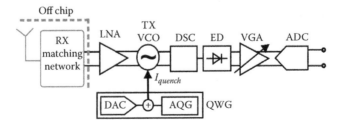

FIGURE 10.7   Block diagram of the implemented super-regenerative receiver front end.

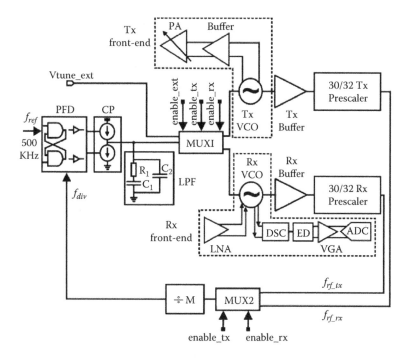

FIGURE 10.8 Block diagram of the implemented integer-N PLL.

between the Tx and the Rx, while each VCO has its own buffer and a 30/32 pre-scaler. Each 30/32 prescaler consists of a divider-by-2 and a divider-by-15/16. With multiplexers MUX1 and MUX2, the PLL can be activated for either the Tx or the Rx. Also, external analog tuning can be applied instead of using the PLL. With a 500 kHz reference frequency (f_{ref}), the PLL locks the VCO frequency from 2.36 to 2.485 GHz with increments of 15 or 16 MHz, dependent on the setting of the divider-by-15/16. The PLL consumes 1.1 mW from a 1.2 V supply.

Because of the quenched operation in the super-regenerative receiver, the VCO is continuously enabled and disabled. For this reason a continuous feedback PLL is not possible. Therefore, during the Rx frequency locking mode, the Rx VCO is biased with a constant current as shown in Figure 10.9. During this time, the PLL is enabled to lock the VCO to the desired frequency. After that, the PLL is disabled, and the Rx starts the receiving mode (t_{RX}). In the transmission mode, the PLL can also be dis-abled to decrease power consumption further. The system-level measurements show that the Rx is able to receive packets with packet error rate (PER) of 10^{-2} in 25 ms time at 1 m line of site (LOS) distance while the PLL is off. Within 25 ms, the Rx is able to receive five packets with a maximum packet size of 5 ms.

10.1.3.5 Receiver Timing

On the Rx side, the receiver timing is important and it is controlled by three clocks from the digital baseband: an_clk_qch, dig_clk_qch, and ADC_clk. The Rx timing is shown in Figure 10.10.

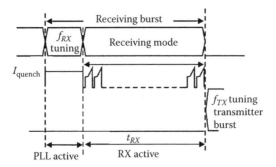

FIGURE 10.9   Rx/PLL timing.

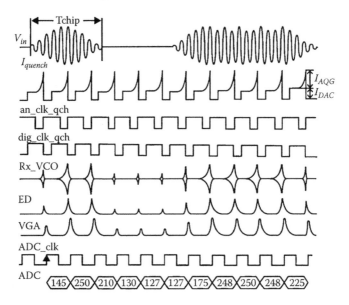

FIGURE 10.10   Receiver timing.

In order to achieve better performance, in our implementation a received signal VIN is 3× oversampled by the quench signal. In this way better peak synchronization will be achieved. By omitting oversampling, the data rate can be increased to the quench rate, but additional synchronization is needed. Further, the shape of I_{quench} is controlled by dig_clk_qch and an_clk_qch. In this way the VCO on-time is controlled. The ADC samples the output of the VGA on the rising edge of the ADC_clk. Each of the control clocks is derived from a 6 MHz baseband clock by a dedicated clock generator. In order to maximize the performances, the phase and duty cycle of each clock is programmable with a resolution of 750 ps [24,25].

10.1.3.6   Digital Baseband

The digital baseband operates at 6 MHz clock and is supplied by a single 1.2 V power supply. Apart from the clock generation, the digital baseband includes the transmitter

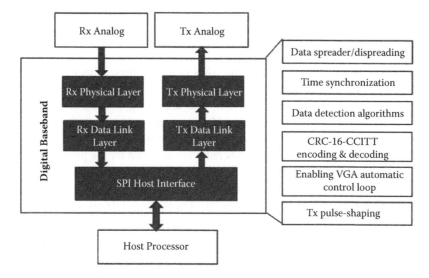

FIGURE 10.11   Digital baseband.

data link layer, the transmitter physical layer, the receiver physical layer, the receiver data link layer, and the SPI host interface, as shown in Figure 10.11. In the physical layer, to achieve the scalability of data rate and robustness, the direct-sequence spread spectrum (DSSS) technology [31] is used. This technology spreads the baseband signal coming from the data link layer by directly multiplying the bit with a pseudonoise (PN) sequence consisting of a number of PN code symbols. The PN code symbol is called the chip and the number of chips in 1 bit can be 1, 2, 4, or 16 in implementation. In the data link layer, the CCITT CRC-16 code [32] is used to check whether there are data errors in packet transmission. The SPI host interface links our designed chip with the external host processor such as the microprocessor or the field-programmable gate array (FPGA) for data and control information communication.

In the transmitter side, the raw data bits from the SPI host interface are assembled in the transmitter data link layer as packets, at the end of which 16 cyclic redundancy check (CRC) bits are generated based on the CCITT CRC-16 code and the raw data bits within the packet. In the transmitter physical layer, the bit is spread over chips. Each chip is further passed through a pulse shaper to make the transmitted signal better suited to the communication channel. The shaped digital signal is passed through the thermometer encoder, which translates the signed 6-bit values into unsigned 15-bit values for the PA in the transmitter analog front end.

In the receiver side, the output data from the ADC is first passed to the receiver physical layer, where the data dispreading, time synchronization, and data detection algorithms are performed. In addition, a VGA updating algorithm, used to increase the sensitivity of the data detection (by avoiding clipping of the ADC samples), has been implemented in the physical layer. The detected bits within one packet will then be verified in the data link layer by comparing the detected CRC bits with the one freshly calculated from the detected data bits. Finally, the detected data bits are passed to the external host processor through the SPI host interface for further processing.

10.1.3.7 Radio Measurements in BAN Environment

The BAN radio has been implemented in 90 nm CMOS. The die photo of the radio is shown in Figure 10.12. It occupies 2.4×1.85 mm^2.

The system-level performance of the BAN radio has been evaluated. From our measurement results, PER of 10^{-2} is achieved within 25 m line-of-sight (LOS) transmission distance wireless communication. This maximum distance is achieved for 1 Mbps chip rate when PN sequence of 16 is applied. This will result in a total of 62.5 kbps achieved data rate. The packet size is 288 bits.

The measurement results of the BAN radio are achieved for the power consumption summarized in Figure 10.13. The measured Tx power consumption is for 50% OOK 0 dBm transmit power. The Tx front end consumes 4.192 mW when transmitting "1" and 0.868 mW when transmitting "0." The measured Rx power consumption is for a 1 Mbps input signal oversampled at 3 MHz quench rate. The digital baseband is running at 6 MHz clock. The peak current of the PLL is 915 µA from 1.2 V. Since

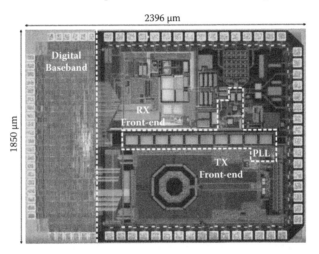

FIGURE 10.12   Die photo of the single-chip transceiver.

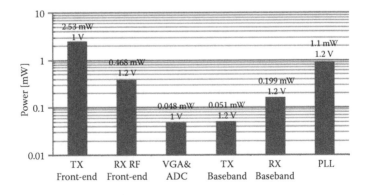

FIGURE 10.13   Power consumption summary.

the duration of the Tx/Rx frequency locking modes is negligible compared to the duration of the transmitting/receiving modes, the contribution of the PLL to the total transceiver power can be neglected.

Further, we have evaluated the system performance of the developed transceiver chip in the BAN environment. The measurement evaluation is based on our proto-typed ECG necklace (see Figure 10.4) with the BAN radio and the corresponding base station as shown in Figure 10.14. The base station consists of our integrated BAN transceiver chip to receive the signal, an MSP430 microcontroller to read data from the transceiver chip, and the UART-USB bridge to transfer the data from the microcontroller to the computer through a USB cable. Through the cable, 5 V power is also provided to the power management circuitry of the base station. In this system we first evaluated the power consumption of the ECG necklace as discussed in Section 10.1.3, where the Nordic radio is replaced with our BAN radio. When replacing the original Nordic radio with the ULP BAN radio, the system power consumption is reduced to 0.366 mW and 2.5× power reduction at system level is achieved. The new power breakdown is shown in Figure 10.15.

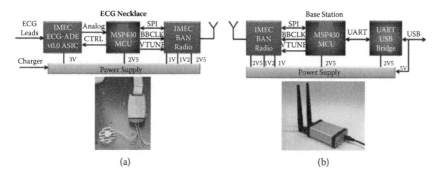

FIGURE 10.14 (a) ECG necklace; (b) base station.

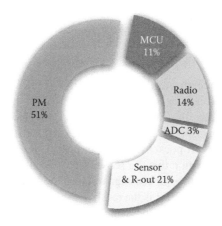

FIGURE 10.15 Power breakdown of optimized ECG necklace with Imec biopotential and Imec BAN radio.

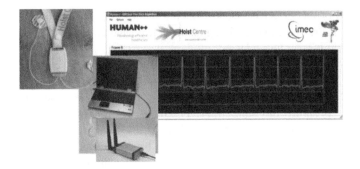

FIGURE 10.16   Heart's electrical activities measured by using the ECG necklace.

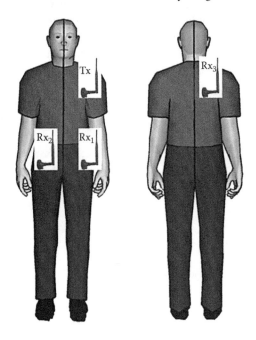

FIGURE 10.17   Antenna positions in BAN measurements.

The ECG necklace together with the base station is used to monitor the heart's electrical activities for fitness or healthcare applications, as shown in Figure 10.16.

In the measurement, as illustrated in Figure 10.17, we placed the Tx (ECG necklace) close to the heart position (i.e., the left side of the chest) and placed the Rx (the base station) on the left pocket (Rx1), right pocket (Rx2), or the back (Rx3) in order to represent the situation that the sensed ECG signal from the sensor node is sent to a relay node for further processing. We used the PER to evaluate the system performance. The considered PER is the result of the synchronization error (both the missed detection and false alarm) and data detection error:

- Misdetection error: the packet is transmitted while not detected by the synchronization algorithm.

TABLE 10.5
PER for Different BAN Scenarios

Rx	Rx1	Rx2	Rx3
PER	<0.1%	0.4995%	1.4%

- False alarm: the start frame delimiter (SFD), which indicates the end of the preamble, is wrongly detected.
- Data detection error: the packet is in error if at least one data bit within the packet is wrongly detected.

To compute the value of each PER, we use 1,000 packets. Thus, when the PER is below 0.001, the results are less reliable and will be shown only as <0.1%. The transmit power is set to 0 dBm.

From Table 10.5, it is clear that the transmission along the front side of the body (i.e., heart to left pocket) results in the best performance, as expected, since the transmission distance is the shortest and it is in the line-of-sight scenario. The transmission from the front to the back of the body (i.e., heart to back) has the worst scenario. However, it can still achieve a PER of around 1%, which is sufficient for many BAN applications.

10.2 SUMMARY AND CONCLUSIONS

In this chapter an ultralow power radio for emerging healthcare applications has been presented. To provide proper context of the research, the chapter first highlights the benefits from the WBAN technology. The technology breakthroughs in WBAN can lead to new application opportunities in the areas of health, wellness, and lifestyle. Some examples of applications are vital signals monitoring such as EEG, ECG, EMG, temperature, respiratory, heart rate, pulse oximeter, blood pressure, oxygen, pH value, glucose, and cardiac arrhythmia. Further, the wireless networks can be used in wireless capsule endoscopes (gastrointestinal), wireless capsules for drug delivery, deep brain stimulators, visual neurostimulators, audio neurostimulators, remote control of medical devices (pacemaker, actuators, insulin pump), hearing aids, retina implants, and disability assistance. The benefits of the wireless ECG system for cardiovascular disease monitoring, the wireless device for sleep monitoring, and the system for emotion monitoring have been discussed in more detail.

From the application study it has been clearly shown that the number of applications where WBAN can be applied is very large. This will require a universal WBAN system with scalable data rates ranging from tens of Kbps to 1 Mbps such that most of the WBAN applications could be supported. Furthermore, the ultralow power wireless system consuming less than 1 mW is expected to enable autonomy and miniaturization of WBAN. For that purpose we have designed and implemented an ultralow power radio that can be used in WBAN applications. The radio can operate in 2.36–2.4 GHz medical BAN and 2.4–2.485 GHz ISM band. The Tx of the BAN radio transmits the OOK signal with 0 dBm peak power. Including the Tx baseband, the complete transmitter consumes 4.243 mW when transmitting "1" and 0.919 mW when transmitting "0," which leads in 2.59 mW with 50% OOK. The Rx front end of the BAN radio achieves −75 dBm sensitivity at 5 Mbps and −78 dBm at 0.5 Mbps. The complete receiver consumes 715 μW at 1 Mbps data rate, oversampled at 3 MHz. The transceiver achieves PER = 10^{-2} at 25 m line-of-site with 62.5 kbps data rate and 288-bit packet size, and it has been incorporated in an ECG necklace to monitor the heart's electrical property. This necklace has a 2.5× power reduction at system level when compared to the original necklace with a best-in-class, off-the-shelf radio.

REFERENCES

1. R. Schmidt, T. Norgall, J. Mörsdorf, J. Bernhard, and T. von der Grün, Body area network, BAN, a key infrastructure element for patient-centered medical applications. *Biomedical Technology,* vol. 47 (s1a), pp. 365–358, Jan. 2002.
2. L. Huang, M. Ashouei, F. Yazicioglu, J. Penders, R. Vullers, G. Dolmans, P. Merken, et al., Ultra-low power sensor design for wireless body area networks: challenges, potential solutions, and applications. *International Journal Digital Content Technology and Applications,* vol. 3, no. 3, pp. 136–148, Sept. 2009.
3. Continua Health Alliance: www.continuaalliance.org
4. http://www.cardionet.com/
5. http://www.corventis.com/us/default.asp
6. http://alivecor.com/
7. http://www.airstriptech.com/
8. N. de Vicq, F. Robert, J. Penders, B. Gyselinckx, T. Torfs, Wireless body area network for sleep staging. *Proceedings International Conference on Biological Circuits and Systems,* 2007.
9. F. Massé, J. Penders, A. Serteyn, M. van Bussel, and J. Arends, Miniaturized wireless ECG-monitor for real-time detection of epileptic seizures. In *Wireless health 2010* (WH '10). New York: ACM, 111–117, 2010.
10. J. Penders, B. Gyselinckx, et al. Human++: From technology to emerging health monitoring concepts. *Proceedings of the 5th International Workshop on Wearable and Implantable Body Sensor Networks,* Chinese University of Hong Kong, China, June 1–3, pp. 94–98, 2008.
11. I. Romero Legarreta, P. Addison, N. Grubb, G. Clegg, C. Robertson, K. Fox, and J. Watson, Continuous wavelet transform modulus maxima analysis of the electrocardiogram: Beat characterization and beat-to-beat measurement. *International Journal of Wavelets, Multiresolution and Information Processing,* vol. 3, no. 1, pp. 19–42, 2005.
12. R. F. Yazicioglu, P. Merken, R. Puers, et al., A 60 μW 60 nV/√Hz readout front end for portable biopotential acquisition systems. *IEEE Journal Solid-State Circuits*, vol. 42, no. 5, pp. 1100–1110, May 2007.
13. A. Rechtschaffen and A. Kales, *A manual of standardized terminology, techniques and scoring system for sleep stages of human subjects.* Washington, DC: US Government Printing Office, National Institute of Health Publication, 1968.
14. L. Brown, B. Grundlehner, J. van de Molengraft, J. Penders, and B. Gyselinckx, Body area networks for monitoring autonomous nervous system responses. *Proceedings of the International Workshop on Wireless Pervasive Healthcare,* London, April 2009.
15. B. Grundlehner, L. Brown, J. Penders, and B. Gyselinckx, The design and analysis of real-time, continuous arousal monitor. *Proceedings of the 6th International Workshop on Wearable and Implantable Body Sensor Networks,* pp. 156–161, 2009.

16. J. Y. Chen et al., A fully integrated auto-calibrated super-regenerative receiver in 0.13 μm CMOS. *IEEE Journal Solid-State Circuits,* vol. 42, pp. 1976–1985, 2007.
17. D. Shi, N. Behdad, J. Y. Chen, and M. P. Flynn, A 5 GHz fully integrated super-regenerative receiver with on-chip slot antenna in 0.13 μm CMOS. *IEEE Symposium on VLSI Circuits,* pp. 34–35, 2008.
18. J. Ayers et al., A 0.4 nJ/b 900 MHz CMOS BFSK super-regenerative receiver. *IEEE CICC,* pp. 591–594, Sept. 2008.
19. K. Y. Yazdandoost and K. Sayrafian-Pour, Channel model for body area network (BAN). (Available at https://mentor.ieee.org/802.15/dcn/08/15-08-0780-09-0006-tg6-channel-model.pdf), April 2009.
20. L. Huang, K. Imamura, P. Harpe, C. Zhou, S. Rampu, M. Vidojkovic, G. Dolmans, and H. de Groot, Performance evaluation of an ultra-low power receiver for body area networks (BAN). *Proceedings IEEE International Symposium Personal, Indoor and Mobile Radio Communications (PIMRC),* Tokyo, Japan, pp. 89–94, Sept. 2010.
21. B. Zhen, M. Patel, S. Lee, E. Won, and A. Astrin, TG6 technical requirements document [Online]. (Available at https://mentor.ieee.org/802.15/dcn/08/15-08-0644-09-0006-tg6-technical-requirements-document.doc), Nov. 2008.
22. M. A. Hanson, H. C. Powell, A. T. Barth, K. Ringgenberg, B. H. Calhoun, J. H., Aylor, and J. Lach, Body area sensor networks: Challenges and opportunities. *Computer,* vol. 42, no. 1, pp. 58–65, Jan. 2009.
23. E. H. Armstrong, Some recent developments of regenerative circuits. *Proceedings IRE,* vol. 10, pp. 244–260, Aug. 1922.
24. M. Vidojkovic, X. Huang, P. Harpe, S. Rampu, C. Zhou, L. Huang, K. Imamura, et al., A 2.4 GHz ULP OOK single-chip transceiver for healthcare applications. *IEEE ISSCC Digest Technical Papers,* pp. 458–459, 2011.
25. M. Vidojkovic, X. Huang, P. Harpe, S. Rampu, C. Zhou, L. Huang, K. Imamura, et al., A 2.4 GHz ULP OOK single-chip transceiver for healthcare applications. *IEEE Transactions on Biomedical Circuits and Systems,* vol. 5, no. 6, pp. 523–534, December 2011.
26. X. Huang, P. Harpe, X. Wang, G. Dolmans, and H. de Groot, A 0 dBm 10 Mbps 2.5 GHz ultra-low power ASK/OOK transmitter with digital pulse-shaping. *IEEE RFIC Symposium,* pp. 263–266, May 2010.
27. Y. H. Chee et al., An ultra-low-power injection locked transmitter for wireless sensor networks. *IEEE Journal Solid-State Circuits,* vol. 41, no. 8, pp. 1740–1748, Aug. 2006.
28. A. Vouilloz, M. Declercq, and C. Dehollain, A low-power CMOS super-regenerative receiver at 1 GHz. *IEEE Journal Solid-State Circuits,* vol. 36, no. 3, pp. 440–451, March 2001.
29. M. Vidojkovic, S. Rampu, K. Imamura, P. Harpe, G. Dolmans, and H. de Groot, A 500 μW 5 Mbps ULP super-regenerative RF front end. *IEEE ESSCIRC,* pp. 462–465, Sept. 2010.
30. P. Harpe, C. Zhou, Y. Bi, N. van der Meijs, X. Wang, K. Philips, G. Dolmans, and H. de Groot, A 26 μW 8 bit 10 MS/s asynchronous SAR ADC for low energy radios. *IEEE Journal Solid-State Circuits,* vol. 46, no. 7, July 2011.
31. J. G. Proakis, *Digital communications,* 4th ed. New York: McGraw–Hill, 2001.
32. P. Koopman and T. Chakravarty, Cyclic redundancy code (CRC) polynomial selection for embedded networks. *Proceedings DSN,* pp. 145–154, 2004.

11 Ultralow Power Techniques in Small Autonomous Implants and Sensor Nodes

Benoit Gosselin, Sébastien Roy,
and Farhad Sheikh Hosseini

CONTENTS

11.1 INTRODUCTION

There is growing interest in the development of autonomous small-scale devices capable of sensing/actuating and performing communication and processing tasks in personnel healthcare technology. Such devices form the basis of many novel diagnostic and prosthetic systems and are often networked to form so-called sensor networks, which constitute a very active field of research. Similarly, autonomous implants within the human body that serve to capture data or to stimulate cells are essentially specialized sensor/actuator nodes, whether or not they are networked.

Such implants are typically powered by a battery or through a transdermal inductive link with a device that is external to the body. It is well known that extremely low power consumption is a critical feature in order to maximize the operational life expectancy of a node, an implant, or an entire network of such devices powered by batteries. Alternatives to batteries generally fall under the energy-scavenging category, with the exception of the inductive link discussed before. In all such cases, the power available is likely to be very low. Multiple power-saving schemes have been proposed and are in usage in several power-conscious devices. Possibly one of the most common is the introduction of inactive or sleep states whenever appropriate. Indeed, an entire electronic device can be shut down when there is no work to perform, or certain components can be turned on and off depending on dynamic demand. The power consumption of devices in sleep or standby modes can be orders of magnitude lower than their active consumption.

We present the following two case studies illustrating the utilization of low power design techniques, both relevant to personal healthcare and medical research. The first one is concerned with the implementation of monitoring devices incorporating several parallel sensing channels; such devices are specifically applicable to brain–computer interfacing technology. The second case study centers on the broad and increasingly active area of wireless sensor networks and, more specifically, on the wake-up radio concept that allows an autonomous device to remain in deep sleep until needed, when it can then be woken up by a specific RF signal received through a passive radio.

Brain–computer interfacing technology aims at collecting bioelectrical signals from hundreds of electrodes to take snapshots of the activity occurring in specific areas of the brain. Indeed, each electrode generates weak analog signals that need to be amplified, further processed, and digitized to be transferable into the world of computing for analysis or storage. Such functions are conducted in dedicated sensory circuits providing low noise amplifiers, filters, and data converters, thus dissipating a considerable amount of power. However, the sensitivity of brain tissues restricts excessive heat dissipation in the nearby circuits, necessitating low power consumption. Indeed, data reduction and data compression schemes that can decrease the overall output data rates of sensory circuits are becoming critical building blocks in such monitoring devices.

On the other hand, different power scheduling schemes have been proposed in order to power up the sensory circuits only when needed. Specifically, activity-based schemes exploit the intermittent nature of biopotential recordings along with their low duty cycles by powering up the sensory building blocks only during occurrences of biopotentials. Such a strategy requires the utilization of accurate signal detectors along with a smart power management approach. Typically, most recording building blocks will remain in an idle state, draining practically no power until biopotentials occur.

The first case study is covering dedicated electronic recording strategies to address the challenge of operating large numbers of recording channels to gather the neural information from several neurons within very low power constraints and an appropriately compact form factor. Specifically, we cover system-level-based strategies for smart power management, and we present dedicated sensory circuit topologies for extracting and separating multiple biopotential modalities with high energy efficiency. A practical implementation is presented to illustrate the application of the described strategies. Such application consists of a discrete-time analog front end, which is leveraging highly energy efficient, low noise sensory circuits and dedicated system-level power saving schemes.

The second case study is in the context of sensor networks where communication itself is often a dominant factor in the power budget. For example, Texas Instruments' low power CC2240 consumes 18.8 mA when in receive mode and 17.4 mA when transmitting a 0 dBm signal. Furthermore, it can be shown that the receive side consumes more power on average than the transmit side, because the receiver is constantly "listening," while transmissions are typically rare and short. This picture is, however, heavily influenced by the communication protocol in use, including the medium access mechanism. Typical solutions include various forms of synchronization where it is possible to turn the receive radios on only during certain predetermined intervals. More sophisticated dynamic evolutions of this concept are termed "rendezvous" protocols.

The holy grail in this realm is the wake-up radio, a receiver that can normally operate in an entirely passive fashion, thereby consuming very little power, until reception of a radio impulse with certain characteristics "wakes" it. In other words, the energy of the received pulse is used to trigger a power-on mode for subsequent reception of a packet. Only through such a mechanism can it be ensured that the receive radio is turned on only when receiving bits. However, it has been estimated in the literature that the power consumption of the wake-up radio should be below 50 µA for this scheme to be effective and competitive with the best rendezvous protocols. Very few realizations of wake-up devices are reported in the literature. This case study presents the first reported design having power dissipation below 40 µW. It consists of a complete wake-up device, including an RF detector and address decoder, with an average power consumption of less than 20 µW. One of the unique features of this design is the use of pulse-width modulation (PWM) instead of the more common on/off keying schemes. This choice has opened up significant power-saving opportunities.

While the sphere of applications of wake-up radios and energy-efficient wireless sensor networks in general is very broad, there are many medical contexts where such technology—allowing an autonomous device to consume significant power only when polled—could be advantageously leveraged. This includes in-body devices with an autonomous limited power source and body-area networks designed to monitor vital signs.

11.2 MICROSYSTEMS FOR BIOPOTENTIAL RECORDING

Nowadays, neural interfacing microsystems capable of continuously monitoring large groups of neurons are being actively researched by leveraging the recent advancements in neuroscience, microelectronics, communications, and microfabrication. Such monitoring microdevices are pursuing two critical objectives for prosthetic applications and advanced research tools: (1) replacing hardwired connections with a wireless link to eliminate cable tethering and infections, and (2) enabling the local processing of neural signals on an as needed basis to improve signal integrity.

A suitable interface to the cortex must enable chronic utilization and high resolution through the simultaneous sampling of the activity of hundreds of neurons.

In prosthetic applications, there are severe limitations on size, weight, and power consumption of monitoring implants in order to limit invasiveness and heat dissipation in surrounding tissues. A major effort has recently been directed toward designing neural recording circuitry consuming very few microwatts per channel by leveraging low power circuit techniques and smart data/power management. Indeed, low power sensory circuits, energy-efficient system-level architectures, and dedicated on-chip management strategies are necessary means for achieving high resolution while addressing stringent power requirements.

11.2.1 MULTICHANNEL SYSTEM ARCHITECTURES

Gathering the sampled neural activity from hundreds of channels, digitizing it, and sending it wirelessly to a base station in an efficient fashion is very challenging and thus requires a dedicated system architecture. A straightforward approach consists in sharing a fast digitizer between several sensory channels. In such a scheme, the sensory channels are directed toward an analog-to-digital converter (ADC) by employing time-division multiplexing (TDM) in the analog domain. The challenge with such an approach consists in minimizing power consumption from the high-speed unity gain buffers, sample-and-hold (SHA) circuit, and ADC. Indeed, unity gain buffers presenting wide bandwidth much higher than the maximum frequency of the incoming signal (f_{max}) are needed to drive the SHA circuit or the ADC within small TDM time intervals, a requirement that results in high power consumption. According to Xiaodan et al. [1], the buffers and the SHA circuit must feature a low pass cutoff frequency (f_{-3dB}) that is at least five times higher than f_{max} in order to achieve a tracking error smaller than 1/2 LSB for a 10-bit representation. Moreover, the analog multiplexers must be designed carefully in order to avoid excessive cross talk and distortion. Examples of such system-level configurations are presented in Bonfanti et al. [2] and Sodagar, Wise, and Najafi [3].

Another approach consists of providing one low power, low rate ADC for each sensory channel and performing TDM in the digital domain. In contrast with the first scheme described earlier, this approach has the advantage of avoiding the need for several power-consuming unity gain buffers and of eliminating interchannel cross talk. However, great care must be taken in the design of a suitable ADC in order to minimize the chip area. An example of this type of implementation is reported in Gosselin et al. [4].

A third approach consists of performing digitization off chip to save power and silicon areas. Digitization is performed in two phases. A first phase consists of converting the multiplexed analog samples into time delays, an operation known as analog-to-time conversion (ATC). Then, after transmitting the ATC signal outside the body where power and size are not highly constrained, a second phase consists of performing time-to-digital conversion (TDC). In addition to saving power, this approach does not require synchronization with a clock signal. However, TDM must be performed in the analog domain, thus leading to cross talk. Such an approach is presented in Seung Bae et al. [5].

11.2.2   Low Power System-Level Strategies

The need for a parallel arrangement comprising several power-hungry modules, like low noise sensory circuits, high speed data converters, and wireless transmitters, motivates the application of dedicated system-level approaches for addressing excessive power consumption in the targeted highly constrained application. Power scheduling consists of powering up the recording circuits only when necessary (Figure 11.1).

Specifically, current-supply modulation [5,6] and duty cycling [5,7,8] are power scheduling techniques consisting of switching specific building blocks between an active and an idle state with low duty cycle. The shorter the duty cycling period is, the lower the power dissipation in the circuits is. The difference between both techniques is that the former uses a low level bias current in the idle state (Figure 11.2), whereas the latter draws no current in the idle state, thus leading to more power savings (Figure 11.3). However, a duty cycle period that is too short can potentially degrade the reading accuracy of a neural recording channel because fast intermittent monitoring requires circuits with larger bandwidth, thus allowing more noise to enter the system.

The battery-powered sensor interface presented in Gosselin and Ghovanloo [8] addresses this trade-off by providing different levels of accuracy and lifetime through the utilization of a programmable low pass filter. Such a filter allows selecting between different input-referred noise levels and duty cycle lengths, which

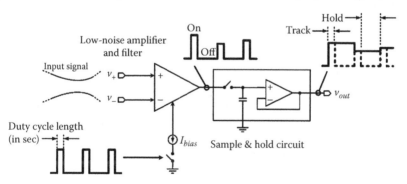

FIGURE 11.1   Simplified representation of a power scheduling strategy.

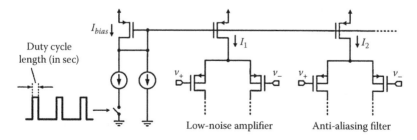

FIGURE 11.2   Simplified schematic of a current supply modulation scheme.

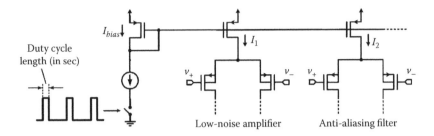

FIGURE 11.3 Simplified schematic of a duty cycling scheme.

determines the overall accuracy and power consumption. The power-scheduling mechanism employed in Seung Bae et al. [5] consists of putting most of the neural amplifiers that are not being sampled in sleep mode, where they draw a fraction of their active current consumption (0.5 µA). Indeed, not turning the neural amplifiers completely off affords time for them to reach their active state more quickly ahead of each sampling instant. This leads to a power reduction of 18% in the analog multichannel front-end block.

Another technique, the multiplexing of several electrodes toward one low noise amplifier (LNA), has been demonstrated to save power and silicon area. Time division multiplexing of four channels toward a single neural amplifier is achieved in Chung-Ching, Zhiming, and Bashirullah [9]. In the proposed topology, a single LNA is shared between four independent input electrodes in order to decrease the power consumption per channel by four. A frequency division multiplexing (FDM) scheme is implemented in Joye, Schmid, and Leblebici [10], where the amplitude of the neural activity seen at several individual electrodes is modulated and directed toward a single wideband neural amplifier. These authors showed that the maximum number of electrodes that can be multiplexed toward a single amplifier is limited by the sum of the thermal noise from each electrode at the input node of the wideband neural amplifier, and it is in the range of 5 to 10 for typical cases.

On the other hand, activity-based schemes are exploiting the transient characteristics of neural signals to maximize efficiency. Adaptive sampling [11] is an activity-based technique that affords a significant decrease of the data rate by dynamically varying the sampling rate of an ADC, based on the input signal activity (Figure 11.4). Reduction factors of seven are reported with this scheme. The algorithm requires the implementation of the second derivative of the signal within dedicated circuits to measure the rate of change of the input. Other activity-based schemes exploit the intermittent nature of neural recordings along with their low duty cycles by powering up the recording building blocks only when neural events occur (Figure 11.5).

Such strategies require the utilization of accurate biopotential detectors along with power scheduling. In such schemes, most recording building blocks remain in an idle state, draining practically no current until neural events occur. In Gosselin and Sawan [12], a low overhead analog detector is employed to detect neural events and trigger the recording circuitry. Indeed, the approach proposed in Gosselin, Sawan, and Kerherve [13] is using low overhead analog delay elements that are implemented

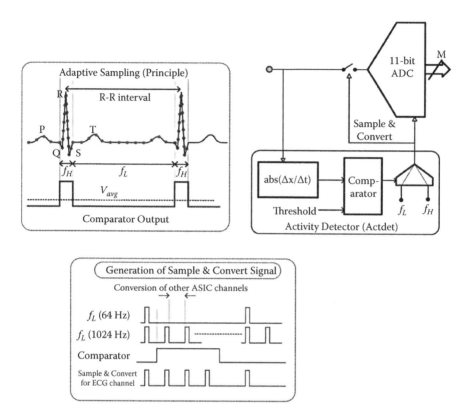

FIGURE 11.4   Architecture of an adaptive sampling analog-to-digital converter [11].

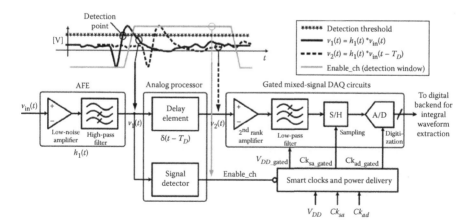

FIGURE 11.5   An activity-based scheme that is powering up the recording building blocks only when necessary. To decrease power consumption drastically, this strategy employs an accurate biopotential detector along with power scheduling [13].

within ultralow power linear-delay filters to wake up the recording circuits "ahead" of biopotential occurrences, a critical behavior to avoid truncated waveforms.

Adaptive mechanisms have also been proposed for achieving high power efficiency. In such approaches, the DC operating points of a circuit are optimally adjusted by employing feedback. A neural amplifier using adaptive biasing is demonstrated in Sarpeshkar et al. [14]. This closed-loop scheme consists of indirectly adjusting the signal-to-noise ratio of an LNA by changing its bias current in order to set the input-referred noise of the amplifier right above the noise floor of the input electrode, thus avoiding any waste of energy.

11.2.3 ENERGY-EFFICIENT SENSORY CIRCUIT TOPOLOGIES

The LNA is the main building block of an analog front end; it must amplify and filter the neural waveforms in order to remove any input DC offset seen across a pair of differential electrodes, thereby maximizing the dynamic range in the recording channel. Indeed, it must provide sufficient gain, appropriate bandwidth, high signal-to-noise ratio (SNR), excellent linearity, and high common mode and high power supply rejection ratios (CMRRs and PSRRs) to provide the expected signal quality. In the case of a multichannel interface, one such sensory circuit per electrode is needed. Therefore, the LNA must consume little power and be of small size as well as scalable to multiple parallel channels. Furthermore, it is essential to optimize the design of the LNA for very low power through a dedicated circuit design methodology, as in Harrison and Charles [15]. The noise efficiency factor (NEF) has been widely adopted as a main figure of merit to assess the performance of LNAs and compare the several existing topologies together.

Designing sensory circuits that can capture multimodal neural information is critical to gather as much neural information as possible. Table 11.1 shows typical values of amplitude and bandwidth for different neural signals. Appropriate amplifiers can discriminate among multiple types of biopotentials by accommodating different frequency ranges [5,16,17]. They can provide different bandwidth settings by (1) tuning the resistive values in a filter, (2) selecting different capacitors values from an array, or (3) changing the operating points of a circuit. In several designs, the high-pass cutoff is changed by varying the gate voltage of a pseudoresistor [17,18] or that of a weakly inverted MOSFET (metal oxide semiconductor field-effect transistor) [19]. In

TABLE 11.1
Characteristics of Various Bioelectric Signals

Signal	Bandwidth (Hz)	Signal range (mVpp)
Electrocardiogram (ECG)	0.05 ~ 256	0.1 ~ 10
Electroencephalogram (EEG)	0.001 ~ 100	0.01 ~ 0.4
Electrocorticogram (ECoG)	0.1 ~ 64	0.02 ~ 0.1
Electromyogram (EMG)	1 ~ 1 K	0.02 ~ 1
Local field potential (LFP)	0.001 ~ 200	0.1 ~ 5
Extracellular action potential (EAP)	100 ~ 10 K	0.04 ~ 0.5

contrast, the low pass cutoff is changed by varying the operating point of the LNA directly [15] or by changing its capacitive load through the selection of different output capacitors [7].

However, tuneable circuits can present significant distortion and exhibit excessive process-dependent variations. Indeed, the resistance of the MOS devices is highly dependent on the voltage level of the output signal [18]. Dedicated linearization circuits have been proposed for performing appropriate biasing of the gate of a MOS device and linearizing the MOS resistor [1,18]. Such approaches have been extended further in Gosselin and Ghovanloo [8] by replacing any inaccurate current mirrors and source followers by precise closed-loop operational amplifiers, thus achieving high linearity above 74 dB within ±200 mV and enabling process-independent frequency cutoff values. This proposed strategy consists in linearizing a pseudoresistor by reporting any variations in the input voltage right at the gates of its constituting pMOS (positive metal oxide conductor) devices with unity gain. This maintains the gate to source voltage of the MOSFETs at constant value, thus canceling any nonlinearity, while providing adequate DC biasing to set the desired cutoff frequency.

A switched-capacitor neural amplifier with tuneable characteristics was recently demonstrated in Jongwoo, Johnson, and Kipke [20] as an alternative to conventional continuous-time circuits. In addition to providing low noise and satisfactory gain, this amplifier can accommodate local field potentials (LFPs) and action potentials (APs) through tuning of its clocking frequency, thus implementing different low- and high pass cutoff frequencies in a straightforward fashion.

11.2.4 A Low Power Discrete-Time Neural Interfacing Front End

An analog front-end topology employing a power scheduling strategy is presented. This approach aims at decreasing the average power consumption by leveraging duty cycling. The schematic of the front end is presented in Figure 11.6. In this scheme, an LNA providing a gain of 100 is powered up only for brief time intervals. The output of the LNA is sampled with a low power switched capacitor (SC) SHA circuit

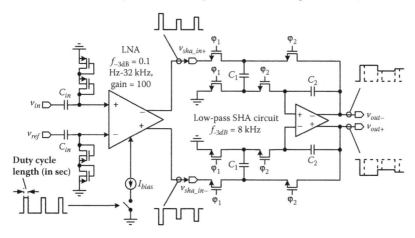

FIGURE 11.6 Analog front end employing duty cycling.

TABLE 11.2

Performance of the Analog Front End

Parameter	Value
Supply voltage	1.8 V
Power consumption	<3 μW
Gain	100 V/V
Input referred noise	31 nV/√Hz
Bandwidth	1 mHz to 8 kHz
Process	CMOS 0.35 μm

Source: B. Gosselin, *The 33rd Annual International Conference of the IEEE Engineering in Medicine and Biology Society (EMBC'11),* Boston, MA, pp. 5855–5859, 2011.

featuring an embedded low pass cutoff frequency for limiting the Johnson noise in the channel. This SHA circuit efficiently combines sampling action with low pass filtering in a compact building block. One challenge in the implementation of this duty cycling strategy is designing an LNA whose output can settle within a short time frame in order to provide high signal quality as well as low power. Indeed, wide bandwidth must be achieved in the LNA to allow fast settling, but this must be done without adding too much power overhead. The cutoff frequency of the LNA is set to 40 kHz, given a typical recording bandwidth of 8 kHz for extracellular AP. Indeed, the SHA circuit provides an embedded low pass cutoff frequency to limit the noise power below 8 kHz. Its simulated NEF was found to be as low as 1.3. The amplifier circuit is simulated using Cadence design tools and CMOS (complementary metal oxide semiconductor) 0.35 μm MOSFET models provided by the foundry. Table 11.2 summarizes the performance of the front end.

Before deriving the cutoff frequency of the SHA circuit, its z-transfer function will be obtained. Then, the continuous transfer function of the circuit will be derived from the z-transfer function, and the cutoff frequency of the SC-SHA will be calculated. We will use the single-ended representation of the circuit for simplicity, the schematic of which is shown in Figure 11.7.

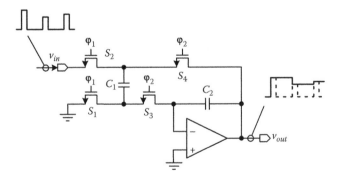

FIGURE 11.7 Single-ended form of the SHA circuit.

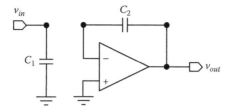

FIGURE 11.8   Equivalent representation of the circuit in sampling mode.

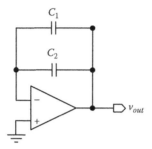

FIGURE 11.9   Equivalent representation of the circuit in holding mode.

First, in sampling mode (nT-T), when φ_1 is high and φ_2 is low, switches S_1 and S_2 are closed and C_1 is charged to v_{in} (Figure 11.8).

After this step, the expressions of the charges on C_1 and C_2 are as follows:

$$Q_1 = C_1 \cdot v_{in}(nT - T), \tag{11.1}$$

$$Q_2 = C_2 \cdot v_{out}(nT - T). \tag{11.2}$$

Then, in the holding mode $(nT - T/2)$, when φ_1 is low and φ_2 is high, switches S_3 and S_4 are closed (Figure 11.9) and the charge on C_1 is combined with the charge that is already present on C_2. The resulting charge can be written as follows:

$$\left(C_1 + C_2\right) \cdot v_{out}\left(nT - \frac{T}{2}\right) = C_2 \cdot v_{out}(nT - T) + C_1 \cdot v_{in}(nT - T). \tag{11.3}$$

We note that once φ_1 turns off, the charge on C_2 will remain the same during the next φ_1, until φ_2 turns on again in the next cycle. Therefore, the charge on C_2 at time (nT), at the end of the next φ_1, is equal to that at time $(nT - T/2)$ or, mathematically,

$$\left(C_1 + C_2\right) \cdot v_{out}\left(nT - \frac{T}{2}\right) = \left(C_1 + C_2\right) \cdot v_{out}(nT). \tag{11.4}$$

Then, v_{out} can be obtained by combining (11.4) and (11.5), i.e.,

$$\left(C_1 + C_2\right) \cdot v_{out}(nT) = C_2 \cdot v_{out}(nT - T) + C_1 \cdot v_{in}(nT - T) \quad (11.5)$$

Thus, v_{out} of the SC-SHA will be

$$v_{out}(nT) = \frac{C_2 \cdot v_{out}(nT - T) + C_1 \cdot v_{in}(nT - T)}{\left(C_1 + C_2\right)} \quad (11.6)$$

The z-transfer function of the circuit is then derived as follows:

$$V_{out}(z) = \frac{C_2 \cdot V_{out}(z) \cdot z^{-1} + C_1 \cdot V_{in}(z) \cdot z^{-1}}{C_1 + C_2} \rightarrow H(z) = \frac{V_{out}(z)}{V_{in}(z)} = \frac{C_1 \cdot z^{-1}}{C_1 + C_2 - C_2 \cdot z^{-1}} \quad (11.7)$$

$$H(z) = \frac{C_1 \cdot z^{-1}}{C_1 + C_2 - C_2 \cdot z^{-1}} = \frac{z^{-1}}{1 + \dfrac{C_2}{C_1} - \dfrac{C_2}{C_1} \cdot z^{-1}} \rightarrow$$

$$H(z) = \frac{z^{-1}}{1 + \dfrac{C_2}{C_1} \cdot \left(1 - z^{-1}\right)}.$$

To find the frequency response, we use $z = e^{j\omega T}$ in (11.7), which gives

$$H(e^{j\omega T}) = \frac{e^{-j\omega T}}{1 + \dfrac{C_2}{C_1} \cdot \left(1 - e^{-j\omega T}\right)}, \quad (11.8)$$

where the period T is the inverse of the clock frequency of the SC circuit f_{clk}. The cutoff frequency of the SC circuit can be obtained from the denominator as follows:

$$1 + \frac{C_2}{C_1} \cdot \left(1 - e^{-j\omega T}\right) = 0 \rightarrow e^{-j\omega T} = 1 + \frac{C_2}{C_1}. \quad (11.9)$$

If the clock frequency is much higher than the highest frequency component of the signal ($\omega T \ll 1$), we can approximate $e^{-j\omega T}$ with the two first terms of its Taylor expansion. Therefore, the cutoff frequency at −3 dB will be approximately

$$1 - j\omega T = 1 + \frac{C_2}{C_1} \rightarrow j\omega T = -\frac{C_2}{C_1} \rightarrow \omega = -\frac{C_2}{C_1} \cdot \frac{1}{T} \rightarrow \quad (11.10)$$

$$f_{-3dB} = -\frac{1}{2\pi} \frac{C_2}{C_1} \cdot f_{clk}.$$

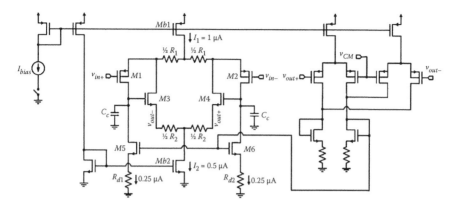

FIGURE 11.10   Schematic of an LNA employing duty-cycling.

The schematic of the wideband low power LNA along with its common-mode feedback (CMFB) circuit is presented in Figure 11.10. This design naturally features wide bandwidth since it employs only seven transistors, which implies minimum parasitic in the signal path. This, in turn, means that the overall power consumption of the LNA is solely determined by the required input-referred noise level. A transconductor (made of M1–M2, M5–M6, and R1) converts $v_{in+} - v_{in-}$ into a current, which is deflected, in part, into a transimpedance amplifier (made of M3–M4 and R2) and then translated into an amplified output voltage $v_{out+} - v_{out-}$.

Then, an accurate and a simplified expression are obtained for the gain of the LNA. After that, the cutoff frequency of the LNA will be derived from this simplified gain expression. The half-circuit model of the LNA, the schematic of which is shown in Figure 11.11(a), is used to derive an expression for the gain.

From Figure 11.11(b), we can obtain the voltages at nodes V_x and V_y by writing the KCL equations at the different nodes. We first obtain i_3 and i_1, as illustrated in Figure 11.11(b).

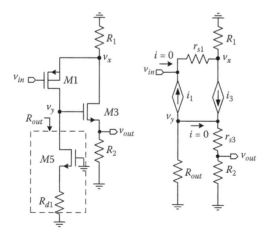

FIGURE 11.11   (a) Half circuit model of the LNA; (b) small signal model of the half circuit.

$$v_{out} = i_3.R_2 \quad \rightarrow \quad i_3 = \frac{v_{out}}{R_2} \tag{11.11}$$

and

$$i_1 = \frac{(v_{in} - v_x)}{r_{s1}} \tag{11.12}$$

where r_{s1} corresponds to the inverse of g_{m1} in the T model of transistor M1. Then, voltage V_y is obtained as follows:

$$v_x = r_{s1}.\left(\frac{v_{in}}{r_{s1}} + \frac{r_{s3}}{R_{out}}.\left(\frac{v_{out}}{R_2} + \frac{v_{out}}{r_{s3}} \right) \right) \tag{11.13}$$

where r_{s3} corresponds to the inverse of g_{m3} in the T model of transistor M3 and R_{out} is the equivalent resistance from the drain of M5. The latter is much larger than R1 and R2. Voltage V_y is obtained as follows:

$$v_y = -R_{out}.i_1. \tag{11.14}$$

Then, we obtain the accurate gain expression of the LNA by writing

$$\frac{v_x}{R_1} + i_3 - i_1 = 0 \rightarrow \frac{r_{s1}.\left(\frac{v_{in}}{r_{s1}} + \frac{r_{s3}}{R_{out}}.\left(\frac{v_{out}}{R_2} + \frac{v_{out}}{r_{s3}} \right) \right)}{R_1} + \frac{v_{out}}{R_2} + \frac{r_{s3}.\left(\frac{v_{out}}{R_2} + \frac{v_{out}}{r_{s3}} \right)}{R_{out}} = 0 \rightarrow \tag{11.15}$$

$$A_v = \frac{v_{out}}{v_{in}} = -\frac{1}{R_1}.\left(\frac{1}{\frac{1}{R_2} + \frac{1}{R_{out}}.\left(\frac{r_{s1}.r_{s3}}{R_1 R_2} + \frac{r_{s1}}{R_1} + \frac{r_{s3}}{R_2} + 1 \right)} \right)$$

Finally, we replace r_{s1} and r_{s3} with $1/g_{m1}$ and $1/g_{m3}$, respectively, and the gain expression of the LNA becomes

$$A_v = \frac{v_{out}}{v_{in}} = -\frac{1}{R_1}.\left(\frac{1}{\frac{1}{R_2} + \frac{1}{R_{out}}.\left(\frac{1}{g_{m1}.g_{m3}.R_1 R_2} + \frac{1}{g_{m1}R_1} + \frac{1}{g_{m3}R_2} + 1 \right)} \right). \tag{11.16}$$

We use the identity $x + y + xy + 1 = (1 + x).(1 + y)$ to simplify (11.16). Assuming that

$$x = \frac{1}{g_{m1}R_1} \quad \text{and} \quad y = \frac{1}{g_{m3}R_2},$$

the gain expression then becomes

$$A_v = -\frac{1}{R_1} \cdot \left(\frac{1}{\frac{1}{R_2} + \frac{1}{R_{out}} \cdot \left(\left(1 + \frac{1}{g_{m1}R_1} \right) \cdot \left(1 + \frac{1}{g_{m3}R_2} \right) \right)} \right) \cdot \tag{11.17}$$

Defining

$$\alpha = \left(1 + \frac{1}{g_{m1}R_1} \right) \cdot \left(1 + \frac{1}{g_{m3}R_2} \right),$$

the gain expression can be rewritten as

$$A_v = -\frac{1}{R_1} \cdot \left(\frac{1}{\frac{1}{R_2} + \frac{\alpha}{R_{out}}} \right) \rightarrow A_v = -\frac{R_2}{R_1} \cdot \left(\frac{1}{1 + \frac{\alpha R_2}{R_{out}}} \right). \tag{11.18}$$

If $\alpha R_2 \ll R_{out}$, then the gain is approximately

$$A_v \approx -\frac{R_2}{R_1}. \tag{11.19}$$

Denoting C_p the capacitor at node v_y and knowing that $\alpha R_2 \ll R_{out}$, we can write

$$\alpha R_2 \cdot C_p \ll R_{out} \cdot C_p \rightarrow \frac{1}{\alpha R_2 \cdot C_p} \gg \frac{1}{R_{out} \cdot C_p},$$

thus yielding the 3 dB cutoff frequency of the LNA according to

$$f_{-3dB} = \frac{1}{2\pi} \cdot \frac{1}{\alpha R_2 \cdot C_p}. \tag{11.20}$$

This design is using source degeneration resistors $R_{d1}–R_{d2}$ in the active load M5–M6 to decrease the contributed noise from these MOS devices, as in Wattanapanitch, Fee, and Sarpeshkar [17]. Moreover, split resistors $R_1–R_2$ are used along with a single current source in both differential pairs to cancel the second harmonic terms efficiently and achieve high linearity. The total current consumption of this optimized design can be very low since it features only three main branches and employs only seven transistors (not counting the CMFB and bias circuits). Furthermore, the utilization of source degeneration resistors in the MOS current mirror load is significantly lowering the overall input-referred noise. Also, the SHA circuit can be designed for very low power since the sampled signal is much less sensitive to noise after having been amplified by the LNA. For power consumption, the current drained from the power supply scales with the duty cycle length.

Indeed, using a duty cycle of 25% in this proposed design reduces the power consumption of the LNA by a factor of four. Similarly to a duty cycling arrangement, the same discrete-time front end can be incorporated into a TDM strategy where one common wideband LNA would be shared between several neural recording electrodes to save power. In such a configuration, the LNA must roughly provide n times the signal bandwidth ($n \times f_{max}$) in order to process n TDM electrodes. For instance, the proposed LNA could enable the multiplexing of four neural recording electrodes with $f_{max} = 8$ kHz since its cutoff frequency is higher than four times the signal bandwidth ($4 \times f_{max} \approx 32$ kHz).

11.2.5 SUMMARY

In this section, we have reviewed energy-efficient system-level architectures, smart power management strategies, and low power sensory circuit topologies that are improving efficiency in power-constrained multichannel recording arrangements. Moreover, the described strategies were illustrated through the design of a practical implementation—namely, a discrete-time neural recording front end employing duty cycling as a means to decrease its power consumption by 75%.

11.3 WAKE-UP RADIO FOR WIRELESS SENSOR NETWORKS

A wireless sensor network (WSN) is a collection of low cost, low energy computing devices, designated *nodes* or *motes;* each is equipped with one or more sensors, with limited computational and memory resources, that operate as a whole to accomplish a specific task. Typical applications include monitoring and data aggregation in fields including military, agricultural, medical, and environmental.

Two outstanding characteristics of WSNs are (1) the ad hoc, self-organized, and self-healing nature of the network, and (2) the extremely low energy budget each sensor node is allocated. The latter constraint is due in part to the small size and cost of the nodes, which leaves little room for bulky batteries, and also to the nature of the applications, which require the nodes to remain operational for months, perhaps years after initial deployment without human intervention. Energy sources include small batteries and energy scavenging mechanisms (solar, mechanical vibrations, heat). In the case of batteries, which is by far the more common solution, the useful life of the network is limited by the first node failure, so network protocols are typically designed to ensure that all nodes are equally active so that their batteries are drained at an equal rate.

The power consumption of the individual nodes is therefore a topic of the highest importance, which tends to be dominated by the RF transmit and receive functions in the vast majority of applications. Any subsystem of a node that is not needed momentarily is typically put in sleep mode, and the RF subsystem is no exception.

Various techniques have been devised to allow a node to "wake up" only when it needs to transmit and/or receive a data packet and then go back to sleep once its task is complete. In typical operation, nodes can acquire data from their sensor at periodic intervals, process this data locally (feature extraction, data compression, etc.), forward the data, relay packets that are meant for another node or for the base/ aggregation station, or receive query/command packets. Multihop packet relaying is favored because it is more energy efficient than transmitting in a single hop over a larger distance.

To allow nodes to be in deep sleep (including the RF section) most of the time, thus consuming minimal power, typically requires protocols that operate in a synchronous or pseudosynchronous manner to ensure that nodes that need to communicate are awake at the same time [22]. In synchronous operation, the whole network wakes up at periodic intervals; performs whatever acquisition, processing, and communication tasks are required; and then goes back to sleep. In pseudosynchronous or *rendezvous* protocols, nodes attempt to predict at what moment they will be needed and should wake up based on past history. In effect, the network as a whole attempts to establish a dynamic optimum schedule in a totally distributed manner. This obviously implies additional processing (and possibly signaling) overhead to determine the continuously changing sleep interval, as well as missed rendezvous and false-alarm wake-ups. In the end, no matter how sophisticated the rendezvous protocol is, the receive sections of nodes have to stay awake significantly longer than the packet duration in order to have a fair chance of making the rendezvous. For this reason, the RF receive function typically dominates the power consumption budget in WSN, even though its instantaneous consumption is less than the transmit function.

11.3.1   THE WAKE-UP RADIO CONCEPT

The preceding discussion highlights how desirable it is to minimize the on-time of the RF receive functions and, if possible, to avoid the complications of rendezvous protocols. The only remaining option—totally asynchronous operation—requires a means to "wake up" a node remotely on demand, using only an RF signal. Forgoing the need to have the network maintain a communication schedule, each node is equipped with an ultra low power wake-up radio (WUR), which continuously monitors the channel. This is in addition to the main RF transceiver unit, which can therefore be turned off most of the time. When a node wishes to communicate with a neighbor, it sends a wake-up signal, containing the wake-up code or address of the target, thus waking only the desired node (see Figure 11.12). After successful reception and address decoding of the wake-up signal, the WUR brings the entire node out of sleep mode if the wake-up code matches one of the node's wake-up codes. An acknowledgment packet is then sent to signal the first node that transmission can proceed. After the data exchange is complete and after any additional tasks are performed, both nodes can go back to sleep, activating their WUR before doing so.

For asynchronous operation to be effective and competitive with rendezvous protocols, the power consumption of the wake-up device must be extremely low. It has been estimated that the wake-up radio must have a power consumption below 50 μW in order to constitute an attractive alternative [23] and should be below 10 μW to extend the range of WSN applications [24]. In a 2010 survey [25], the feasibility requirements of wake-up radios were listed: low cost, low power usage, low latency, low interference,

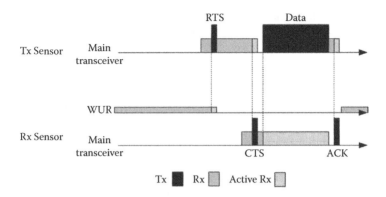

FIGURE 11.12 Example of asynchronous operation presenting a wake-up, acknowledge, and receive wake-up behavior. (a) Transmission of the wake-up code. (b) Reception and processing of the wake-up code and possible wake-up of the node. (c) Acknowledgment of wake-up. (d) Data transmission and return to sleep mode.

low missed wake-up rate, and appropriate range (with a typical goal of 10 m). As of 2010, no reported implementation meets all these requirements and very few complete, stand-alone, and working implementations have been reported to date.

Restricting our attention to complete, fully functional sensor nodes, the first reported wake-up radio (in 2007) is described in Van der Doorn, Kavelaars, and Langendoen [26]. Since it is built from off-the-shelf components (including a microcontroller) and consists of a wake-up radio piggyback board in conjunction with Delft University's T-node platform, the power consumption in practice is quite high—above 600 µW, although the design could, in principle, have operated at 170 µW after resolution of certain practical issues. Furthermore, the range was only 2–3 m. It follows that while this is a conceptually compelling first effort, its interest is mostly academic.

Examples of the current state of the art include the design reported by Ansari, Pankin, and Mähönen in 2008 [27] and that of Hambeck, Mahlknecht, and Herndl in 2011 [28]. Based on the Telos node, the design in Ansari et al. [27] uses a secondary radio in the 868 MHz band to transmit the wake-up signal, thus shifting some of the complexity and power burden to the transmitter. The receiver employs impedance matching and a five-stage voltage multiplier, resulting in a wake-up circuit that consumes only 876 nA. The entire node in sleep mode consumes 12.5 µW, which is significantly below the 50 µW barrier. However, the observed range is of the order of 2.5 m, thus falling short of the 10 m objective.

Hambeck et al. [28] report a wake-up radio chip implementation in 130 nm CMOS leveraging various low power techniques to achieve a consumption of 2.4 µW (wake-up unit only) and a sensitivity of −71 dBm at 868 MHz. However, it seems that this extraordinary sensitivity is obtained in part by correlating over very long sequences, thus trading off wake-up time for sensitivity and possibly shifting part of the power burden to the transmitter. In fact, the paper cites a wake-up time of 40 to 110 ms, which is enormous, and the extraordinary receiver sensitivity is based on correlation over a 7 ms period (thus accumulating energy) with an external surface acoustic wave (SAW) filter.

11.3.2 A LOW POWER WUR DESIGN BASED ON PWM

It has been established previously that in assessing the quality of a WUR design, multiple variables must be taken into consideration in addition to receiver sensitivity and power consumption, such as wake-up time, bit rate, robustness to interference, range, band of operation, the completeness of the design, etc. As a case study, we examine a design dating from 2008 [29,30] that presents some unusual and innovative characteristics.

The vast majority of reported wake-up radio designs operate at a frequency of 868 MHz, including the latest state-of-the-art ones [27,28]. This allows better range and/or lower power consumption compared with higher frequency bands such as the popular 2.4 GHz unlicensed band. However, it should be noted that 868 MHz does not constitute a globally available band. It is an unlicensed band in Europe, but not in North America (where the ISM band is at 908 MHz instead). Furthermore, while path loss is lower at 868 MHz, antenna gain is also lower and/or antenna size is larger than at 2.4 GHz. It follows that for compact nodes with decent antenna gains *and* for designs that can operate globally, 2.4 GHz is a better option.

11.3.3 PWM MODULATION

Aside from operating in the 2.4 GHz band, the design under study also features a highly unusual modulation format. Indeed, nearly all surveyed WUR implementations leverage on/off keying (OOK) for simplicity of operation, while the design under study is based on pulse-width modulation (PWM). This is arguably the central innovation of this design, as this choice leads to many benefits. To the best of our knowledge, this modulation has not been used before in single-chip wake-up radio implementations, although an early effort from off-the-shelf components has been reported [31], where PWM was used specifically to alleviate complex synchronization requirements.

First, we observe that both OOK and PWM are forms of amplitude modulations that are amenable to mostly passive, extremely simple receiver structures based on, for example, an envelope detector and an integrator. Second, unlike OOK, the PWM is self-synchronizing (since there is a pulse for every transmitted bit) and does not require sophisticated clock recovery circuitry. Third, PWM allows arbitrary control of the duty cycle. For example, choosing to transmit zeros as 3 μs pulses and ones as 12 μs pulses, and assuming a symbol period of 20 μs, the mean duty cycle (with equiprobable bits) is 37.5%. This leads to transmit energy savings with respect to constant envelope (frequency and/or phase modulation, duty cycle of 100%) and OOK (duty cycle of 50%).

The fourth and possibly most important benefit of PWM is its natural robustness to both noise and interference. Indeed, because the information is encoded in the duration of the pulses, the effect of additive white Gaussian noise is different, thus leading to higher resistance [32,33]. Additionally, because this modulation is radically different from that of other modulation types used by interfering signals and its detection is based on pulse duration, it is also robust to interferers.

It is noteworthy that PWM has been forgotten for a long time as a wireless form of modulation because it is inherently not bandwidth efficient. However, bandwidth efficiency is not a primary concern in the unusual context of wireless sensor networks.

11.3.4 ARCHITECTURE

The WUR circuit architecture is depicted in Figure 11.13. The RF detector is a form of envelope detector—more specifically, a zero-bias Schottky diode voltage doubler. Impedance matching is integrated into this component with lumped reactive elements in order to minimize the form factor. This matching is optimized for a subset of the 2.4 GHz ISM band, with a worst-case return loss of –10 dB for the first six 802.11 channels (out of 14).

The amplifier is by far the most power-hungry block of the WUR reception chain. It consists of a two-stage structure (see Figure 11.14) providing an approximate gain of 86 dB for a bandwidth of 180 kHz. A Schmitt trigger inverter at the output provides digitization prior to entering the PWM demodulator.

PWM modulation is performed by an asymmetrical inverter followed by a capacitor, the pair acting as a resettable integrator. Transistors are sized to determine integration time constants such that pulses shorter than 10 μs yield a logical zero and

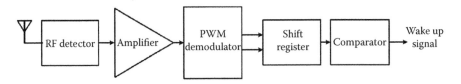

FIGURE 11.13 Architecture of PWM-based wake-up radio. The antenna is not necessarily dedicated to the wake-up device; it can be shared with the main transceiver. The RF detector provides a baseband signal to the demodulator. The amplitude modulation used is PWM since it provides clocking information. The received information is fed to a shift register and its content is compared to the node's address, generating a wake-up signal if there is a match.

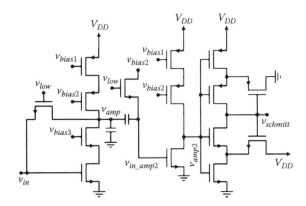

FIGURE 11.14 Two-stage amplifier structure with Schmitt trigger output.

longer pulses yield a logical one. A symbol-rate clock is generated along with the main output of the integrator, and both signals are passed to the shift register.

The shift register is 8 bits long and, together with the comparator, allows matching with a local unique 8-bit address. Thus, only wake-up radio signals targeted at this specific node will be recognized and activate the wake-up line to the core of the node.

11.3.5 Performance

The proposed WUR circuit, consisting of a layout in 130 nm CMOS using an IBM process, was validated through SPICE simulation given a 50 kHz PWM signal having a 15% duty cycle for zeros and a 60% duty cycle for ones. It was seen that the device functions adequately with a sensitivity of −55 dBm. This is deemed to correspond to a practical range of between 4 and 5 m for propagation exponents between two and three.

The breakdown of the power consumption per component is given in Table 11.3. It can be seen that all components except the amplifier consume negligible power, thus providing a clear path for further improvement in this respect. With a 1 V power supply, the entire circuit (in simulation) consumes 10 µA.

It is noteworthy that addresses longer than 8 bits could be used with minimal impact on the power budget. Only the current drawn by the comparator stage would change, in a manner that is linear with the address length.

The range could also be substantially increased by making use of directional antennas. One possible scenario involves equipping a node with four directional antennas of moderate gain pointing 90° away from each other with patterns that are wide enough to collectively cover 360° in azimuth adequately. This could provide a gain of approximately 6 dB or more, depending on how directive the antennas are in elevation. A similar gain could be obtained at the transmitter (for a total gain of 12–15 dB) if the transmitter "knows" in what direction the target node is (a likely hypothesis, provided the network is capable of geographic self-discovery and self-organization).

Two possible schemes can be devised to integrate such a four-antenna layout with a wake-up radio. In the first scheme, a single WUR circuit is shared by all four antennas. A switching mechanism connects it to a single selected antenna out of the set of

TABLE 11.3

Per-Component Breakdown of the Average Power Consumption of the WUR Circuit with 15%/60% Duty Cycle 50 kHz PWM Signal

Building block	Average consumption with 50 kHz signal and noise (µW)	Average consumption with noise only (µW)
RF detector	0	0
Amplifier	17.80	18.29
PWM demodulator	0.8	0.06
Address decoder	0.02	<0.01
Total	18.61	18.35

four. A continuously running timer would trigger a switching event at regular intervals so that all four antennas are listened to in turn. This has the benefit of requiring only one complete WUR circuit, albeit at the cost of slightly increased complexity and power consumption (the timer and switching mechanism) and a higher chance of missed wake-ups.

In the second scheme, a full WUR circuit is implemented for each antenna so that wake-up signals from any direction can be intercepted at any time. Obviously, the power consumption of this scheme would be much higher. However, it has been observed that most of the power consumption (60%) is in the amplifier bias circuitry. Since this circuitry can be shared among all four WUR components, it can be assumed that the overall power consumption would be $(60\% + 4 \times 40\%) \times 19\ \mu W = 2.2 \times 19\ \mu W = 41.8\ \mu W$, which is still below the 50 μW threshold.

This demonstrates that the multiple antenna context is feasible, especially in light of potential improvements in the amplification section to make it more power efficient.

11.3.6 SUMMARY

Throughout this section, we have surveyed the state of the art in wake-up radio circuit design as a means to achieve fully asynchronous ultralow power wireless sensor networks. We have also presented a complete stand-alone wake-up radio architecture that, in the form of a layout in 130 nm CMOS, provides an overall power consumption of less than 19 μW in the 2.4 GHz band and with a sensitivity on the order of –55 dBm. This demonstrates that the wake-up radio concept is both feasible and useful, since the power consumption is far below the 50 μW usefulness threshold agreed upon in the literature. Furthermore, ideas could be borrowed from recently reported designs with even lower power consumptions (although they operate at 868 MHz instead of 2.4 GHz) to improve the reviewed architecture. The notion of using multiple directive antennas in order to achieve directive gain and thus greater range and/ or RF energy savings was also discussed.

REFERENCES

1. Z. Xiaodan, X. Xiaoyuan, Y. Libin, and L. Yong, A 1 V 450 nW fully integrated programmable biomedical sensor interface chip. *IEEE Journal Solid-State Circuits,* vol. 44, pp. 1067–1077, 2009.
2. A. Bonfanti et al., A multi-channel low-power system-on-chip for single-unit recording and narrowband wireless transmission of neural signal. *Annual International Conference of the IEEE Engineering in Medicine and Biology Society (EMBC'10),* pp. 1555–1560, 2010.
3. A. M. Sodagar, K. D. Wise, and K. Najafi, A wireless implantable microsystem for multichannel neural recording. *IEEE Transactions on Microwave Theory and Techniques,* vol. 57, pp. 2565–2573, 2009.
4. B. Gosselin et al., A mixed-signal multichip neural recording interface with bandwidth reduction. *IEEE Transactions on Biomedical Circuits and Systems,* vol. 3, pp. 129–141, 2009.

5. L. Seung Bae, L. Hyung-Min, M. Kiani, J. Uei-Ming, and M. Ghovanloo, An inductively powered scalable 32-channel wireless neural recording system-on-a-chip for neuroscience applications. *IEEE Transactions on Biomedical Circuits and Systems,* vol. 4, pp. 360–371, 2010.
6. X. Zhiming, T. Chun-Ming, C. M. Dougherty, and R. Bashirullah, A 20 µW neural recording tag with supply-current-modulated AFE in 0.13 µm CMOS. *IEEE International Solid-State Circuits Conference (ISSCC'10),* pp. 122–123, 2010.
7. M. S. Chae, Z. Yang, M. R. Yuce, L. Hoang, and W. Liu, A 128-channel 6 mW wireless neural recording IC with spike feature extraction and µWB transmitter. *IEEE Transactions Neural Systems Rehabilitation Engineering,* vol. 17, pp. 312–321, 2009.
8. B. Gosselin and M. Ghovanloo, A high-performance analog front-end for an intraoral tongue-operated assistive technology. *IEEE International Symposium on Circuits and Systems (ISCAS),* pp. 2613–2616, 2011.
9. P. Chung-Ching, X. Zhiming, and R. Bashirullah, Toward energy efficient neural interfaces. *IEEE Transactions Biomedical Engineering,* vol. 56, pp. 2697–2700, 2009.
10. N. Joye, A. Schmid, and Y. Leblebici, Extracellular recording system based on amplitude modulation for CMOS microelectrode arrays. *IEEE Biomedical Circuits and Systems Conference,* pp. 102–105, 2010.
11. R. F. Yazicioglu, K. Sunyoung, T. Torfs, K. Hyejung, and C. Van Hoof, A 30 µW analog signal processor ASIC for portable biopotential signal monitoring. *IEEE Journal Solid-State Circuits,* vol. 46, pp. 209–223, 2011.
12. B. Gosselin and M. Sawan, Circuits techniques and microsystems assembly for intracortical multichannel ENG recording. *IEEE Custom Integrated Circuits Conference (CICC'09),* pp. 97–104, 2009.
13. B. Gosselin, M. Sawan, and E. Kerherve, Linear-phase delay filters for ultra-low-power signal processing in neural recording implants. *IEEE Transactions on Biomedical Circuits and Systems,* vol. 4, pp. 171–180, 2010.
14. R. Sarpeshkar et al., Low-power circuits for brain–machine interfaces. *IEEE Transactions on Biomedical Circuits and Systems,* vol. 2, pp. 173–183, 2008.
15. R. R. Harrison and C. Charles, A low-power low-noise CMOS amplifier for neural recording applications. *IEEE Journal Solid-State Circuits,* vol. 38, pp. 958–965, 2003.
16. M. Mollazadeh, K. Murari, G. CauWenberghs, and N. Thakor, Micropower CMOS integrated low-noise amplification, filtering, and digitization of multimodal neuropotentials. *IEEE Transactions on Biomedical Circuits and Systems,* vol. 3, pp. 1–10, 2009.
17. W. Wattanapanitch, M. Fee, and R. Sarpeshkar, An energy-efficient micropower neural recording amplifier. *IEEE Transactions on Biomedical Circuits and Systems,* vol. 1, pp. 136–147, 2007.
18. Y. Ming and M. Ghovanloo, A low-noise preamplifier with adjustable gain and bandwidth for biopotential recording applications. *IEEE International Symposium on Circuits and Systems (ISCAS'07),* pp. 321–324, 2007.
19. P. Mohseni and K. Najafi, A fully integrated neural recording amplifier with DC input stabilization. *IEEE Transactions on Biomedical Engineering,* vol. 51, pp. 832–837, 2004.
20. L. Jongwoo, M. D. Johnson, and D. R. Kipke, A tunable biquad switched-capacitor amplifier-filter for neural recording. *IEEE Transactions on Biomedical Circuits and Systems,* vol. 4, pp. 295–300, 2010.
21. B. Gosselin, Approaches for the efficient extraction and processing of neural signals in implantable neural interfacing microsystems. *The 33rd Annual International Conference of the IEEE Engineering in Medicine and Biology Society (EMBC'11),* Boston, MA, pp. 5855–5859, 2011.
22. H. Carl and A. Willig, *Protocols and architectures for wireless sensor networks.* Chichester, England: Wiley, 2007.

23. E.-Y. Lin, J. Rabaey, and A. Wolisz, Power-efficient rendezvous schemes for dense wireless sensor networks. *Proceedings ICC (IEEE International Conference Communications)*, pp. 3769–3776, 2004.

24. M. Spinola Durante, Wakeup receiver for wireless sensor networks, PhD dissertation, Institute of Computer Technology, Vienna, University of Technology, Vienna, Austria, 2009.

25. B. Kersten and C. G. U. Okwudire, MAC layer energy concerns in WSNs: Solved? Unpublished research paper, SAN (System Architecture and Networking) Group Seminar, TU Eindhoven, Feb. 2010.

26. B. Van der Doorn, W. Kavelaars, and K. Langendoen, A prototype low cost wakeup radio for the 868 MHz band. *International Journal Sensor Networks*, vol. 5, no. 1, pp. 22–32, 2009.

27. J. Ansari, D. Pankin, and P. Mähönen, Radio-triggered wake-ups with addressing capabilities for extremely low power sensor network applications. *Proceedings PIMRC (IEEE International Symposium on Personal, Indoor, Mobile and Radio Communications)*, pp. 1–5, 2008.

28. C. Hambeck, S. Mahlknecht, and T. Herndl, A 2.4 µW wake-up receiver for wireless sensor nodes with −71 dBm sensitivity. *Proceedings ISCAS (IEEE International Symposium on Circuits and Systems)*, pp. 534–537, 2011.

29. P. Le-Huy and S. Roy, Low-power 2.4 GHz wake-up radio for wireless sensors. *Proceedings WiMob (Wireless Mobility Conference)*, pp. 13–18, 2008.

30. P. Le-Huy and S. Roy, Low-power wake-up radio for wireless sensor networks. *Mobile Networks and Applications*, vol. 15, no. 2, pp. 226–236, April 2010.

31. S. von der Mark et al., Three stage wakeup scheme for sensor networks, *Proceedings International Conference on Microwave and Optoelectronics*, pp. 205–208, 2005.

32. H. S. Black, *Modulation theory.* New York: Van Nostrand, 1953.

33. A. B. Carlson, *Communication systems,* 4th ed. New York: McGraw–Hill, 2002.

12 An Open-Loop Multiphase Local-Oscillator Generation Technique

Kai-Fai Un, Pui-In Mak, and Rui P. Martins

CONTENTS

12.1 INTRODUCTION

Multiphase local-oscillator (LO) generators are widely employed in modern wireless/wireline communication systems. This chapter describes how an inverter-based open-loop four-/eight-phase LO generator can be utilized to cover a wide spectrum and relax the speed requirement of the phase-locked loop (PLL) and voltage-controlled oscillator (VCO), when comparing with the conventional frequency-divider-based method. The mathematical models of open-loop four-phase and eight-phase LO generators are developed. A design example targeting the mobile TV applications is implemented in 65 nm CMOS (complementary metal oxide semiconductor) technology to prove the feasibility of the technique. It covers the full band (VHF-III, UHF, and L) of mobile TV with two operating modes (i.e., the eight-phase mode supports image reject down-conversion whereas the four-phase mode supports typical quadrature down-conversion).

12.2 OVERVIEW OF MULTIPHASE LO GENERATION

Many different types of multiphase LO/clock generators have been reported for wireless and wireline applications. A ring VCO based on CMOS inverters (Figure 12.1a) is a compact solution to realize a multiphase LO with a large frequency range [1,2]. However, its phase noise performance is normally unacceptable for high-tier wireless systems. Although the phase noise can be substantially reduced by replacing all inverters with LC VCOs (Figure 12.1b), the associated inductors occupy a significant part of the chip area [3,4]. Delay-locked loop (DLL) using numerous delay units can also be a multiphase clock generator (Figure 12.1c) [5,6]. The key drawback is that the phase-noise performance is heavily dependent on the number of output phases required.

Alternatively, a multiphase LO signal can be generated in an open-loop way by using a low noise LC VCO followed by a frequency divider (Figure 12.1d). Elementarily, a four-phase LO can be generated via a div-by-2 circuit. Such a division factor implies that the associated PLL and VCO have to operate at a doubled

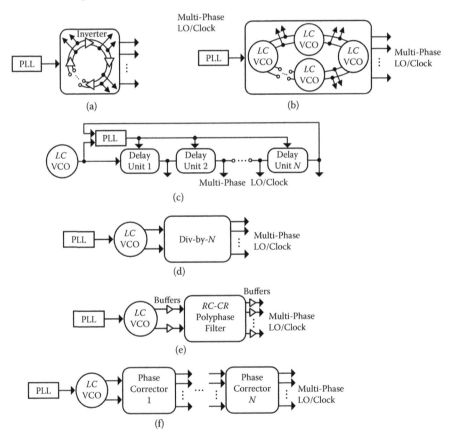

FIGURE 12.1 Multiphase LO/clock generation methods: (a) ring with inverters, (b) ring with LC VCOs, (c) DLL, (d) frequency divider, (e) RC-CR polyphase filter, and (f) phase correctors in cascade.

frequency. The design complexity, however, rises dramatically with the number of phases required. For instance, to generate an eight-phase LO, a div-by-4 circuit is necessary; the PLL and VCO have to operate at four times the output frequency. A higher operating frequency unavoidably calls for more power to lower the phase noise and phase error. Moreover, the PLL and VCO will be more sensitive to parasitic capacitances, implying narrower locking and tuning ranges, respectively. To surmount these constraints, the frequency divider can be replaced by a passive *RC-CR* polyphase filter (Figure 12.1e) [7,8], but the performance can be strongly affected by the temperature and process variations, while power-hungry buffers are required for proper interface.

Recently, an open-loop four-phase clock generator was proposed for wireline applications [9,10]. Multiple phase correctors are cascaded in an open-loop way to improve the phase precision (Figure 12.1f). The prime advantages of this method are its simplicity (i.e., open loop and inverter only), no power-hungry buffer, and that the number of output phases is independent of the operation frequency of the circuit itself and its driving source. The achieved frequency range in Kim et al. [9] is 0.37 to 2.5 GHz and the phase precision is ±5° for a four-phase output.

In this content, we extend the concept of such open-loop architecture for wireless applications with different requirements on the phase precision and the number of output phases [11,12]. The targeted phase error is ±1° and both four- and eight-phase LO generators will be designed and analyzed.

12.3 MATHEMATICAL MODEL OF OPEN-LOOP MULTIPHASE LO GENERATOR

12.3.1 FOUR-PHASE LO GENERATION

12.3.1.1 Architecture

The block diagram of a four-phase LO generator is depicted in Figure 12.2. It is structured by putting numerous phase correctors in cascade to interpolate a four-phase LO from a two-phase differential input (v_{clk} and v_{clkp}). From left to right,

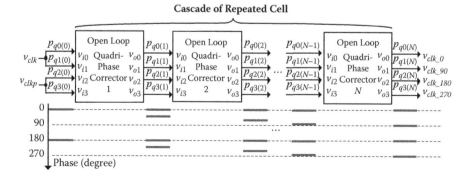

FIGURE 12.2 Block diagram of a four-phase LO generator. Phase error is reduced down progressively from the correctors 1 to N.

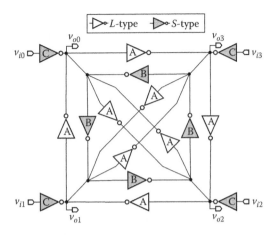

FIGURE 12.3 Architecture schematic of an inverter-based four-phase corrector.

the phase correctors improve the phase precision progressively until reaching the desired accuracy. The schematic of each inverter-based quadriphase corrector is shown in Figure 12.3. Every corrector is composed of 16 inverters (CMOS) classified according to two different device sizes as *L*-type or *S*-type. *L*-type inverters feature a larger geometrical size than the *S*-type to optimize the phase precision in a specific frequency range. The inverters can be divided into three groups according to their functionality:

- Set **A** is for phase correction. With three inverters in a loop, it is able to oscillate and interpolate the intermediate phases.
- Set **B** is for natural-frequency suppression. It leads to a larger operating frequency range [16].
- Set **C** is for signal injection. It allows multiple phase correctors to be cascaded directly to improve the output phase precision.

12.3.1.2 Mathematical Model

In order to determine the optimum conditions in terms of frequency range and phase precision, the four-phase corrector is modeled by a signal flow graph (SFG) as shown in Figure 12.4. For simplicity, a linear model is assumed [13,14]. Each inverter is modeled as a single-pole amplifier with a transfer function of $h(f)$ as given by

$$h(f) = -\frac{G}{1 + \dfrac{jf}{f_C}}, \tag{12.1}$$

where G is the normalized DC gain and f_C is the −3 dB cutoff frequency. The constants a and b in Figure 12.5 represent the driving capability of *L*-type and *S*-type inverters, respectively.

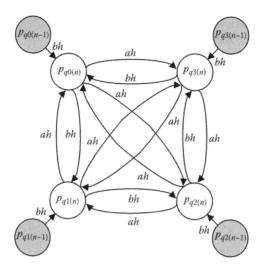

FIGURE 12.4 SFG of a four-phase corrector.

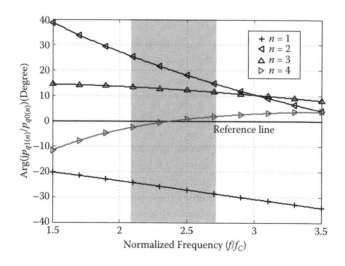

FIGURE 12.5 Static phase error of four-phase corrector with $a/b = 3.4$ and $G = 10$ (linear model simulation).

Phasor-domain analysis is applied to obtain the phase correction transformation of the nth phase corrector as expressed by

$$
\begin{cases}
A_q P_{q(n)} = P_{q(n-1)} \\[4pt]
A_q = \begin{bmatrix}
1 & -ah & -ah & -bh \\
-bh & 1 & -ah & -ah \\
-ah & -bh & 1 & -ah \\
-ah & -ah & -bh & 1
\end{bmatrix}, \\[4pt]
P_{q(n)} = \begin{bmatrix} P_{q0(n)} & P_{q1(n)} & P_{q2(n)} & P_{q3(n)} \end{bmatrix}^T
\end{cases}
\tag{12.2}
$$

where A_q is the phase transformation matrix and $p_q(n)$ is the output phase vector. Rearranging (12.2) yields

$$
P_{q(n)} = A_q^{-1} P_{q(n-1)},
\tag{12.3}
$$

For N quadriphase cascaded correctors, $p_{q(N)}$ becomes

$$
P_{q(N)} = A_q^{-N} P_{q(0)},
\tag{12.4}
$$

where $p_{q(0)}$ is the input phase vector represented by

$$
P_{q(0)} = \begin{bmatrix} 1 & 1 & -1 & -1 \end{bmatrix}^T bh,
\tag{12.5}
$$

since the input phase can be either $0°$ or $180°$. In the phasor domain only 1 or -1 is available for the input vector. Figure 12.5 shows the steady-state phase error function defined by (12.6) with a different number of stages in cascade

$$
\Phi_{eq(n)} = Arg\left(j \frac{P_{q1(n)}}{P_{q0(n)}} \right),
\tag{12.6}
$$

where a/b is chosen to be 3.4 to provide an acceptable frequency range and G is selected as 10 (practical DC gain value of a CMOS inverter in nanometer technologies). The steady-state phase error depends on the ratio of the output frequency (f) to the corner frequency of an inverter (f_c), as well as the number (n) of phase correctors in cascade. With $n = 4$, the phase error is minimized over a wide range of f/f_c (between 2.1 and 2.7). Figure 12.5 also shows that the steady-state phase error can be minimized by cascading additional stages of four-phase correctors for certain frequency ranges.

Based on the preceding definitions and following a particular context, the transfer function of the linearized quadriphase LO generator will be derived next. According to it, the optimal conditions for minimizing the phase error are obtained to build up the device-sizing strategy. From (12.2), we can simply prove that

$$
\begin{cases}
p_{q0(n)} = -p_{q2(n)} \\
p_{q1(n)} = -p_{q3(n)}
\end{cases}.
\tag{12.7}
$$

Thus, (12.2) and (12.3) can be simplified as

$$
\begin{bmatrix}
1+ah & -(a-b)h \\
(a-b)h & 1+ah
\end{bmatrix}
\begin{bmatrix}
p_{q0(n)} \\
p_{q1(n)}
\end{bmatrix}
=
\begin{bmatrix}
p_{q0(n-1)} \\
p_{q1(n-1)}
\end{bmatrix}.
\tag{12.8}
$$

And also for a simplification of the notations, A, B, and C are introduced as follows:

$$
A = 1 + ah
\tag{12.9}
$$

$$
B = (a-b)h
$$

$$
C =
\begin{bmatrix}
A & -B \\
B & A
\end{bmatrix}.
$$

Similarly to (12.4), we can obtain

$$
\begin{bmatrix}
p_{q0(N)} \\
p_{q1(N)}
\end{bmatrix}
= C^{-N}
\begin{bmatrix}
p_{q0(0)} \\
p_{q1(0)}
\end{bmatrix}
= C^{-N}
\begin{bmatrix}
1 \\
1
\end{bmatrix}.
\tag{12.10}
$$

Since the matrix C can be diagonalized as

$$
C = \left(
\frac{1}{\sqrt{2}}
\begin{bmatrix}
j & -j \\
1 & 1
\end{bmatrix}
\right)
\begin{bmatrix}
A+jB & 0 \\
0 & A-jB
\end{bmatrix}
\left(
\frac{1}{\sqrt{2}}
\begin{bmatrix}
-j & 1 \\
j & 1
\end{bmatrix}
\right),
\tag{12.11}
$$

substituting it in (12.10) will finally lead to

$$
\begin{bmatrix}
p_{q0(N)} \\
p_{q1(N)}
\end{bmatrix}
= 2^{N}
\begin{bmatrix}
(1+j)(A+jB)^{-N} + (1-j)(A-jB)^{-N} \\
(1-j)(A+jB)^{-N} + (1+j)(A-jB)^{-N}
\end{bmatrix}.
\tag{12.12}
$$

Finally, the transfer function of $p_{q1(N)}$ divided by $p_{q0(N)}$ can be obtained as

$$
\frac{p_{q1(N)}}{p_{q0(N)}} =
\frac{
j + \left(\dfrac{1+ah-j(a-b)h}{1+ah+j(a-b)h} \right)^{N}
}{
1 + j \left(\dfrac{1+ah-j(a-b)h}{1+ah+j(a-b)h} \right)^{N}
}.
\tag{12.13}
$$

The criteria for phase error minimization are equivalent to set

$$\frac{P_{q1(1)}}{P_{q0(1)}} = -j, \tag{12.14}$$

which implies a 90° phase shift. Substituting (12.14) into (12.13) leads to

$$\begin{cases} 1 - 2Ga + Gb + \dfrac{f}{f_c} = 0 \\[2mm] 1 - Gb - \dfrac{f}{f_c} = 0 \end{cases}. \tag{12.15}$$

By solving (12.15) the optimal conditions for phase error minimization can be obtained as

$$\begin{cases} a = G^{-1} \\ f = G(a-b)f_c \end{cases}. \tag{12.16}$$

This implies that an ideal 90° phase shift happens at the natural frequency of the circuit: $f_n = G(a-b)f_c$. Also, a larger a/b ratio can provide a stronger phase-correcting ability.

12.3.2 Eight-Phase LO Generation

12.3.2.1 Architecture

Eight-phase LO generation can be obtained by further extending the four-phase LO concept and architecture previously outlined. As such, the block diagram of an eight-phase LO generator can be drawn as illustrated by Figure 12.6. From left to right, an

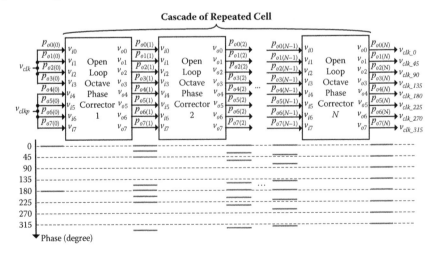

FIGURE 12.6 Block diagram of an eight-phase LO generator.

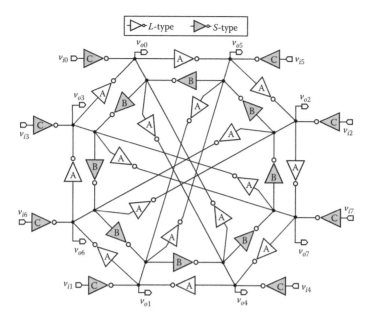

FIGURE 12.7   Architecture schematic of an eight-phase corrector.

eight-phase LO can be composed of multiple octave-phase correctors. The number of stages in cascade will directly depend on the final phase-precision requirement. The architecture schematic of an eight-phase corrector is shown in Figure 12.7, which is composed of 32 inverters (CMOS). Similarly to the four-phase design, the inverters are also classified as L-type or S-type and the three sets (**A, B,** and **C**) previously mentioned are also maintained.

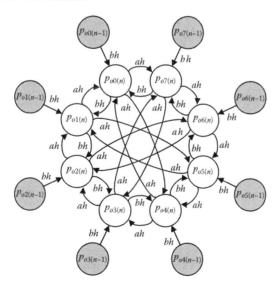

FIGURE 12.8   SFG of an eight-phase corrector.

12.3.2.2 Mathematical Model

The SFG of the eight-phase corrector is presented in Figure 12.8, where a linear model is also assumed and each inverter is modeled, similarly, as a single-pole amplifier. Again, phasor-domain analysis is applied to obtain the phase transformation of the nth octave-phase corrector as given by

$$
\begin{cases}
A_o p_{o(n)} = p_{o(n-1)} \\
A_o = \begin{bmatrix}
1 & -ah & 0 & 0 & -ah & 0 & 0 & -bh \\
-bh & 1 & -ah & 0 & 0 & -ah & 0 & 0 \\
0 & -bh & 1 & -ah & 0 & 0 & -ah & 0 \\
0 & 0 & -bh & 1 & -ah & 0 & 0 & -ah \\
-ah & 0 & 0 & -bh & 1 & -ah & 0 & 0 \\
0 & -ah & 0 & 0 & -bh & 1 & -ah & 0 \\
0 & 0 & -ah & 0 & 0 & -bh & 1 & -ah \\
-ah & 0 & 0 & -ah & 0 & 0 & -bh & 1
\end{bmatrix}, \\
p_{o(n)} = \begin{bmatrix} p_{o0(n)} & p_{o1(n)} & \cdots & p_{o7(n)} \end{bmatrix}^T
\end{cases}
\tag{12.17}
$$

where A_o is the octave-phase transformation matrix and $p_{o(n)}$ is the output phase vector of the nth octave-phase corrector. Rearranging (12.17) will yield

$$
p_{o(n)} = A_o^{-1} p_{o(n-1)}.
\tag{12.18}
$$

For N octave-phase correctors in cascade, $p_{o(N)}$ is given by

$$
p_{o(N)} = A_o^{-N} p_{o(0)},
\tag{12.19}
$$

where $p_{o(0)}$ is the input phase vector

$$
p_{o(0)} = \begin{bmatrix} 1 & 1 & 1 & 1 & -1 & -1 & -1 & -1 \end{bmatrix}^T bh,
\tag{12.20}
$$

since the input phase can be either $0°$ or $180°$. In the phasor domain, only 1 or -1 is available for the input phase vector. The steady-state phase error function can be defined by

$$
\Phi_{eo(n)} = Arg\left(e^{\frac{j\pi}{4}} \frac{p_{o1(n)}}{p_{o0(n)}} \right).
\tag{12.21}
$$

As illustrated by Figure 12.9, the steady-state phase error can be minimized by cascading additional stages of octave-phase correctors for a certain frequency range from 0.7 to 1.3 (normalized frequency: f/f_c). The a/b ratio is set to be 3.5 and G is 10.

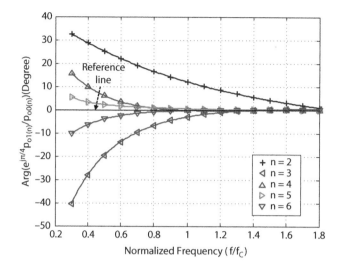

FIGURE 12.9   Static phase error of octave-phase corrector with $a/b = 3.5$ and $G = 10$ (linear model simulation).

Although the optimum conditions in terms of frequency range and phase precision can be derived from the linear model, the final phase permutation can only be determined from transistor-level simulation because not all the transistors can be operated, simultaneously, in the saturation region.

12.4   PRACTICAL DESIGN CONSIDERATIONS

12.4.1   SIZING

Both quadri- and octave-phase correctors have a limited operating frequency range. The channel length of the inverter's transistors is correlated to the upper frequency limit (i.e., a smaller channel length allows a higher operating frequency). When the channel length is fixed, the a/b ratio is correlated to the range and the phase-correcting ability of the phase corrector. A smaller a/b ratio can increase its operating range at the expense of a weaker phase-correcting ability. Thus, subject to different applications, the optimum a/b ratio should be chosen such that the desired frequency range can be covered. On the other hand, the optimum number of phase correctors needed in cascade can be determined according to the required phase precision.

12.4.2   PVT VARIATIONS

Similarly to the ring oscillator, the frequency range covered by the phase corrector can be sensitive to process, voltage, and temperature (PVT) variations. For fast-fast (FF)/slow-slow (SS) process corner with temperature and voltage variations, the covered frequency range is shifted up or down significantly. For a reliable design, the channel length of the inverters is determined at "SS corner + low supply voltage + high temperature" for the highest operating frequency to be larger than the desired frequency. Then, a suitable a/b at "FF corner + high supply voltage + low temperature" is chosen for the lowest operating frequency, which is also lower than the desired frequency. Increasing the width of the transistors can only lead to a better variability control at the expense of power. The operating principle is not dependent on the transistors' width.

12.4.3 DESIGN AND VERIFICATION FLOW

The design and verification flow is graphically summarized in Figure 12.10(a) and (b). For simplicity, only the quadriphase LO generator is considered. Based on the developed linear model and equations, the phase error vector can be determined with initial values of G, a/b ratio, and n. These values can be adjusted to minimize the phase error over the desired frequency range with fast simulation speed. The obtained circuit parameters are then transferred to the transistor-level design. Since the circuit is dynamic, the optimization involves mostly transient simulations—except for the particular case of the phase noise, which was checked through periodic noise (pnoise) simulations. Although the linear model can provide a set of parameters that are close to the optimum values, transistor-level fine tuning is still necessary to account for PVT variations. Circuit nonlinearity may also affect the final phase permutation and it must be confirmed at transistor level. Finally, the optimized circuit can be transferred to the layout design phase. Postlayout verification with parasitic effects is needed to reconfirm all performance metrics and it will be repeated until all specifications are met.

12.5 DESIGN EXAMPLE: A FOUR-/EIGHT-PHASE LOG FOR A MOBILE TV RECEIVER IN 65 NM CMOS

A four-/eight-phase LOG was fabricated with a mobile TV receiver in 65 nm CMOS [15,16]. The entire scheme is shown in Figure 12.11. It is based on two chains of eight-phase corrector (8PC) for achieving harmonic rejection (HR) in the VHF-III and UHF bands, and one chain of four-phase corrector (4PC) for simple quadrature down-conversion in the L-band. A high-level algorithm has been developed in MATLAB to optimize the device size, number of stages, and frequency range using the developed linearized model. This dedicated coverage lowers the tuning range of each phase corrector chain, thereby lowering the phase error. The entire LOG is implemented with thin-oxide MOSFETs (metal-oxide semiconductor field-effect transistors) operating at a 1.2 V supply. Selectors with a logic arrangement feature assign the correct LO phase to each mixer in different modes.

The injection signal LO_{in} might cover the desired bands by using a 1.27 to 1.92 GHz PLL + VCO (not integrated in this work) with selectable output division ratios (1, 2, 3, 6, and 8). The required division ratios can be of *any* number as they are unrelated to the phases of LO_{out}. In the L-band, the PLL + VCO works at the same RF frequency. The phase precision of the LOG is mainly limited by the routing in the layout, where the transistor intrinsic RC value is sensitive to the parasitics.

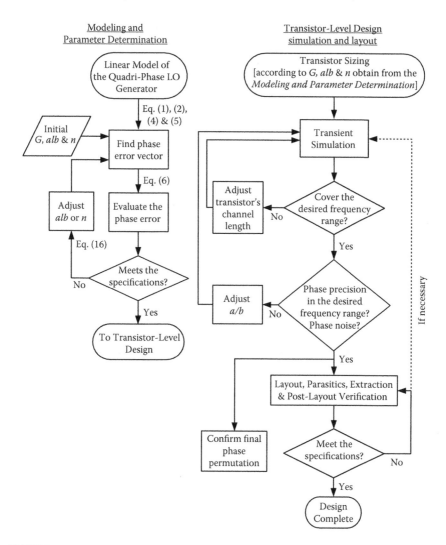

FIGURE 12.10   Design and verification flow: (a) modeling and parameter determination and (b) transistor-level design, simulation, and layout.

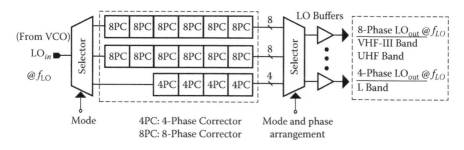

FIGURE 12.11   A direct injection-locked four-/eight-phase LOG.

The LOG was laid out and extracted to tune out this effect in several iterations. In the postlayout simulation (PLS), the phase error for eight-phase output is optimized to be <1° for the VHF-III and UHF bands. This precision fairly meets our target of HR ratio of 33 dB, even under a possible gain mismatch up to 5%. For the four-phase output, the phase error is controlled to be <1.5° for the L-band, which corresponds to an image rejection ratio of around 38 dB for a possible gain mismatch up to 2.5%.

Similarly to the ring oscillator, the robustness of the phase corrector can be improved by adopting a supply regulator and a bandgap reference to cope with the voltage and temperature variations, respectively. Simulations show that the regulator should stabilize the power supply with less than 80 mV fluctuation such that the phase noise of the multiphased LO will not be degraded by more than 1 dB at 1 MHz frequency offset. Those schemes are under development and have not been included in this work.

In the receiver measurements, the third- and fifth-order HR ratios are 35 and 39 dB, respectively, confirming the high phase precision of the LOG. Note that, to increase the accuracy, the data are averaged values of 12 samples with σ of 1.8 and 2.4 dB, respectively.

The phase noise performance of the LOG was characterized using a signal generator as LO_{in}, as shown in Figure 12.12(a). The measured phase noise (Figure 12.12b) is well below that achieved in Vassilios et al. [17], which uses a practical PLL + VCO as LO_{in}. Thus, the phase noise induced by the LOG should be tolerable when the PLL + VCO is presented.

The injection-locking characteristic of the LOG is of interest. When LO_{in} is deactivated (Figure 12.13a), LO_{out} is free running at its natural oscillation frequency of around 240 MHz and noticeable spurious tones appear throughout the spectrum. With LO_{in} activated with a frequency of 205 MHz, LO_{out} tracks LO_{in} at the same frequency and the dynamic range is >60 dB (Figure 12.13b). On the other hand, the second to sixth harmonics will be suppressed by the differential I/Q mixer with HR. These results demonstrate that a high-purity multiphase LO can be achieved without frequency division.

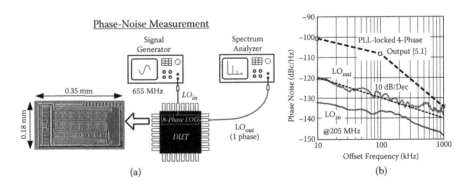

(a)   (b)

FIGURE 12.12   (a) Test setup for phase noise measurement and (b) measured phase noise of the LOG with one of the eight-phase outputs at 205 MHz.

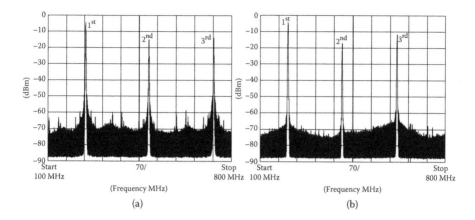

FIGURE 12.13   Measured injection-locking characteristic of the LOG: (a) free running at ~240 MHz; (b) injection locked at the desired 205 MHz.

12.6   CONCLUSIONS

Multiphase LO generators are critical blocks in modern wireless/wireline communication systems. This chapter has covered the design of an open-loop, inverter-only, four-/eight-phase LO generator. The corresponding mathematical models are developed so as to be easy to synthesize at a higher level, optimizing the device sizing with respect to phase error, power, and locking range. A four-/eight-phase LOG fabricated with a wireless receiver in 65 nm CMOS combines two chains of eight-phase correctors and one chain of four-phase corrector to cover the full-band mobile TV. Experimental results with the receiver prove that the third- and fifth-order HR ratios are 35 and 39 dB, respectively, confirming the high phase precision of the LOG, while adding relatively low phase error. The direct injection locking significantly reduces the spurious tones appearing throughout the spectrum.

REFERENCES

1. D. Y. Jeong, S. H. Chai, W. C. Song, and G. H. Cho, CMOS current-controlled oscillators using multiple-feedback-loop ring architectures. *IEEE International Solid-State Circuit Conference (ISSCC) Digest,* pp. 386–387, 491, Feb. 1997.
2. J. Y. Chang, C. W. Fan, C. F. Liang, and S. I. Liu, A single-PLL UWB frequency synthesizer using multiphase coupled ring oscillator and current-reused multiplier. *IEEE Transactions on Circuits and Systems—II: Express Briefs,* vol. 56, no. 2, pp. 107–111, Feb. 2009.
3. L. C. Cho, C. Lee, and S. I. Liu, A 1.2 V 37–38.5 GHz eight-phase clock generator in 0.13 mm CMOS technology. *IEEE Journal of Solid-State Circuits (JSSC),* vol. 42, no. 6, pp. 1261–1270, June 2007.
4. A. Rofougaran, J. Rael, M. Rofougaran, and A. Abidi, A 900 MHz CMOS LC-oscillator with quadrature outputs. *IEEE International Solid-State Circuit Conference (ISSCC) Digest,* pp. 392–393, Feb. 1996.

5. X. Gao, E. A. M. Klumperink, and B. Nauta, Advantages of shift registers over DLLs for flexible low jitter multiphase clock generation. *IEEE Transactions on Circuits and Systems—II: Express Briefs,* vol. 55, no. 2, pp. 244–248, March 2009.

6. J. M. Chou, Y. T. Hsieh, and J. T. Wu, Phase averaging and interpolation using resistor strings or resistor rings for multi-phase clock generation. *IEEE Transactions on Circuits and Systems—I: Regular Papers,* vol. 53, no. 5, pp. 984–991, May 2006.

7. M. J. Gingell, Single-sideband modulation using asymmetric polyphase networks. *Electronic Communications,* vol. 48, pp. 21–25, 1977.

8. F. Behbahani, Y. Kishigami, J. Leete, and A. A. Abidi, CMOS mixers and polyphase filters for large image rejection. *IEEE Journal of Solid-State Circuits (JSSC),* vol. 36, no. 6, pp. 873–887, June 2001.

9. K. H. Kim, P. W. Coteus, D. Dreps, S. Kim, S. V. Rylov, and D. J. Friedman, A 2.6 mW 370 MHz to 2.5 GHz open-loop quadrature clock generator. *IEEE International Solid-State Circuit Conference (ISSCC) Digest,* pp. 458–627, Feb. 2008.

10. K. H Kim, D. M. Dreps, F. D. Ferraiolo, P. W. Coteus, S. Kim, S. V. Rylov, and D. J. Friedman, A 5.4 mW 0.0035 mm^2 0.48 psrms-jitter 0.8 to 5 GHz non-PLL/DLL all-digital phase generator/rotator in 45 nm SOI CMOS. *IEEE International Solid-State Circuit Conference (ISSCC) Digest,* pp. 98–99, 99a, Feb. 2009.

11. K.-F. Un, P.-I. Mak, and R. P. Martins, An open-loop octave-phase local-oscillator generator with high-precision correlated phases for VHF/UHF mobile-TV tuners. *Proceedings of IEEE International Symposium on Circuits and Systems (ISCAS),* pp. 433–436, May 2009.

12. K.-F. Un, P.-I. Mak, and R. P. Martins, Analysis and design of open-loop multi-phase local-oscillator generator for wireless applications. *IEEE Transactions on Circuits and Systems—I: Regular Papers,* vol. 57, no. 5, pp. 970–981, May 2010.

13. A. Rezayee and K. Martin, A three-stage coupled ring oscillator with quadrature outputs. *Proceedings IEEE International Symposium on Circuits and Systems (ISCAS),* pp. 484–487, May 2001.

14. P. R. Gray and R. G. Meyer, *Analysis and design of analog integrated circuits,* 3rd. ed. New York: Wiley, 1993.

15. P.-I. Mak and R. P. Martins, A 0.46 mm^2 4 dB NF unified receiver front-end for full-band mobile TV in 65 nm CMOS. *IEEE Journal of Solid-State Circuits,* pp. 1970–1984, Sept. 2011.

16. P.-I. Mak and R. P. Martins, A 0.46 mm^2 4 dB NF unified receiver front-end for full-band mobile TV in 65 nm CMOS. *IEEE International Solid-State Circuit Conference (ISSCC) Digest,* pp. 172–173, Feb. 2011.

17. I. Vassilios, K. Vavelidis, N. Haralabidis, et al., A 65 nm CMOS multistandard, multiband TV tuner for mobile and multi-media applications. *IEEE Journal Solid-State Circuits,* vol. 43, pp. 1522–1533, July 2008.

13 On-Chip Accelerometers Using Bondwire Inertial Sensing

Yu-Te Liao, William Biederman, and Brian Otis

CONTENTS

13.1 INTRODUCTION

Inertial sensors are widely used to measure physical velocity, acceleration, tilt, and vibration, allowing applications in a diverse range of industries, such as automotive, consumer electronics, and healthcare monitoring. Consequently, accelerometers are one of the fastest growing markets in inertial sensing. One of the principal driving forces for the accelerometer market is the automotive industry, which has now incorporated accelerometers in air bags, electronic stability control systems, and navigation systems in virtually every vehicle. Accelerometers have also started to see wide use in handheld devices for orientation, tilt, and shock detection. This has opened the application space to smart phones, video game controllers, and health-care instrumentation. Applications for accelerometers in the healthcare sector are still emerging, but are currently used to monitor the characteristics of physical movements to improve the quality of life—for example, fall detection for the elderly [1,2] and tremor detection for patients with Parkinson's disease [3].

Each of the aforementioned market applications for accelerometers has very different design requirements. Automotive applications usually require large dynamic ranges (crash detection), large bandwidth, and/or high accuracy (navigation) while operating in a rigid environment. Consumer electronics and healthcare applications typically require more moderate performance. For example, most human physical activity is within acceleration amplitudes of ±12 g and at frequencies less than 20 Hz [4]. Table 13.1 summarizes these observations and shows the acceleration and frequency ranges for various applications [4–7].

Many different types of accelerometers exist today, such as piezoelectric, piezoresistive, and capacitive. Each type of accelerometer has different performance characteristics and applications. For example, piezoresistive and piezoelectric accelerometers are usually used as vibration and shock sensing devices due to their high acceleration detection range (>100,000 g) with a bandwidth over 20 kHz [8]. Furthermore, capacitive microelectro-mechanical (MEMS) inertial sensor technology was introduced in the 1980s, allowing a reduction in the sensor size. Analog

TABLE 13.1

Summary of the Acceleration Amplitude and Frequency Ranges for Different Applications

Applications	Acceleration amplitude	Frequency
Automotive [5]		
Navigation	±2 g	<20 Hz
Air bag	±50 g	~400 Hz
Ride control	±2 g	DC-10 Hz
Antilock brake system	±1 g	0.5–50 Hz
Human movement [4,6]		
Head movement	0.5–9 g	3.5–8 Hz
Hand movement	0.5–9 g	< 12 Hz
Finger movement	0.04–1 g	< 12 Hz
Walk	±2 g	< 20 Hz
Jump/run	±12 g	< 20 Hz
Biomedical activity hand [3]		
Tremor	±5 g	Normal: 9–25 Hz
		Essential: 4–12 Hz
		Parkinson's: 3–8 Hz (at rest)

Devices was the first group to develop a commercial MEMS accelerometer in 1985, which was used for an air bag system. Recently, MEMS accelerometers have demonstrated microgravity resolution, high stability, and linearity through force feedback [9–11]. Existing MEMS accelerometers provide high-quality inertial sensing but require complicated sensor fabrication (bulk/surface machining) and packaging at the wafer level. The complex postprocessing required to integrate the MEMS sensor and electronic IC on the same silicon substrate is the major cost limitation of accelerometer manufacturing and constrains the flexibility and the size of the proof mass.

The ability to fabricate a low cost sensor is the key factor in the widespread use of deploying ubiquitous wireless sensor networks. With advances in silicon technology, traditional electronics with data processing, wireless technology, and memory can be integrated into a single square millimeter-sized IC, with a fabrication cost of a few cents. Thus, integrating a sensor with these active circuits, on a CMOS (complementary metal oxide semiconductor) process, is mandatory for low cost sensor node implementation. However, it is impossible to fabricate a free-moving structure in silicon using currently available IC design foundry services. Therefore, to implement an accelerometer that can be fabricated in a standard CMOS technology, we have proposed sensing acceleration using bondwires [12], which are widely used to connect a silicon chip to its package.

In Section 13.2, we will explore the mechanical and electrical characteristics of bondwires and present a bondwire model that reveals the performance implications on a bondwire inertial sensor. The corresponding bondwire model is verified using finite element method simulations and the results are presented. The bondwire inertial sensing theory and circuitry are shown in Sections 13.3 and 13.4, respectively. A single-axis bondwire accelerometer prototype is shown in Section 13.5. Finally, a conclusion and future research directions are discussed in Section 13.6.

13.2 MECHANICAL AND ELECTRICAL PROPERTIES OF A BONDWIRE INERTIAL SENSOR

Wirebonding, shown in Figure 13.1, is a chip-to-package/board interconnection technique where fine metal wires (gold, aluminum, or copper) are connected from a silicon chip's I/O pads to the associated package pins. This technique provides many advantages of high flexibility, low defect rate (less than 100 ppm), programmable bonding cycles, and established instrument support. Consequently, wirebonding is widely used in low cost and large volume IC assembly. In this section we will review the various properties of bondwires and explain how they influence the metrics for a bondwire inertial sensor, including: mechanical sensitivity, bandwidth, linearity, isolation between axes, and resolution.

13.2.1 INERTIAL MECHANICAL SYSTEM

The differential force equation of a mechanical system can be derived from the following:

$$f = ma = m\ddot{x} + b\dot{x} + kx \tag{13.1}$$

FIGURE 13.1   An example of chip-on-a-board assembly using gold bondwires.

where m is the mass, a is the acceleration, k is the spring constant, and b is the damping constant of the system. The displacement transfer function can be solved by using the Laplace transform and can be shown to be

$$\frac{X(s)}{A(s)} = \frac{1}{s^2 + \dfrac{\omega_r}{Q} s + \omega_r^2} \tag{13.2}$$

where ω_r is the natural frequency and Q is the quality factor of the system. The resonant frequency is equal to $\sqrt{k/m}$ and Q is equal to $\omega_r \dot{m}/b$. When the acceleration frequency is well below the resonant frequency, the displacement is $1/\omega_r^2$, which is proportional to acceleration. For example, if a mechanical system has a 1 kHz mechanical resonant frequency, this yields a displacement sensitivity of 248 nm/g (where g is the acceleration due to gravity, 9.8 m/s²), approximately. Further, (13.2) also reveals a fundamental trade-off between bandwidth and displacement in the accelerometer design: A lower resonant frequency yields a higher acceleration sensitivity but reduces the sensor bandwidth.

13.2.2   BONDWIRE MATERIALS

Gold, aluminum, and copper are the most common materials for bondwires. Table 13.2 shows the relevant mechanical and electrical property for these materials. Gold has

TABLE 13.2
Bondwire Materials

Materials	Density $(g \cdot cm^{-3})$	Young's modulus (GPa)	Electrical conductivity $(S \cdot m^{-1})$
Aluminum	2.7	70	3.7×10^7
Gold	19.3	79	4.5×10^7
Copper	8.94	128	5.9×10^7

the highest density and is the most robust when exposed to environmental variables. Aluminum has the lowest density and electrical conductivity and is widely used in low cost, low pressure ultrasonic wirebonding. Copper has the highest electrical conductivity and stiffness; however, it oxidizes easily, which may affect reliability.

13.2.3 SENSITIVITY OF A BONDWIRE ACCELEROMETER

A bondwire typically has a length of 1–5 mm, a diameter of 0.7–2 mil, and traces an approximately parabolic arc between the chip and package. To simplify calculations, the bondwires can first be modeled as a semicircular arch. The peak displacement at the apex of a semicircular trace can be calculated by

$$\Delta X = \frac{\rho R^4}{E r^2} \cdot (\pi^2 + 2\pi) \tag{13.3}$$

where R is the radius of the semicircle, r is the radius of the bondwire, E is the Young's modulus, and ρ is the density.

Figure 13.2(a) shows the calculated results of a semicircular gold bondwire from (13.2), which is verified by the FEM simulation results. From 13.3, the displacement sensitivity is proportional to the density of materials and inversely proportional to the square of the radius. Therefore, by utilizing a different material or configuration (radius, length, and height) between two bondwires, we can create a relative displacement between the bondwires during acceleration. To create a large relative displacement between two bondwires, the material for one bondwire was chosen to be aluminum and the other gold. Gold is about seven times denser than aluminum, which yields deflections up to seven times larger for the same bondwire size and acceleration. Figure 13.2(b) shows the bondwire model and the FEM simulation results for the relative displacement between gold and aluminum bondwires (3.5 mm length, 12.5 µm radius, and 0.5 mm height). The simulated sensitivity of the gold-aluminum bondwire acceleration sensor is 32 nm/g in the major axis (X).

13.2.4 CROSS-AXIS SENSITIVITY

Cross-axis sensitivity defines the coupling from one axis to the other orthogonal axes while applying acceleration signals. The cross-axis sensitivity can be calculated by

$$S = \frac{\sqrt{S_{y-x}^2 + S_{z-x}^2}}{S_x} \tag{13.4}$$

where S_{y-x} is the coupling sensitivity from the Y-axis to the X-axis. Intuitively, the structure is most compliant in the X-axis and much stiffer in the Y- and Z-axes. FEM simulation models the coupled displacement at 1 g acceleration in the Y- and Z-axes to be 0.003 and 0.0005 nm for a 3 mm long bondwire, respectively. The displacement in the X-axis is 14.36 nm per 1 g acceleration, yielding a coupling sensitivity of less than –36 dB.

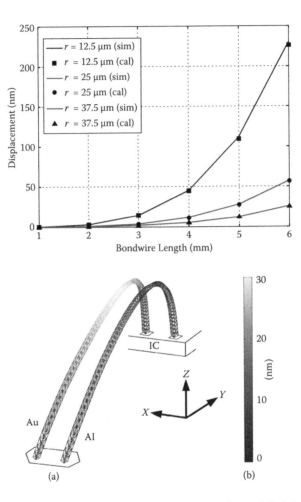

FIGURE 13.2 (a) Calculation and FEM simulation results of a semicircle bondwire struc-ture; (b) FEM model of bondwire sensor.

13.2.5 BANDWIDTH

In an open loop inertial sensor system, the mechanical resonant frequency (ω_r) of the proof mass determines the bandwidth of the acceleration sensor. To simplify the calculation for the mechanical resonant frequency of a bondwire, the bondwire can be modeled as a parabolic arc. Using this approach, the resonant frequency of a bondwire was derived [13]:

$$\omega_r = C_n \frac{1}{l^2 r} \sqrt{\frac{EI}{\rho}} \tag{13.5}$$

where I is the second moment of inertia, l is wire length, E is the Young's modulus, ρ is material density, and C_n is a constant for a given vibration mode. Note that C_n is related to l/h, where h is the height of the bondwire arc. For a circular cross section, the second moment of inertia is

$$I = \frac{\pi r^4}{4} \tag{13.6}$$

Therefore, the mechanical resonant frequency is proportional to the bondwire radius and inversely proportional to the bondwire length. Figure 13.3(a) shows the FEM simulated frequency response of a pair of gold and aluminum bondwires (3.5 mm length, 12.5 μm radius, and 0.5 mm height). The calculated mechanical resonant frequency is ~2.9 and ~8.5 kHz for gold and aluminum bondwires,

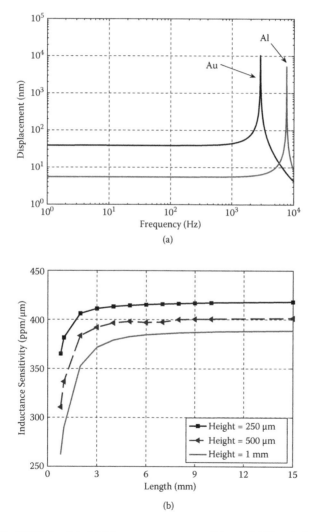

FIGURE 13.3 FEM simulated (a) frequency response of Au and Al bondwires; (b) inductance sensitivity versus length.

respectively. The bandwidth of the sensor system is therefore determined by the lower resonant frequency of the denser gold bondwire.

13.2.6 RESOLUTION AND NOISE FLOOR

There are two major noise sources in an accelerometer: (1) mechanical noise and (2) electronic interface noise. Brownian motion creates the mechanical noise of a structure, which is a type of mechanical-thermal noise caused by molecular collisions from the surrounding environment. The noise source represents the fundamental noise limit of an inertial sensor. The Brownian noise equivalent acceleration (BNEA) can be calculated using [14]

$$BNEA = \frac{\sqrt{4K_BTD}}{M_a} = \sqrt{\frac{4K_BT\omega_r}{Q_{Mech}M}} \tag{13.7}$$

where K_B is Boltzmann constant, T is the absolute temperature, D is damping constant, M_a is proof mass, and Q_{Mech} is the mechanical quality factor. To reduce the mechanical noise in an accelerometer system, a large proof mass, a high mechanical quality factor, and a low mechanical resonant frequency are needed. For instance, a 3.5 mm length and 25 μm diameter gold bondwire has a mass of 30 μg, a resonant frequency of 3 kHz, and a Q_{Mech} of 200, which leads to a BNEA noise density of $0.73\mu g / \sqrt{(Hz)}$ at room temperature. Since this noise density is so small, electrical interface noise usually dominates the resolution of the acceleration sensor system. Therefore, minimizing electronic noise is critical. An analysis of electrical interface noise will be presented in Section 13.3.

13.2.7 BONDWIRE INDUCTANCE

Bondwires can appear as explicit or parasitic inductors. They have a larger surface area per unit length compared to planar spiral inductors and are further from the ground plane. This increased distance from the ground plane reduces parasitic capacitance and substrate losses, resulting in a high Q inductor. The electrical properties of bondwires depend on their physical dimensions: the height above the die plane, the horizontal length, and the distance between other adjacent bondwires. When we ignore the nearby conductor effects, the self-inductance of a bondwire is [15]

$$L = \frac{\mu_o l}{2\pi}\left[ln\left(\frac{2l}{r}\right) - 0.75 \right] \tag{13.8}$$

The mutual inductance of two parallel bondwires with equal length is approximately

$$M = \frac{\mu_o l}{2\pi}\left[ln\left(\frac{2l}{d}\right) - 1 + \frac{d}{l} \right] \tag{13.9}$$

where d is the distance between them. For a pair of bondwires carrying a differential current, the total inductance is

$$L_t = L - M \approx \frac{\mu_o l}{2\pi}\left[ln\left(\frac{d}{r}\right) + 0.25 - \frac{d}{l} \right] \qquad (13.10)$$

The inductance sensitivity to displacement is defined by

$$S_{ind} = \frac{L_t(d + \Delta d)}{L_t} \qquad (13.11)$$

and is plotted in Figure 13.3(b). Both the total inductance and mutual inductance increase with the length of bondwire; however, when the length becomes excessively large

$$ln\left(\frac{d}{r}\right) + 0.25 >> \frac{d}{l} \qquad (13.12)$$

the inductance sensitivity becomes independent of the length. Thus, there is an optimal choice for length of the bondwire sensor, which will maximize sensitivity and minimize the sensor size.

Note that this calculation is used to illustrate an idea and models a pair of parallel bondwires that undergo the same fixed relative displacement across their entire length. In practice, the peak displacement only occurs at the apex of the bondwire since the two ends are fixed. Thus, the FEM simulation results of two parabolic bondwires with fixed ends shows a much smaller inductance sensitivity (Figure 13.3b) than what is predicted from calculation. The bondwire spacing is set by the bondpad pitch used in the CMOS fabrication process, so it is typically not a design parameter. In our process (0.13 µm), the on-chip bond pad spacing is about 90 µm, which results in a 390 ppm/µm inductance sensitivity (FEM simulation) for a 3 mm length bondwire.

13.2.8 TEMPERATURE AND FABRICATION VARIATION

Bondwire inductance varies with temperature and package assembly (due to the bonding strength and machine tolerances). The temperature coefficient (TC) of a bondwire inductor comes from (1) the linear expansion of the wire with increasing temperature and (2) the change in the contribution of the internal flux to the total inductance [15]. The internal flux is inversely proportional to frequency due to the skin effect reducing the effective conductor area. Usually, the total inductance of a bondwire inductor has a TC of roughly 50–70 ppm/°C at 1 GHz. In addition, the TCs for the resistivity of gold and aluminum are 3400 ppm/°C and 3900 ppm/°C, respectively.

In addition to temperature variations, the bonding process can affect bondwire inductance since a bondwire's inductance depends on its structure and geometry. To enhance the precision of bonding and eliminate board/package stress, the sensing gold and aluminum wires can be bonded between pads on the same chip (instead of chip to package), which more precisely controls the length and separation of the bondwires. Chip-to-chip wire bonding can reduce inductance deviation to ~6% [16], which is comparable to the variation of on-chip inductors and capacitors in a standard IC technology design (>10%). This value is tolerable in standard IC design since on-chip capacitors and inductors typically have a variation larger than 10%.

13.3 ELECTRICAL READOUT INTERFACE FOR BONDWIRE ACCELEROMETER

13.3.1 System Architecture for Bondwire Inertial Sensing

Figure 13.4 shows the diagram of the proposed sensing scheme using a bondwire inertial sensor. A dense and relatively elastic gold (Au) wire is used in conjunction with a stiff and less dense aluminum (Al) wire to create an inertial sensor. The difference in material properties creates a relative deflection between the two bondwires during acceleration. Despite the fact that Au and Al have a similar Young's modulus, Au is 7.14 times denser than Al, resulting in a much greater bondwire displacement for the same applied acceleration. The relative displacement of adjacent bondwires changes their mutual inductance, which is then converted to a frequency deviation using an oscillator. Due to the small displacements of the bondwire and the corresponding small changes in inductance under acceleration, the oscillator frequency can be linearly approximated as

$$\omega = \frac{1}{\sqrt{C(L_t + \Delta L)}} = \frac{1}{\sqrt{L_t C}} (1 - \frac{\Delta L}{2L_t} \dots) \qquad (13.13)$$

where C is an on-chip capacitance that is independent of acceleration. For example, for a 2 GHz bondwire oscillator with an inductance sensitivity of 350 ppm/μm and a

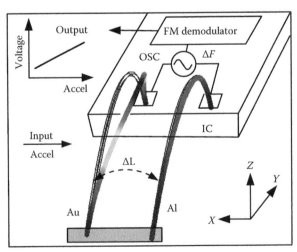

FIGURE 13.4 System diagram of the proposed bondwire inertial sensor.

30 nm relative displacement due to a 1 g acceleration, the center frequency will shift by 10.5 kHz, which is detectable by an on-chip demodulator. In addition, the oscillator has a large amplitude output (hundreds of millivolts), which is less sensitive to induced noise in the following stages. Further, the frequency modulated signal also offers immunity to amplitude noise.

13.3.2 ELECTRICAL NOISE IN READOUT CIRCUITRY

In a frequency modulated (FM) system, frequency uncertainty can be described by phase noise, Allan variance, and residual frequency noise. Phase noise represents the short-term frequency stability, whereas the Allan variance signifies the long-term frequency stability in the time domain. Residual frequency noise represents the RMS frequency fluctuation, which can be derived from the phase noise measurement.

13.3.3 PHASE NOISE

In an LC oscillator design, losses in the resonant tank, periodic variations of the active device parameters, and nonlinearity of devices can all contribute the amplitude and phase fluctuations. To simplify the analysis, the phase noise of an oscillator can be approximately modeled by Leeson's phase noise formula [17]:

$$\pounds(f_m) = \left[1 + \left(\frac{1}{2Q} \frac{f_{osc}}{f_m} \right)^2 \right] \frac{2FK_bT}{P_{rf}} \tag{13.14}$$

where F is the oscillator noise factor, Q is the quality factor of the resonant tank, T is the absolute temperature, and P_{rf} is the average power at the output of the oscillator. Phase noise improves linearly as carrier power increases and quadratically as Q increases. Bondwire inductors typically have a Q factor of 30–70 at frequencies in the low gigahertz, resulting in excellent oscillator noise performance.

13.3.4 ALLAN VARIANCE

Biomedical signals usually require a low sampling rate; therefore, long-term frequency stability is also important to sensor performance. The time domain frequency stability measurement is based on the statistical analysis of the phase/frequency fluctuation as a function of time. The Allan variance can be described as

$$\sigma^2(\tau) = \frac{1}{2} \overline{\left(\overline{y_{n+1}(\tau)} - \overline{y_n(\tau)} \right)^2} \tag{13.15}$$

where $\overline{y_n}$ is the nth average over the observation time τ. Figure 13.5 shows an example Allan variance plot and is displayed in a log-log scale. Five basic phase and frequency noise sources contribute noise in three regions of the Allan variance plot. The regions define acceleration sensor noise performance in terms of resolution, bias stability, and drifts from environmental effects, such as temperature and aging.

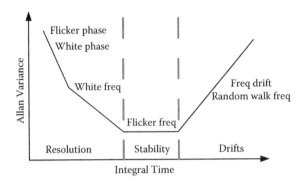

FIGURE 13.5 Allan variance plot.

Frequency/phase domain and time domain conversions can be expressed between the slopes of phase noise and Allan variance [18,19]. With respect to sensor design, the Allan variance plot shows the optimized integral sampling time for achieving minimum resolutions by minimizing the flicker frequency noise.

13.3.5 RESIDUAL FM NOISE

Residual frequency is an important specification that defines the RMS value of the frequency fluctuation within a frequency range. It can be derived from the phase noise of an oscillator as

$$\Delta f = \sqrt{2 \int_{f_1}^{f_2} \pounds(f_m) f_m^2 \, df_m} \tag{13.16}$$

where $\pounds(f_m)$ is the single sideband (SSB) noise power spectral density, and f_m is the offset from the carrier frequency. The integration in (13.16) can be performed using a known phase noise profile within the sensor bandwidth. Assuming the gain of the accelerometer is 10 kHz/g and a phase noise slope of −20 dB/decade, calculations show that the oscillator requires a phase noise of approximately −50 dBc/Hz at 10 kHz offset frequency to achieve a full bandwidth (5 kHz) resolution of 0.1 g. Increasing the power consumption of the oscillator, using a higher Q resonant tank, or limiting the bandwidth of the sensor signal can further improve the resolution.

13.4 CIRCUIT IMPLEMENTATION

Figure 13.6 shows the overall circuit block diagram of the implemented inertial sensor readout system. The sensor readout circuit consists of a 2 GHz sensing oscillator (VCO1), a phase locked loop (PLL), mixers, baseband amplifier, a bandgap temperature-stabilized bias circuitry, and a low dropout (LDO) supply regulator. One major difference between the sensor readout architecture and a traditional FM radio receiver is the input signal strength. Our sensor architecture needs to handle a fixed large signal input, where an FM radio requires wide dynamic range and the signal

Bondwire inertial sensor

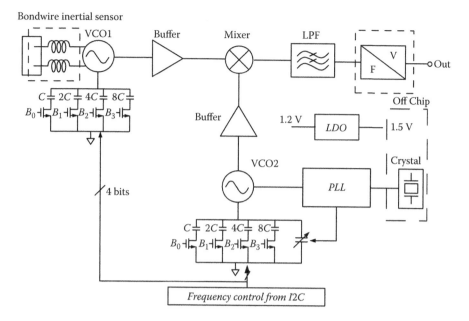

FIGURE 13.6   System blocks of readout circuitry.

is usually small. The large input signal amplitude makes the sensor architecture less susceptible to noise from the down-conversion stages. In addition, using an FM detection architecture provides more relaxed requirements for the linearity of the mixer and IF amplifiers. However, one of the major design concerns with this architecture is signal locking between oscillators. To eliminate this effect, the parasitics in the signal path and substrate coupling must be minimized and the two oscillator frequencies must be sufficiently different. Therefore, the frequency-modulated signal (2 GHz ± 10 kHz) was set 20 MHz away from the stable reference frequency from PLL. Once the signals are down-converted (20 MHz ± 10 kHz), they can be easily detected and digitalized for further signal processing.

A cross-coupled differential oscillator is chosen to suppress common-mode noise. To reduce noise up-conversion due to nonlinearity, varactors were avoided. In order to compensate for variations in the wirebonding process and fabrication, the oscillator is tunable from 1.5 to 2.5 GHz by utilizing a 4-bit discrete capacitor bank. To further enhance the power supply rejection ratio (PSRR) and reduce the temperature dependency of the sensing oscillator's frequency, an on-chip bandgap reference and low dropout regulator (LDO) are used to convert a 1.5 V external battery supply to an on-chip 1.2 V supply. The regulator can accommodate a load current up to 16 mA with only a 10 mV drop from the desired 1.2 V output. The maximum voltage deviation is 7 mV across temperature, and the measured low frequency PSRR is 32 dB from 1.35 to 1.55 V.

The clock signal is generated by a third-order type II PLL. The PLL oscillator design is similar to the bondwire sensing oscillator design. A MOS varactor was added to tune oscillator frequency continuously (by about 100 MHz/V) within the sensor frequency band (1.5–2.5 GHz), which is set by a 4-bit capacitor bank. The in-band noise induced by the nonlinearity of the varactor is suppressed by the loop when the oscillator is locked to the crystal frequency reference. Thus, the noise of the entire system is dominated by the free-running sensing oscillator (VCO1). The phase-frequency detector was designed using true single-phase clock logic (TSPC), which has higher speed than static logic and provides less in-band phase noise. A differential charge pump circuit is employed to reduce the common mode noise and minimize the effects of charge injection and signal feed-through. All schematic components are fully integrated except the crystal resonator. It should be noted that, since an accurate time reference is not critical, even the crystal resonator can be eliminated if necessary.

The RF mixer down-converts the modulated RF signal to a low frequency to facilitate demodulation. Unlike conventional FM receivers, the mixer in this design is driven by two large signals. Though this provides noise immunity, it also leads to signal distortion and LO pulling that can overwhelm the input signal. CMOS passive mixers have been demonstrated to produce better linearity and low flicker noise [20]. A two-stage cascaded buffer inserted between the oscillator and the mixer reduces the effect of LO pulling in the sensing oscillator. The third-order active RC low pass filter amplifies the signals and consequently reduces the effect of amplitude-induced errors in the baseband demodulator. The detailed circuit analysis and measurement results can be found in Liao, Biederman, and Otis [12].

13.5   EXPERIMENTAL RESULTS OF THE SINGLE-AXIS BONDWIRE ACCELEROMETER

The bondwire accelerometer, fabricated in a 0.13 μm CMOS process, is completely integrated in a 1×1.1 mm^2 area, except for an 8 MHz quartz crystal reference and

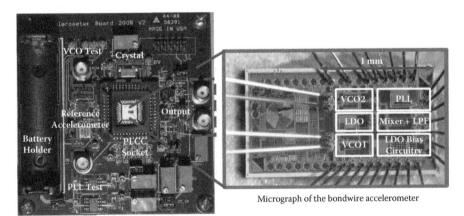

Assembled PCB prototype

Micrograph of the bondwire accelerometer

FIGURE 13.7   Assembled PCB prototype and micrograph of the accelerometer IC. The active area is outlined in white boxes.

bypass capacitors. Figure 13.7 shows the assembled PCB and micrograph of the prototype accelerometer. To reduce the PCB strain or other undesired external forces, the accelerometer chip was mounted in a 44-pin plastic leaded chip carrier (PLCC) package using aluminum/gold wedge wire bonding without encapsulation and was powered by one AA battery on the PCB. Two commercial accelerometers for three-axis and large acceleration measurement were mounted on the PCB near the chip to ensure accurate calibration of the applied acceleration from a shaker table. The prototype PCB was mounted on a custom machined aluminum platform to maximize the mechanical energy transferred from the shaker table to the accelerometers. An oscilloscope and a spectrum analyzer were used to monitor the accelerometer outputs.

Using the shaker table, an acceleration of 1g at 40 Hz was applied along the sensitive axis (X) of the bondwire accelerometer. The output was compared to the commercial accelerometers after normalizing for differences in gain. The comparison result is presented in Figure 13.8(a) and shows that the bondwire accelerometer is capable of real time acceleration tracking. Figure 13.8(b) shows the measured accelerometer frequency response, exhibiting a gain of 10 kHz/g within a 700 Hz bandwidth. The mechanical resonances of the Au and Al bondwires are visible at approximately 3.1 and 8.7 kHz, respectively, which is consistent with the FEM simulation. Figure 13.8(c) shows the measurement results of a three-axis acceleration test on the bondwire accelerometer, revealing a linear gain of 10 kHz/g (matching calculations) and greater than 20 dB isolation between the sensitive axis (X) and nonsensitive axes (Y and Z).

As a control, the same test was performed on a chip that was completely encapsulated in nonconductive epoxy (*Stycast 1266*). The goal of this experiment was to show decisively that the frequency deviations under acceleration were caused by bondwire deflection. The control package exhibited no detectable output from acceleration along any axis. Finally, a 90° rotation test with acceleration only from gravity was performed on non-encapsulated sensor, and the result is plotted alongside the expected sinusoidal response in Figure 13.8(d).

Figure 13.9(a) shows the measured phase noise of the sensing oscillator, PLL, and IF stage, confirming that the accelerometer noise floor is dominated by the free-running sensor oscillator. The bondwires exhibit a high electrical quality factor (Q ~ 40), allowing a low sensor oscillator phase noise of −64 and −121 dBc/Hz at 10 kHz and 1 MHz offsets, respectively, while consuming 2 mA.

The Allan variance is often used to define a clock system and resonant sensor performance [19]. Figure 13.9(b) shows the measured Allan variance of the outputs of the sensing oscillator, PLL, and IF with a sample rate of 1 Hz. The Allan variance gradually flattens out as average time increases. The drift is most likely caused by the temperature and environment fluctuations, which can be removed with feedback control to VCO1. Furthermore, this plot reveals that the accelerometer has a resolution of 80 mg and a bias stability of 35 mg for a 10 s integration window. The noise floor of this accelerometer is limited by the phase noise of the sensing oscillator, not by mechanical noise sources. Accelerometer performance is summarized in Table 13.3.

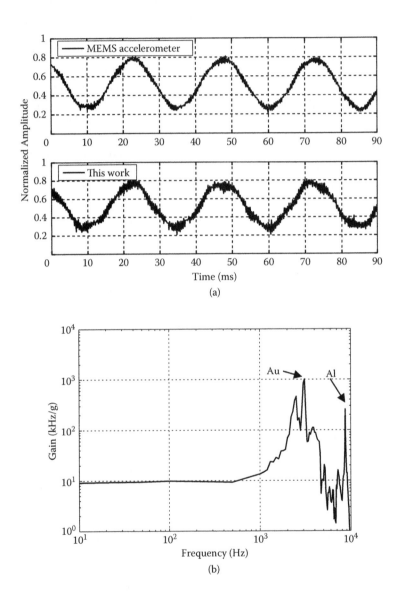

FIGURE 13.8 Measured results of the bondwire accelerometer: (a) waveforms; (b) frequency response

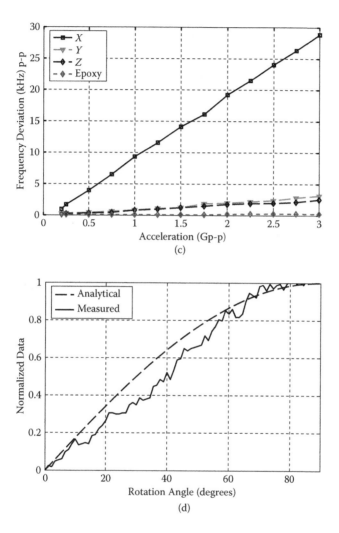

FIGURE 13.8 (CONTINUED) Measured results of the bondwire accelerometer: (c) a three-axis acceleration test; (d) the rotation test.

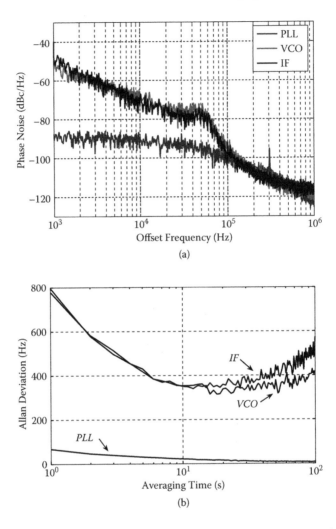

FIGURE 13.9 Measured (a) phase noise and (b) Allan variance at the output of the sensing oscillator, PLL, and IF stages.

TABLE 13.3
Performance Summary

Technology	0.13 μm CMOS
Chip area	1×1.1 mm^2
Sensing oscillator frequency	2.1 GHz
PN at 10 kHz	–64 dBc/Hz
PN at 1 MHz	–121 dBc/Hz
Sensitivity	10 kHz/g
Bandwidth	700 Hz
Axis isolation	>20 dB
Resolution	80 mg
Bias stability	35 mg
Power consumption	13.5 mW

13.6 CONCLUSIONS AND DISCUSSION

Successful deployment of wireless sensors to provide ambient intelligence for healthcare will require a confluence of progress in materials, sensor integration, low power electronics, RF communication, and packaging. CMOS-based sensor systems are promising for many healthcare and biomedical applications due to their robustness and low cost. By combining advanced microelectronics and sensors on a single chip, the system offers unique opportunities to perform high-resolution measurements in a variety of applications. In this chapter, we focused on the design and implementation of the first inertial sensor to use bondwires, which can be readily integrated into any IC process. Bondwires can be incorporated onto chips with existing fabrication techniques without complicated MEMS processes and can potentially be mass manufactured at low cost.

The principle of operation, as well as the mechanical and electrical properties of bondwire accelerometers, were explained in detail in Section 13.2. By exploiting advanced IC technology, a bondwire accelerometer has potential to integrate a wireless transmitter and digital processing, all on a single chip. A single-chip bondwire accelerometer has demonstrated performance suitable for many applications, such as fall detection, body movement monitoring, and orientation display. In the future, further data processing/functions can be added into a square millimeter-sized silicon chip (e.g., real-time functionality configuration and calibration such as bandwidth, resolution, and acceleration sensitivity). With further advances in bondwire design, coating isolation, and packaging, the system could perform many measurement functions, such as mechanical, inertial, or magnetic sensing, at a low fabrication complexity and cost.

REFERENCES

1. A. Lombardi, M. Ferri, G. Rescio, M. Grassi, and P. Malcovati, Wearable wireless accelerometer with embedded fall-detection logic for multi-sensor ambient assisted living applications, in *IEEE Sensors Conference Proceedings,* pp. 1967–1970 Oct. 2009.

2. C.-F. Lai, S.-Y. Chang, H.-C. Chao, and Y.-M. Huang, Detection of cognitive injured body region using multiple triaxial accelerometers for elderly falling, *IEEE Sensors Journal*, vol. 11, no. 3, pp. 763–770, March 2011.

3. S. Patel, K. Lorincz, R. Hughes, N. Huggins, J. Growdon, D. Standaert, M. Akay, J. Dy, M. Welsh, and P. Bonato, Monitoring motor fluctuations in patients with Parkinson's disease using wearable sensors, *IEEE Transactions on Information Technology in Biomedicine*, vol. 13, no. 6, pp. 864–873, Nov. 2009.

4. C. Bouten, K. Koekkoek, M. Verduin, R. Kodde, and J. Janssen, A triaxial accelerometer and portable data processing unit for the assessment of daily physical activity, *IEEE Transactions on Biomedical Engineering*, vol. 44, no. 3, pp. 136–147, March 1997.

5. D. Crescini, D. Marioli, and A. Taroni, Low-cost accelerometers: Two examples in thick-film technology, *Sensors and Actuators A: Physical*, vol. 55, no. 2–3, pp. 79–85, 1996.

6. C. Verplaetse, Inertial proprioceptive devices: Self-motion-sensing toys and tools, *IBM Systems Journal*, vol. 35, no. 3.4, pp. 639–650, 1996.

7. R. N. Stiles, Frequency and displacement amplitude relations for normal hand tremor, *Journal Applied Physiology*, vol. 1, no. 40, pp. 44–54, Jan. 1976.

8. Y. Wang, J. Fan, P. Xu, J. Zu, and Z. Zhang, Shock calibration of the high-g triaxial accelerometer, *IEEE Instrumentation and Measurement Technology Conference Proceedings*, pp. 741–745, May 2008.

9. R. Olsson, K. Wojciechowski, M. Baker, M. Tuck, and J. Fleming, Post-CMOS-compatible aluminum nitride resonant MEMS accelerometers, *Journal of Microelectromechanical Systems*, vol. 18, no. 3, pp. 671–678, June 2009.

10. L. He, Y. P. Xu, and M. Palaniapan, A CMOS readout circuit for SOI resonant accelerometer with 4 μg bias stability and 20-μg / $\sqrt{Hz}$ resolution, *IEEE Journal of Solid-State Circuits*, vol. 43, no. 6, pp. 1480–1490, June 2008.

11. J. Wu, G. Fedder, and L. Carley, A low-noise low-offset capacitive sensing amplifier for a 50 μg/√Hz monolithic CMOS MEMS accelerometer, *IEEE Journal of Solid-State Circuits*, vol. 39, no. 5, pp. 722–730, May 2004.

12. Y.-T. Liao, W. Biederman, and B. Otis, A fully integrated CMOS accelerometer using bondwire inertial sensing, *IEEE Sensors Journal*, vol. 11, no. 1, pp. 114–122, Jan. 2011.

13. H. A. Schafft, Testing and fabrication of wire-bond electrical connections: A comprehensive survey, US National Bureau of Standards, 1973.

14. B. Boser and R. Howe, Surface micromachined accelerometers, *IEEE Journal of Solid-State Circuits*, vol. 31, no. 3, pp. 366–375, March 1996.

15. T. H. Lee, *The design of CMOS radio-frequency integrated circuits*, Cambridge University Press, 2004.

16. J. Craninckx and M. Steyaert, A 1.8 GHz CMOS low-phase-noise voltage-controlled oscillator with prescaler, *IEEE Journal of Solid-State Circuits*, vol. 30, no. 12, pp. 1474–1482, Dec. 1995.

17. D. Leeson, A simple model of feedback oscillator noise spectrum, *Proceedings of the IEEE*, vol. 54, no. 2, pp. 329–330, Feb. 1966.

18. O. Baran and M. Kasal, Allan variances calculation and simulation, *Radioelektronika, 19th International Conference*, pp. 187–190, 2009.

19. N. El-Sheimy, H. Hou, and X. Niu, Analysis and modeling of inertial sensors using Allan variance, *IEEE Transactions on Instrumentation and Measurement*, vol. 57, no. 1, pp. 140–149, Jan. 2008.

20. S. Zhou and M.-C. Chang, A CMOS passive mixer with low flicker noise for low-power direct-conversion receiver, *IEEE Journal of Solid-State Circuits*, vol. 40, no. 5, pp. 1084–1093, May 2005.

14 Design of a Frequency-Shift-Based CMOS Magnetic Sensor Array for Point-of-Care (PoC) Biomolecular Diagnosis Applications

Hua Wang and Ali Hajimiri

CONTENTS

14.1 INTRODUCTION

Future point-of-care (PoC) molecular-level diagnostic systems require advanced biosensors that can offer high sensitivity, ultraportability, and a low price tag. Targeting on-site detection of biomolecules, such as DNAs, RNAs, or proteins, this type of system is believed to play a crucial role in a variety of emerging applications such as in-field medical diagnostics, epidemic disease control, and biohazard detection [1,2].

Conventionally, microarray technology is used to provide sensing information for biomolecules [3]. However, traditional microarray systems rely on optical detection setups for fluorescent molecular tags. This requires bulky and expensive optical devices including multiwavelength fluorescent microscopes, which limit the usage of optical microarray for PoC applications.

Another type of sensor modality, electrochemical biosensors, detects target molecules based on their extra electrical charges or dielectric properties. This includes detection schemes, such as impedance spectroscopy [4], amperometric analysis, redox cycling [5], and cyclic voltammetry [6]. However, electrochemical biosensors are subject to excessive noise at the electrode–electrolyte interface induced by drift and diffusion effects [7]. Moreover, this type of modality is sensitive to the offset and background perturbations, which are exacerbated by in-field measurement environments. These issues limit the typical detection sensitivity to several tens or hundreds of nanomolars [4–6]—orders of magnitude higher than the analyte concentrations in typical biochemical tests.

On the other hand, sensor platforms based on magnetic micro- or nanoparticle labels have been proposed as a promising biosensing scheme to augment or replace these sensing modalities for PoC applications. Affinity-based sandwich assays, such as enzyme-linked immunosorbent assay (ELISA) [8], are typically used in magnetic sensing processes (shown in Figure 14.1). The sensor surface is first coated with the desired molecular probes, which have high affinities with the target molecules. Then the test samples are introduced, and the target molecules are captured by the predeposited molecular probes through the surface chemistry. Finally, the surface-activated micro- or nanomagnetic labels are fed into the system and immobilized onto the sensor surface by the captured target molecules. Therefore, by detecting the

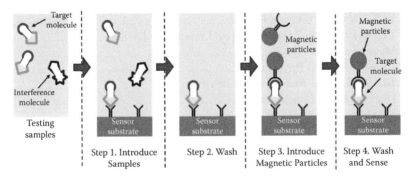

FIGURE 14.1 Typical sensing procedures of a magnetic biosensor using affinity-based sandwich assay.

existence of those magnetic labels, one can infer the presence of the target molecules in the incoming test sample.

In comparison to optical microarrays and electrochemical sensors, magnetic biosensors provide the following advantages: First, they directly eliminate bulky and expensive optical devices, making a low form factor and a low system cost possible. Second, magnetic labels do not have the signal quenching or decaying problems often encountered in fluorescence-based optical detection systems, making the magnetic sensing scheme more robust. Moreover, since most biosamples produce negligible magnetic signals compared to the magnetic labels, magnetic biosensing can achieve a very high signal-to-background-noise ratio. In addition, magnetic particle labels provide a powerful and versatile way of micromanipulation on both the cellular and molecular levels. This can be realized by designing and distributing the excitation electrical currents on-chip [9]; they create a magnetic field and apply forces on the nearby magnetic labels. Based on the high-integration level and the complex digital control capabilities supported in CMOS (complementary metal oxide semiconductor) processes, this on-chip magnetic manipulation concept can potentially be extended to a reconfigurable microfluidic platform particularly useful for tissue engineering.

Currently reported integrated magnetic sensor schemes include giant magneto-resistance (GMR) sensors [10–12], Hall effect sensors [13,14], and nuclear magnetic resonance (NMR) relaxometers [15]. However, these magnetic detection schemes require externally generated magnetic fields to bias the immobilized magnetic labels during their sensing processes. Moreover, expensive fabrication processes such as multilayer metal sputtering and deep dry etching on the passivation layers are demanded in those sensor fabrications. Consequently, these issues affect those systems' ultimate form factors, total power consumption, and manufacturing cost.

To address these impediments, we propose an ultrasensitive frequency-shift-based magnetic biosensing scheme that is fully compatible with standard CMOS processes with no need for costly postprocessing steps or any electrical or permanent external biasing magnets [16]. The core of the sensor is an on-chip low noise LC oscillator, whose oscillation frequency will experience a downshift due to the presence of the magnetic labels on the sensor surface during the detection process. This sensor scheme is conducive to implementing a very large scale magnetic microarray without suffering from significant design complexity and manufacturing cost penalty. This can be achieved by integrating more sensor units onto a single chip and tiling the chips to form the entire sensor system [17]. Therefore, the proposed frequency-shift-based magnetic biosensing scheme presents itself as an ideal solution for PoC molecular-level diagnostic applications.

This book chapter is organized as follows. Section 14.2 focuses on the sensor mechanism. Theoretical modeling on the sensor transducer gain and the fundamental sensor noise floor is demonstrated to yield the sensor signal-to-noise ratio (SNR). The line-width narrowing effect is also presented to justify the advantage of oscillator-based frequency-shift detection. Section 14.3 demonstrates the design optimization to maximize the sensor SNR under various practical implementation constraints. As a design example, an eight-cell frequency-shift-based sensor array realized in a 130 nm CMOS process is demonstrated [18]. The system architecture and the key building blocks are covered in detail in Sections 14.4 and 14.5, respectively. Section

14.6 presents the measurement results of the sensor system for both electrical performance and magnetic sensing. To the authors' best knowledge, this sensor example achieves the best sensitivity among the CMOS magnetic biosensors reported so far.

14.2 SENSOR MECHANISM

14.2.1 SENSOR OPERATION PRINCIPLE AND OSCILLATOR-BASED FREQUENCY-SHIFT DETECTION

The core of our proposed sensing scheme is an on-chip LC resonator (Figure 14.2). The current through the on-chip inductor generates a magnetic field and polarizes the magnetic particles present. This polarization leads to an increase of the total magnetic energy in space and thereby the effective inductance of the inductor. The corresponding resonant frequency downshift of the LC tank is given as

$$f = 12\pi LC = 12\pi L0 + \Delta LC0 \approx f01 - \Delta L2L0, \tag{14.1}$$

where and are the nominal inductance and capacitance, while ΔL is the inductance increase due to the magnetic particles. Therefore, this downshift indicates the existence of the immobilized magnetic particles on the sensor surface, based on which one can infer the presence of the target molecules in the incoming test sample (Figure 14.1).

To detect this frequency downshift, one approach is to measure the LC resonator's impedance in its amplitude and/or phase directly by precision circuits, such as a Wheatstone bridge. However, the line width of the tank impedance is fundamentally determined by its quality factor. For a standard CMOS process, this quality factor is often limited by the on-chip inductor quality factor Q (typically 10 to 20). In contrast, for an on-chip spiral sensing inductor with D_{out} of 100 ~ 200 μm, a single micron-size magnetic particle label typical only induces a part per million (ppm, i.e., 10^{-6}) or even sub-ppm level relative frequency shift. Moreover, to provide a sensitive on-chip

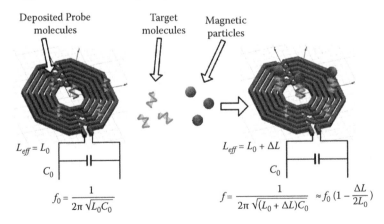

FIGURE 14.2 Proposed frequency-shift-based magnetic sensor scheme.

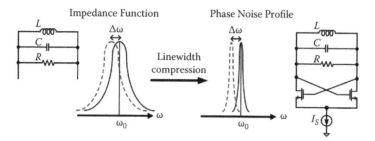

FIGURE 14.3 Line-width compression effect of oscillator phase noise profile compared with passive tank impedance function, assuming that the same LC tank is used.

impedance measurement solution, the system requires fully integrated high-quality test tone generation, frequency sweeping, and analog-to-digital conversions, which add significantly to the design complexity.

On the other hand, if the same on-chip LC tank is implemented as an oscillator's resonator, based on the virtual damping phenomena, the line width of the oscillator's phase noise profile will experience a significant line-width compression effect compared with that of the LC tank impedance function (Figure 14.3) [19]. For a gigahertz range CMOS oscillator, this line-width compression ratio is typically between 10^{-8} and 10^{-9}. Therefore, oscillator-based frequency detection provides an ultrasensitive measurement platform, which can be readily used to discern the sub-ppm level relative frequency shift in sensing the micron or nanometer size magnetic labels. In addition, a simple frequency counter can be implemented locally to the sensor oscillator and directly provides the digital measurement output of the oscillation frequency. Compared with the aforementioned impedance measurement approach, the oscillator-based frequency sensing provides a much more compact and simplified solution conducive to very large scale sensor array implementations with high pixel density.

14.2.2 Sensor Signal-to-Noise Ratio Characterization

To characterize the proposed oscillator-based frequency-shift magnetic sensing platform fully, both its signal response (transducer gain) and the measurement noise floor will be discussed in this section. Subsequently, boldface letters denote vectors and italics denote scalars in the following context.

14.2.2.1 Sensor Transducer Gain

Most off-the-shelf magnetic particle labels are superparamagnetic; induced magnetization vector **M** under external polarization magnetic field **H** can be expressed by a Langevin function and further approximated at room temperature as

$$\mathbf{M} \approx Msat \mu 0 mp 3kT\mathbf{H} = \chi eff\mathbf{H}, \tag{14.3}$$

where *Msat* is the saturation magnetization, $\mu 0$ is the magnetic permeability in vacuum, *mp* is the magnetization factor, and χeff is the effective susceptibility of the superparamagnetic material [20]. The quantity k stands for the Boltzman constant and T for the temperature. Considering the demagnetization effect [21], based on the χeff and the demagnetization factor D, the apparent magnetic susceptibility χapp of the label particles can be calculated; this characterizes how their magnetizations respond to the external polarization field. Since both D and χapp depend on the exact shape of the polarized magnetic objects, they are typically three-dimensional vectors even if the object is made of isotropic magnetic material. But, with a spherical shape assumption for the magnetic particles, both D and χapp can be simplified as scalar quantities [21].

Assume that the electrical current I conducting through the sensing inductor generates an excitation magnetic field of $\textbf{\textit{Hext}}$. If there are N immobilized magnetic particles, each with an equal volume of Vm and located at $\textbf{\textit{ri}} = xi, yi, zi, i \in [1,N]$ in the vicinity of the sensor surface, the total magnetic energy increase in the space due to the particles' magnetic polarizations by $\textbf{\textit{Hext}}$ is given by

$$\Delta Em \approx i = 1N12Vm\chi app\mu 0\textbf{\textit{Hext}}(\textbf{\textit{ri}})2dv \approx i = 1N12\chi app\mu 0\textbf{\textit{Hext}}(\textbf{\textit{ri}})2Vm, \quad (14.4)$$

if the interactions among the adjacent magnetic particles are negligible and $\textbf{\textit{Hext}}(\textbf{\textit{ri}})$ can be treated as a uniform polarization field across the volume Vm for the ith magnetic particle. Therefore, the effective inductance change ΔL is given as

$$\Delta L = 2\Delta Em I2 \approx i = 1N\chi app\mu 0\textbf{\textit{Hext}}(\textbf{\textit{ri}})2Vm I2 = \chi app\textbf{\textit{Bext}}2Vm \times N\mu 0I2, \quad (14.5)$$

where $\textbf{\textit{Bext}}2$ is the spatially averaged excitation magnetic flux density for the N magnetic particles. Consequently, the averaged transducer gain can be defined as

$$\text{Transducer Gain} = \text{Frequency Shift}(\Delta f \,/\, f0) \,\#\, \text{of Magnetic Particles} \quad (14.6)$$

$$= \Delta L2L \cdot 1N = \chi app2\mu 0 \cdot \textbf{\textit{Bext}}2I2L \cdot Vm$$

The preceding equation shows that the sensor signal is composed of three factors multiplied together. The first factor, $\chi app/2\mu 0$, is related only to the magnetic property and the particle shape, while the last factor, Vm, is determined by the particle size. Both of them can be treated as constant values for a given type of magnetic label, assuming the local magnetic field strength does not saturate the particle's magnetic susceptibility. However, the middle factor stands for an excitation magnetic field factor, which is proportional to the magnitude square of the polarization magnetic field per unit current and normalized by the sensing inductance value. Therefore, in order to achieve a large sensor transducer gain, the sensor inductor design should maximize the average excitation magnetic field strength per unit current and at the same time minimize its nominal inductance value.

In addition, the spatial "spot" sensor transducer gain at the location $r = x, y, z$ is given as

$$\text{Transducer Gain } r = \chi app2\mu 0 \cdot \textbf{\textit{Bextr}}2I2L \cdot Vm, \quad (14.7)$$

where $\textbf{\textit{Bextr}}$ is the excitation magnetic flux density at the location $\textbf{\textit{r}}$.

14.2.2.2 Sensor Noise Modeling

Frequency counting can be used to determine the sensing oscillator's frequency shift induced by the immobilized magnetic particle labels. During frequency counting, the total number of transitions M for the oscillator transient voltage waveform is measured within a given counting time window T. The measured frequency f is then given by

$$f = M/T. \tag{14.8}$$

The oscillator's accumulated jitter within the window T presents transition errors of its waveform, which lead to total phase error ϕ_T and the frequency counting error Δf as

$$\Delta f = \phi_T / 2\pi \cdot 1/T = M_{err,T}/T \tag{14.9}$$

where $M_{err,T} = \phi_T/2\pi$ stands for the measurement uncertainty for the number of transitions within time T. Therefore, the noise floor for frequency counting can be formulated based on the sensing oscillator's accumulated jitter σ_T^2 within the counting window T as

$$\sigma_{\Delta f}/f_0^2 = \Delta f^2/f_0^2 = (\phi_T)^2/(2\pi)^2 T^2 \cdot 1/f_0^2 = (\phi_T)^2/\omega_0^2 \cdot 1/T^2 = \sigma_T^2/T^2 \tag{14.10}$$

where Δf is the uncertainty in frequency and f_0 is the nominal frequency. In general, the window T is assumed to be derived from a stable off-chip frequency reference such as an oven-controlled crystal oscillator (OCXO), whose frequency uncertainty is negligible compared to that of the sensing oscillator.

Assuming that ϕ_t is the excessive noisy phase of the oscillator, the accumulated jitter σ_T^2 is fundamentally determined by the sensing oscillator's phase noise as

$$\sigma_T^2 = 1/\omega_0^2 E|\phi_{t+T} - \phi_t|^2 = 4/\pi\omega_0^2 \int_0^\infty S_{\phi,DSB}\omega \sin^2 \omega T/2 \, d\omega \tag{14.11}$$

$$= 2/\pi\omega_0^2 \int_0^\infty S_{\phi,SSB}\omega \sin^2 \omega T/2 \, d\omega$$

where $S_{\phi,DSB}\omega$ and $S_{\phi,DSB}\omega$ are the double side band (DSB) and single side band (SSB) phase noise power spectrum densities, respectively [22]. The quantity ω is the phase noise offset frequency, and ω_0 is the nominal oscillation frequency. In the following discussions, $S_\phi\omega$ will be used to denote the SSB phase noise $S_{\phi,SSB}\omega$. Therefore, based on (14.10) and (14.11), the relative frequency measurement error $\sigma_{\Delta f}/f_0^2$ can be related to the phase noise as

$$\sigma_{\Delta f}/f_0^2 = \sigma_T^2/T^2 = 2/\pi\omega_0^2 T^2 \int_0^\infty S_\phi\omega \sin^2 \omega T/2 \, d\omega. \tag{14.12}$$

The phase noise $S_\phi\omega$ for a CMOS-based oscillator typically contains both $1/f$ and $1/f$ phase noise components. The former is due to up-conversion and folding of the thermal noise in the oscillator circuit, while the latter is caused by up-conversion of the flicker noise mainly from the oscillator's active devices (Figure 14.4a). This frequency measurement uncertainty $\sigma_{\Delta f}/f_0^2$ can be shown to be a function of the actual counting window T as follows.

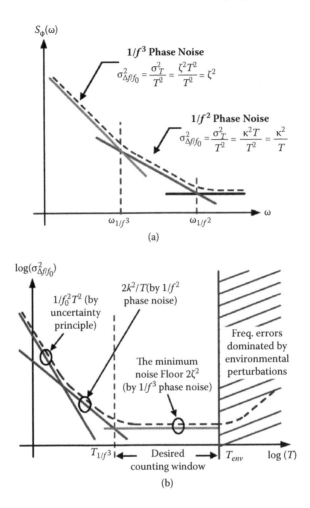

FIGURE 14.4 (a) Typical phase noise profile of an oscillator. (b) Frequency measurement uncertainty versus counting window T in differential operation. The curve is the overall frequency counting uncertainty for a given T with the minimum noise floor of 2.

For a small T—that is, $T \ll 2\pi/\omega 1f3$, where $\omega 1f3$ is the $1/f$ corner frequency for the phase noise profile $S\phi\omega$—the jitter due to $1/f$ phase noise dominates the relative frequency error $\sigma\Delta f/f02$, which becomes inversely proportional to T as

$$\sigma\Delta f \,/\, f02 = \sigma T2T2 = k2TT2 = k2T. \tag{14.13}$$

Here, k is the $1/f$ jitter coefficient for the sensing oscillator [22]. Note that $\sigma\Delta f/f02$ decreases when the window T is increased for a longer counting window. This is because the $1/f$ phase noise behaves like white frequency noise [23], where the uncertainties in edge transition times are uncorrelated and, as a result, the accumulated jitter power $\sigma T2$ is only linearly proportional to T, shown in (14.13). Thus, the frequency measurement error $\sigma\Delta f/f02$ due to the $1/f$ phase noise, as its normalized accumulated jitter (14.12), can always be reduced using a longer counting time window T.

However, when T is large enough $(T \gg 2\pi/\omega_1 f_3)$, phase noise dominates and results in the frequency error $\sigma_{\Delta f}/f_{02}$ actually independent of T as

$$\sigma_{\Delta f} / f_{02} = \sigma_T 2 T 2 = \zeta 2 T 2 T 2 = \zeta 2, \qquad (14.14)$$

where ζ is the $1/f$ jitter coefficient for the sensing oscillator [22]. This is due to the fact that $1/f$ phase noise is flicker frequency noise [23], whose edge transition uncertainties are strongly correlated, leading to $\sigma_T 2$ being proportional to $T2$ (14.14). Since $\sigma_{\Delta f}/f_{02}$ is independent of the window length T for this case, this $\zeta 2$ factor therefore determines the fundamental noise floor for the oscillator-based frequency measurement.

In addition, a measurement error proportional to $1/fT$ due to the uncertainty principle should be superimposed onto the aforementioned two frequency errors. This $1/fT$ error suggests that a frequency resolution of 1 Hz can only be achieved with a counting window longer than 1 second. The total measurement noise with respect to window T is plotted and reveals that the $1/f$ phase noise ($\zeta 2$) indeed determines the ultimate sensor noise floor (Figure 14.4b).

In practical implementations, differential sensing can be used to further reject the environmental perturbations, such as temperature drifting and supply variations. This scheme can be implemented as two identical oscillators—one used for sensing and one used as a reference both physically placed close to each other and sharing the same electrical supply and biasing circuits (shown as an example later in Figure 14.7). Therefore, the environmental variations present themselves as common-mode perturbations to the two oscillators and can be readily suppressed by differential operations (i.e., taking the frequency difference of the sensing and reference oscillators as the sensor output). Note that in this case, since the phase noise is uncorrelated between the two oscillators, the total noise power after the differential sensing actually doubles, if the two oscillators are assumed to have the same phase noise power spectrum density (PSD) (Figure 14.4b). The extra factor of two accounts for the noise power doubling.

Therefore, the overall sensor SNR in the differential sensing scheme can be formulated as

$$SNR = \text{Relative Frequency Shift Relative Frequency Error} \qquad (14.15)$$

$$= \Delta f f_{02} \sigma_{\Delta f} / f_0 = \Delta f f_{02} \zeta$$

where the frequency shift $\Delta f f_0$ and error $2\sigma_{\Delta f}/f_0$ for a given sensing oscillator and magnetic particle distribution can be calculated based on (14.6) and (14.12), respectively. The factor of two is to account for the noise power doubling due to differential sensing.

14.3 SENSOR DESIGN SCALING LAW

The theoretically derived SNR expression enables sensor design optimization, which will be discussed in detail in this section.

First, for the sensor signal response, the transducer gain equation (14.6) demonstrates that for a given type of magnetic particle label, the factors related to the magnetic material and the particle volume are both constants. The field factor $Bext2/2L$, determined by the sensing inductor geometry, is therefore the only factor subjected to design optimization.

The effect of the sensor noise floor $\sigma \Delta f / f02$ with respect to the sensing inductor design (inductance L and quality factor Q) is discussed next. Since the fundamental noise floor is determined by the $1/f3$ phase noise, only flicker noise will be considered for the following derivation.

For a given process technology, assuming a fixed biasing current density for the oscillator active-core devices and a constant tank amplitude (limited by the supply V_{DD}), the transistors' DC biasing current and width W are related to the tank inductor design as

$$Id \propto W \propto VDDRtank = VDD\omega 0LQ, \tag{14.16}$$

where $Rtank$ is the tank's parallel resistance and Q is its quality factor of the tank (typically dominant by the tank inductance Q in CMOS processes), both at the resonant frequency $\omega 0$. The result of $Id \propto VDD/Rtank$ assumes that the oscillator is biased for its optimum operation (i.e., the boundary region between the current-limiting and voltage-limiting regime) [22]. Moreover, for a fixed biasing current density, the device DC current is proportional to its drain output flicker current noise PSD, $1/f2\omega = \beta/\omega$, as

$$Id \propto W \propto \beta \tag{14.17}$$

The $1/f3$ jitter coefficient ζ can be derived using the linear time-varying (LTV) phase noise model at a given offset frequency ω [22], and further simplified based on (14.16) and (14.17) to

$$\zeta 2 \propto S\phi\omega \cdot \omega 3 = A02qmax2 \cdot in, 1 / f2\omega 2\omega 2 \cdot \omega 3 \tag{14.18}$$

$$= A02qmax2 \cdot \beta 2 \propto A02VDD2C2 \cdot VDD\omega 0LQ \propto LQ$$

where $A0$ is the DC term of the oscillator's impulse sensitivity function (ISF) $\Gamma(t)$ and $qmax$ is the tank maximum charge swing [22]. Therefore, assuming differential operation, the relationship between the sensor noise power $\sigma \Delta f / f02$ and the inductor L and Q is given by

$$\sigma \Delta f / f02 = 2 \times 2\pi\omega 02T20 + \infty S\phi\omega \sin 2\omega T2d\omega = 2\zeta 2 \propto LQ. \tag{14.19}$$

This result suggests that the sensor noise floor also depends on the sensor inductor design. Consequently, based on (14.6) and (14.19), the averaged sensor SNR of a single magnetic particle can be obtained as

$$SNR = \Delta ff0\sigma \Delta f / f0 \propto Bext2/2L \cdot QL = Bext2/2 \cdot QL3. \tag{14.20}$$

This equation shows that the sensor inductor design is the key to optimizing the magnetic sensor performance and maximizing its SNR during magnetic sensing.

As an example, the normalized averaged SNR for a six-turn symmetric inductor detecting a single magnetic bead ($D = 1$ µm) at 1 GHz is shown in Figure 14.5. In this example, both the outer diameter (D) and the inductor trace width are swept while the trace separation is kept constant at 3.5 µm. For a given D, the SNR does not vary significantly with the width. This is because, for a constant D, both the inductance L and the averaged **Bext2/I2** scale the same with the inductor width. This leads to a relatively constant SNR for a given outer diameter.

On the other hand, for the same trace width, a smaller D gives a much higher averaged SNR. This is due to a larger **Bext2/I2** (i.e., a more concentrated excitation magnetic field) and a smaller inductance value at a smaller inductor footprint (i.e., a smaller D). However, an inductor with an exceedingly small size is practically undesirable for several reasons. First, the LC tank's metal interconnections start to contribute non-negligible resistances and parasitic inductances comparable to the sensor inductance, which degrades the sensor SNR by lowering both Q and $\Delta L/L0$ during sensing. Moreover, to maintain the same voltage amplitude at the operating frequency, a low impedance tank (for a small L) needs to conduct a larger current, which is susceptible to causing magnetic saturation of the particles and yields the decreased detected magnetic signal due to susceptibility χapp degradation as mentioned previously. Furthermore, the frequency-dependent magnetic loss, which degrades the magnetic signal, limits the maximum operation frequency to around 1 GHz for magnetic beads formed by nanometer size magnetic particles suspended in nonmagnetic structures [25,26]. Thus, the inductor size cannot be too small in order to achieve an acceptable Q at the desired operation frequency. In addition, the design space for the sensing inductor is also limited by other constraints, such as the sensor pixel size (footprint) and power consumption.

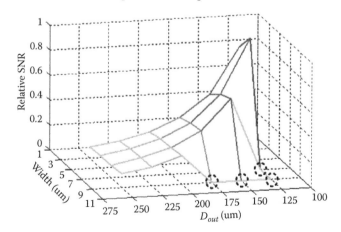

FIGURE 14.5 The simulated averaged SNR for a six-turn differential symmetric inductor with different geometric sizes at 1 GHz operating frequency. The dotted circles indicate inductor geometries unrealizable due to layout rules.

14.4 A CMOS IMPLEMENTATION EXAMPLE (SYSTEM ARCHITECTURE)

In this section, an eight-cell sensor array implemented in a standard 130 nm CMOS process will be demonstrated as a design example for the proposed frequency-shift magnetic sensing scheme. The system architecture is shown in Figure 14.6.

The differential sensing scheme is implemented for each sensor cell. It is composed of two well matched sensor oscillators for sensing and reference operations, both operating at a nominal frequency of 1 GHz (Figure 14.7). In the layout, the active cores of the two oscillators are placed in proximity to improve matching and to minimize local temperature differences. The oscillator pair also shares the same power supply, biasing, and ground lines. As discussed before, these implementation techniques ensure that the environmentally related low frequency noise and drifting appear as the common-mode perturbations and are subsequently suppressed by the differential operation. The active and the reference oscillators are turned on alternately in time to avoid parasitic injection locking or oscillator pulling.

On-chip temperature controllers are implemented in this design. They regulate the local temperature for the oscillator active cores through a thermal-electrical feedback loop. This further reduces frequency drifting induced by ambient temperature changes. Instead of regulating the entire chip, the temperature controllers are placed locally at every sensor cell, regulating the differential sensing oscillators' active core temperature to achieve more efficient thermal control with minimal power consumption overhead.

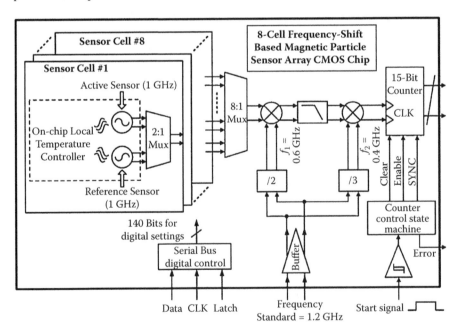

FIGURE 14.6 The eight-cell frequency-shift-based CMOS magnetic biosensor array system architecture.

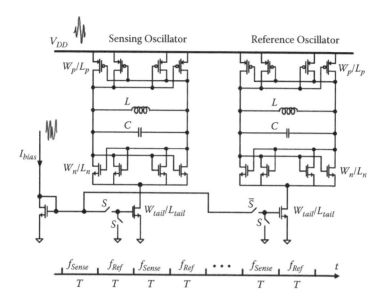

FIGURE 14.7 Schematic of the differential sensing oscillators. The two sensor oscillators are alternatively turned on to perform differential sensing schemes.

Frequency counters are used to detect the sensing signal's frequency shift and provide direct digital output. To facilitate resolving a ppm or sub-ppm level frequency shift at 1 GHz, a two-step down-conversion architecture is used to shift the 1 GHz sensor output to a tunable kilohertz range baseband frequency. Unlike direct down-conversion, this architecture guarantees that the two LO signals (0.6 and 0.4 GHz) are not close to the sensor free-running frequency or its dominant harmonics and, hence, inherently prevents pulling or injection locking on the sensing oscillator pair. By using a 15-bit baseband frequency counter, a maximum counting resolution of better than 3×10^{-4} ppm (0.3 Hz at 1 GHz) can be achieved. Note that digital nature of the sensor system's output signal facilitates tiling multiple sensor array chips to form a very large scale magnetic microarray for high throughput applications.

14.5 A CMOS IMPLEMENTATION EXAMPLE (CIRCUIT BLOCKS)

In this section, circuit designs for the major building blocks in the example sensor array system are presented in detail.

14.5.1 SENSING OSCILLATOR

A differential complementary cross-coupled topology is used for the sensing oscillator, as shown in Figure 14.7. Switches at the tail current sources control the turning on and off of the oscillators, whose bias current is derived from the same master current source. Common centroid layout is used to improve matching and design robustness against process gradients. Since the $1/f3$ phase noise determines the ultimate noise floor of the sensor, a gate length of 240 μm is used for both the negative (N)

MOS and positive (P)MOS transistors in the cross-coupled pair to lower the device flicker noise corner. Moreover, the relative weighting between the NMOS active pair and the PMOS active pair is optimized to shape the oscillation transient voltage waveform and minimize the flicker noise up-conversion from the NMOS current tail [27]. Furthermore, to provide frequency tuneability of the sensing oscillator, a switched capacitor bank has been adopted to set the desired operating frequency.

Based on the noise analysis described in Section 14.2.2.2, a design goal for the $1/f3$ phase noise of the sensor oscillator can be obtained, which achieves a given frequency-shift detection sensitivity. For the differential sensing scheme, the frequency counting noise floor $\sigma\Delta f/f0$, $diff$ has been shown to be 2ζ, with ζ as the $1/f3$ jitter coefficient of the oscillator. Numerically, this ζ factor can be obtained from the oscillator's SSB $1/f3$ phase noise profile, $S\phi\omega = c/\omega3$, based on a close-in corner frequency approximation in Liu and McNeil [24] as

$$\sigma\Delta f / f02 = 2 \times 2\pi\omega02T20 + \infty S\phi\omega \sin 2\omega T 2d\omega = 2\zeta2 \approx 25c2\pi\omega02. \qquad (14.21)$$

Therefore, in order to achieve a frequency-shift noise floor of less than 1 ppm in a differential operation, the SSB $1/f3$ phase noise for the sensing oscillator should be below −43.9 dBc/Hz at a 1 kHz frequency offset for a center tone of 1 GHz. For a 0.1 ppm frequency resolution, the phase noise at an offset of 1 kHz should be −63.9 dBc/Hz.

14.5.2 Temperature Regulator

On-chip temperature regulators are implemented locally for each individual differential sensing cell. A temperature controlling system is typically composed of an electrical-thermal feedback loop. This loop senses the difference between the on-chip and the target temperature and converts it to an electrical signal, which drives an on-chip heater for thermal controlling.

The simplified schematic of our temperature regulator is shown in Figure 14.8. The temperature sensor is implemented as a PTAT (proportional to absolute temperature) voltage, and a bandgap voltage is used as the temperature reference. The PTAT voltage is programmable with 12-bit control on its output resistors; this sets the target temperature for thermal regulation. The voltage difference between the two temperature signals is then amplified by a two-stage buffer to drive a heater transistor array. A feedback op-amp is used to lock the common-mode output voltage of the driver to the threshold voltage of a dummy heater transistor. This provides a reliable stand-by driver output voltage to prevent any false turning on of the heater transistors due to process variations or modeling inaccuracy. The thermal response of the heater array is simulated using Ansoft® Ephysics [28]. Overall, this thermal controller forms a first-order electrical-thermal feedback loop with a typical DC loop gain of 20.5 dB and is stabilized by a dominant pole in the kilohertz range.

The layout configuration of the temperature regulator is shown in Figure 14.9. The bandgap core, including the reference and the PTAT generation, is placed close to the two oscillators' active devices for accurate temperature sensing. The heater array, composed of power PMOS transistors, forms a ring structure surrounding

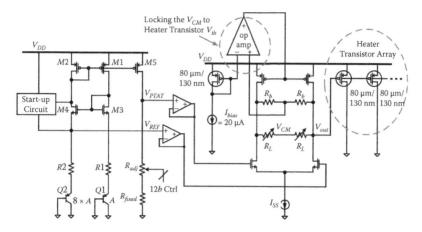

FIGURE 14.8 Simplified schematic for the on-chip temperature regulator.

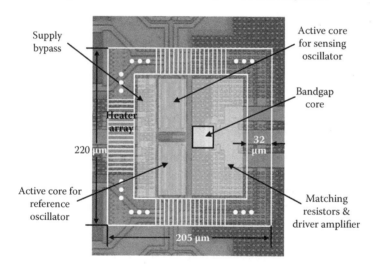

FIGURE 14.9 The layout configuration of the heater structure for one differential sensing cell.

the oscillator active cores to minimize the spatial temperature difference within the regulator.

14.5.3 MULTIPLEXERS AND BUFFERS

Multiplexers and buffer amplifiers together with distribution lines deliver the output signals from the selected sensor cells to a down-conversion block (Figure 14.10). Resistive degeneration is used to linearize the distribution circuit gain. To ensure an adequate bandwidth without peaking inductors or excessive power consumption, signals are distributed in current mode through a modified cascode topology [29]. The low impedance at the current summing nodes minimizes the bandwidth penalty due to parasitic capacitances. Since only buffers for the selected branch need to be

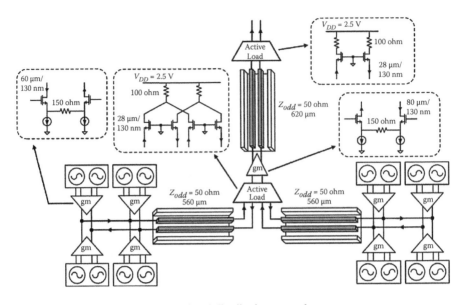

FIGURE 14.10   Schematic for the signal distribution network.

turned on at any given time, the distribution system overall consumes 10 mA from a 2.5 V supply and has a –3 dB bandwidth of 2.5 GHz.

14.5.4   DIVIDE-BY-2 AND DIVIDE-BY-3 FREQUENCY DIVIDERS

To synthesize the down-conversion LO signals, static divide-by-2 and divide-by-3 circuits are implemented in current mode logic (CML) as high-speed digital dividers. Both dividers are composed of feedback D-flip-flops (Figure 14.11) and provide divided signals with a 50% duty cycle to suppress unwanted harmonics. With these dividers, an off-chip 1.2 GHz frequency reference signal is divided into 600 and 400 MHz LO signals that down-convert the sensor oscillators' nominal 1 GHz signal. This frequency plan ensures that both LO signals are separated from the fundamental tone and the dominant harmonics of the sensor free-running frequency to prevent oscillator pulling or injection locking on the sensing oscillator pair. Note that although the 1.2 GHz reference signal is fed from off-chip in this implementation, it can also be synthesized on-chip and locked to a megahertz crystal frequency standard. In this case, the presented down-conversion frequency plan also ensures that this 1.2 GHz tone is away from the 1 GHz sensor oscillators. The dividers consume 8 mA from a 1.2 V supply and add negligible noise to the 1.2 GHz reference signal.

14.6   MEASUREMENT RESULTS

The chip microphotograph for the example eight-cell magnetic sensor array is shown in Figure 14.12. The entire chip is implemented in a standard 130 nm 1P8M CMOS process and occupies an area of 3.0 × 5.2 mm². All the critical building blocks are highlighted. Note that the bonding pads are placed along the upper chip edge. This

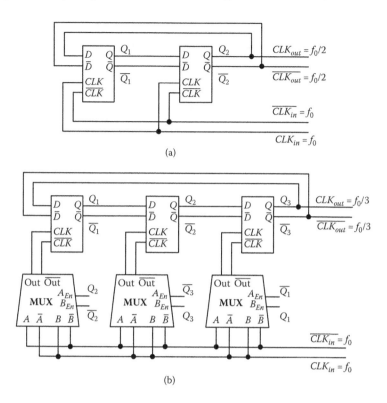

FIGURE 14.11 Schematic for the frequency dividers with 50% duty cycle: (a) divide-by-2; (b) divide-by-3.

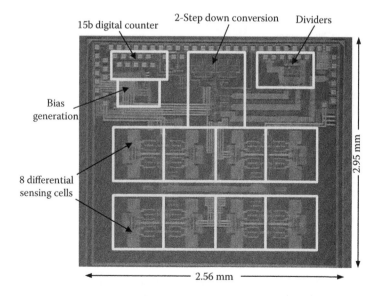

FIGURE 14.12 Chip microphotograph of the CMOS sensor array.

configuration accommodates the integration of the PDMS (polydimethylsiloxan) microfluidic structures for biosample delivery without perturbing the wire bonds. The wire bonds can be eventually sealed and protected by the PDMS solution [30]. A minimum distance of 250 μm between adjacent sensing inductors is used, limited by the minimum achievable microfluidic channel width/separation in our in-house PDMS fabrication facilities. This spacing between sensor cells can be substantially reduced by employing more advanced microfluidic processes. The CMOS sensor chip integrated with the PDMS structure is shown in Figure 14.13.

In this section, the electrical performance of the sensor array is first presented. Magnetic sensing measurements are then shown for detecting micron size magnetic

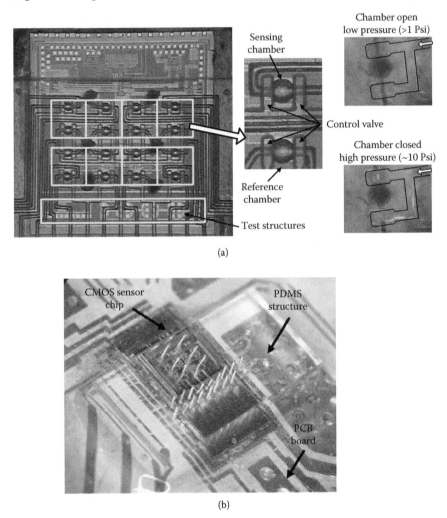

(a)

(b)

FIGURE 14.13   (a) Chip microphotograph of the CMOS sensor array with PDMS microfluidic structure. (b) Integration of the CMOS sensor chip, PDMS structure, and PCB board for the testing module.

particles. The results demonstrate the full functionality of our proposed frequency-shift-based magnetic sensing scheme.

14.6.1 ELECTRICAL PERFORMANCE

The measured phase noise performance of the sensing oscillator is shown in Figure 14.14. Consuming 4 mA from a 1.2 V supply, the oscillator has its $1/f3$ phase noise (at 1 kHz frequency offset) from −59 to −61.8 dBc/Hz for different biasing scenarios with a center tone of 1.04 GHz. Based on the theoretical analysis, this phase noise level provides a frequency measurement uncertainty $\sigma\Delta f/f0$ between 170.6 and 123.6 ppb (parts per billion, i.e., 10^{-9}) in the differential operation scheme, which achieves a sub-ppm sensor sensitivity using frequency-shift detections.

The rejection of environmentally related drifting through differential counting is demonstrated in Figure 14.15. The frequency counting results for the sensor and

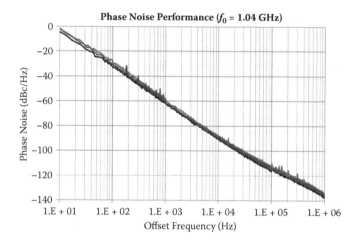

FIGURE 14.14 Measured and simulated phase noise performance of single sensing oscillator at different biasings.

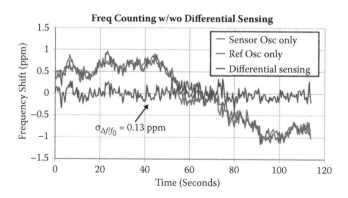

FIGURE 14.15 Suppression of common-mode frequency drifting through differential sensing.

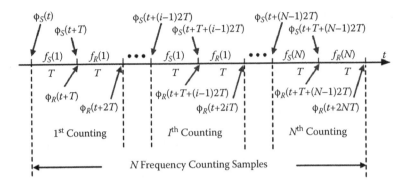

FIGURE 14.16 N-sample averaging on differential frequency counting measurements.

reference oscillators are shown with and without differential operation. A slowly varying common-mode drift of the oscillation frequencies can be observed and is greatly suppressed by the differential operation. Consequently, in the differential operation mode, the measured relative frequency uncertainty $\sigma \Delta f / f0$ is 130 ppb before averaging, which stays within the theoretical calculated noise range based on the phase noise measurement.

Moreover, the differential frequency counting samples can be averaged to improve the sensor SNR further (Figure 14.16). Let fSi and fRi stand for the frequency measurement in the ith differential counting interval. The total N-sample averaged differential frequency result is

$$\Delta fdiff, N = 1Ni = 1N\Delta fi = 1Ni = 1NfSi - fRi. \tag{14.22}$$

Therefore, using the preceding noise model and assuming that the sensor and reference oscillators present the same, yet uncorrelated power spectrum density, the frequency measurement noise power after averaging N differential samples can be formulated as

$$\sigma \Delta f / f0, \tag{14.23}$$

$$diff, N2 = 1T2\omega02E1Ni = 1N\phi St + T + i - 12T - \phi St + i - 12T - \phi Rt + 2T$$

$$+ i - 12T - \phi Rt + T + i - 12T2$$

$$= 1T2\omega02E1Ni = 1N\phi St + T + i - 12T - \phi St + i - 12T2 + 1T2\omega02E1Ni$$

$$= 1N\phi Rt + 2T + i - 12T - \phi Rt + T + i - 12T2$$

$$= 2T2\omega02E1Ni = 1N\phi St + T + i - 12T - \phi St + i - 12T2$$

$$= 2 \times 2\pi\omega02T20 + \infty S\phi\omega \cdot \sin 2\omega T2 \cdot \sin 2N\omega T \sin 2\omega Td\omega'$$

$$= NPRRN \cdot 2 \times 2\pi\omega02T20 + \infty S\phi\omega \sin 2\omega T2d\omega$$

$$= NRFN \cdot \sigma \Delta f / f0, diff2$$

where ϕS and ϕR are the excessive noisy phases for the sensor and reference oscillators, which are uncorrelated to each other. The ratio between the noise power of the N-sample-averaged differential frequency counting result over the unaveraged one can be defined as a noise power reduction ratio NPRR. The simulated NPRR with respect to the averaging sample number N is plotted in Figure 14.17. Note that the noise power for the N-sample averaging is not inversely proportional to the sample number N as the classical averaging scenario for sampled white noise (uncorrelated samples). This is because the waveform transition uncertainties due to $1/f3$ jitter in the frequency measurement exhibit strong correlations, resulting in a much lower suppression after averaging.

The thermal regulator response is depicted in Figure 14.18 as the total heater current versus on-chip temperature variations. The on-chip temperature is measured through the PTAT voltage. When the temperature deviates from the target setting (i.e., 29°C in this measurement), the PMOS heater ring starts to draw a DC

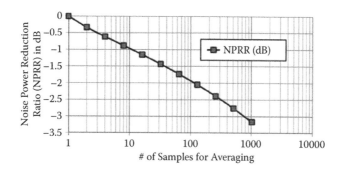

FIGURE 14.17 Simulated noise power reduction ratio, $NPRR(N)$, based on sample averaging, when the phase noise is the dominant noise for the sensing oscillators. The x-axis stands for the number of differential frequency counting measurements.

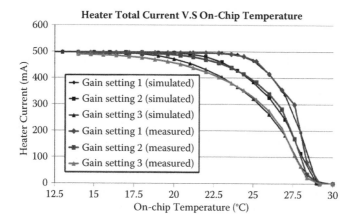

FIGURE 14.18 Simulated and measured heater total current versus on-chip temperature when the target temperature is set at 29°C.

TABLE 14.1

Measured Electrical Performance of the Sensor Array

Power consumption

Sensor oscillator[a]	4.0 mA at 1.2 V
Down-conversion chain, multiplexer[a] and counter	39.6 mA at 2.5 V
On-chip temperature controller[a,b]	24. 3 mA at 2.5 V
Total power consumption	165 mW
Thermal loop gain (nominal setting)	20 dB
Number of differential sensing cells	8
Technology	130 nm CMOS
Chip area	2.95 mm × 2.56 mm
Sensor inductor size (D_{out})	140 µm

[a]   At any given time, only one sensor oscillator with its temperature controller and buffer will be turned on to avoid pulling and/or injection locking between oscillators.

[b]   Here the target chip temperature is set at 29°C with the ambient temperature of 27°C. The 20 dB loop gain ensures the actual on-chip temperature is at 28.8°C.

current from the supply and heat up the enclosed differential sensing oscillators' active cores. The measured heater responses for three different loop-gain settings are shown, closely matching the simulated responses.

The electrical performance of the example CMOS sensor array is summarized in Table 14.1.

14.6.2   MAGNETIC SENSING MEASUREMENTS

To characterize the sensor's magnetic sensing functionalities, two sets of experiments are performed on detecting three types of micron size magnetic particles (i.e., Dynabeads® M-450 Epoxy, Dynabeads® Protein G, and Dynabeads® MyOne™ [31]). The first set of experiments is used to verify the sensor's capability of detecting a single, micron size magnetic particle. The second one is used to characterize the sensor's dynamic range with a wide range of magnetic particle numbers.

In the first experiment, a single magnetic particle is introduced onto the sensor surface for all three bead types. The measurement results are summarized in Table 14.2. The frequency counting window is 0.1 s, and averaging on the sensing data is performed to improve the SNR further. With 160 s used for averaging, an SNR of 8.0 dB is achieved for detecting a single 1 µm magnetic particle (Dynabeads MyOne). After 90 s of averaging, the achieved SNRs for a single 2.4 and 4.5 µm magnetic particle (Dynabeads Protein G or Dynabeads M-450 Epoxy) are 28.6 and 40.0 dB, respectively. As a control sample, polystyrene beads ($D = 1$ µm) are tested. Polystyrene is the nonmagnetic structure material used to hold the nanomagnetic grains and form most off-the-shelf micron size magnetic bead products. The polystyrene beads show significantly lower frequency-shift responses compared with the 1 µm magnetic beads. This verifies that our sensor's frequency shifts for the presented magnetic beads are mainly due to the inductance (magnetic) rather than the

TABLE 14.2

Typical Sensor Response to Off-the-Shelf Magnetic Bead

Bead type	Bead size (diameter)	Recorded Δf/f per bead	Sensitivity (no. of beads)	SNR	Averaging time (seconds)[a]
DynaBeads M-450 Epoxy	4.5 µm	9.6 ppm	1	40.0 dB	90
DynaBeads Protein	2.4 µm	2.6 ppm	1	28.6 dB	90
DynaBeads MyOne carboxylic acid	1.0 µm	0.23 ppm	1	8.0 dB	160
Polystyrene bead (nonmagnetic)	1.0 µm	0.0035 ppm[b]	—	—	—

[a] The counting window T is 0.1 s for these measurements.

[b] This Δf/f per bead is a calculated average value for a measured frequency shift of 100 polystyrene beads sensed at the same time.

capacitance (dielectric) changes. Typical measurement results for a single magnetic bead of 2.4 and 1 µm are shown in Figure 14.19. The blue curves represent the data traces with cumulative averaging.

In the second experiment, different numbers of magnetic beads for all three bead types are separately applied onto the sensor surface, and their corresponding

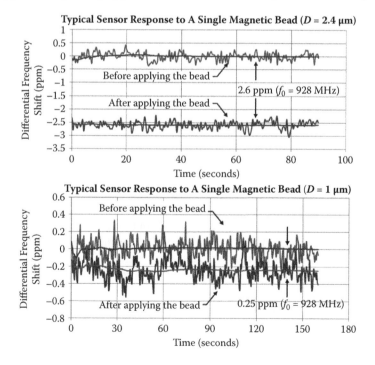

FIGURE 14.19 Typical sensor response of a single Dynabeads protein G (D = 2.4 mm) and MyOne (D = 1 mm).

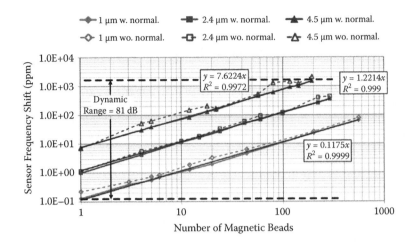

FIGURE 14.20 Measured sensor response with and without spatial transducer gain normalization. The counting window T is 0.1 s.

frequency shifts are recorded respectively. Since the excitation magnetic field amplitude ***Bext(r)*** is generally not a constant across the sensor surface for a six-turn symmetric sensing inductor layout, the local transducer gain will consequently be spatially nonuniform based on (14.7). Therefore, in this experiment the distribution of the beads is first recorded by a high-magnification camera. Then their measured frequency shifts are normalized by their corresponding spatially averaged transducer gain factors, which are obtained through theoretical calculations based on the transducer gain model in (14.7). Note that the normalized frequency-shift results assume that all the beads' responses are normalized to the center of the sensing inductor. The normalized and raw frequency-shift results versus bead numbers are plotted in Figure 14.20. The total bead count is limited to around 200 ~ 400 because, beyond this number, the beads start to aggregate on the sensor surface, which significantly prevents accurate bead identifications under the microscope. Nevertheless, the linear sensor responses after normalization prove the validity of our spatial sensor transducer gain modeling in (14.6) and (14.7). Moreover, this measurement demonstrates a sensor dynamic range of greater than 81 dB.

A sensor performance comparison among state-of-the-art CMOS-based biosensors is summarized in Table 14.3. Sensor testing results based on actual DNA samples are reported in Wang et al. [18]. The frequency-shift magnetic sensor array example presents the best sensitivity among the reported magnetic biosensors in CMOS and achieves a competitive performance compared to the reported CMOS biosensors.

14.7 DISCUSSION AND CONCLUSION

As an extension to the presented oscillator-based magnetic sensor array, a correlated double counting (CDC) frequency-shift sensor scheme is introduced in Wang, Hajimiri, and Kosai [33] and Wang et al. [34] to further decrease the minimum sensor noise floor below $2\delta2$ for the basic differential sensing case shown in (14.20).

TABLE 14.3

Performance Comparison with Published CMOS Biosensors

Sensor type	Sensitivity (magnetic beads)	Sensitivity (molecules)	External magnets?	Costly post-processing?	Optical device?
Magnetic sensor					
Frequency-shift-based sensor (this work)	1 ($D = 1$ μm)	1 nMolar [18]	No	No	No
GMR sensor [8]	1 ($D = 2.8$ μm)	—	Yes	Yes	No
GMR sensor [10]	—	10 nMolar	Yes	Yes	No
Hall effect sensor [12]	1 ($D = 2.8$ μm)	—	Yes	Yes	No
NMR relaxometer [13]	—	4 nMolar	Yes	No	No
Optical sensor					
CMOS image sensor [32]	—	0.125 nMolar	No	Yes	Yes
Electrochemical sensor					
Electrochemical sensor [4]	—	100 nMolar	No	No	No
Electrochemical sensor [5]	—	1 μMolar	No	No	No
Electrochemical sensor [6]	—	100 nMolar	No	Yes	No

Through oscillator active core sharing, this scheme physically establishes $1/f3$ phase noise correlation between the sensing and the reference oscillators. Similarly to the common-mode environmental noise for the oscillator pair, the correlated $1/f3$ phase noise is then largely suppressed via subsequent differential sensing. This significantly reduces the overall sensor noise for the frequency-shift measurements without requiring either low $1/f3$ phase noise or power overhead. Moreover, averaging the CDC frequency counting data can be shown to present a much more effective noise suppression effect—that is, a lower NPRR(N)—compared to a standard differential scheme shown in (14.19), which further boosts the sensor SNR.

In parallel, the issue of transducer gain uniformity is addressed in Wang and Hajimiri [35] and Wang, Sideris, and Hajimiri [36]. In this study, a two-layer stacked coil is proposed as the sensing inductor; this provides an equalized excitation magnetic field (in magnitude) across the sensor surface. In addition, floating shimming metal pieces are introduced to enhance the transducer spatial gain uniformity further. Therefore, the frequency-shift magnetic sensor achieves a linear sensor response versus the particle numbers without any postnormalization steps.

In summary, a novel frequency-shift-based magnetic biosensing scheme is introduced for future point-of-care (PoC) molecular diagnosis applications. Fully compatible with standard CMOS processes, the sensor scheme achieves high sensitivity, handheld portability, and low power consumption without using any external magnetic biasing fields or expensive postprocessing steps. Theoretical limits on sensor SNR and design optimization techniques have been presented. An eight-cell sensor array implemented in a standard 130 nm CMOS process has been demonstrated as a design example. Both electrical and magnetic sensing measurement results have been presented to verify the sensing scheme's functionality. To the authors' best knowledge, this presented magnetic sensor array example achieves the best sensitivity among the CMOS magnetic sensors reported so far.

ACKNOWLEDGMENTS

The authors acknowledge Dr. Yan Chen, Dr. David Wu, and Mr. Constantine Sideris for their technical support during the sensor testing. The authors would also like to thank Professor Axel Scherer, Sander Weinreb, Azita Emami, and the members of the Caltech High-Speed Integrated Circuit (CHIC) Group for their helpful discussions.

REFERENCES

1. N. K. Tran and G. J. Kost, Worldwide point-of-care testing: Compendiums of POCT for mobile, emergency, critical, and primary care and of infectious diseases tests. *Journal Near-Patient Testing & Technology,* vol. 5, no. 2, pp. 84–92, June 2006.
2. G. J. Kost, *Principles and practice of point-of-care testing.* Philadelphia: Lippincott Williams & Wilkins, 2002.
3. D. Stekel, *Microarray bioinformatics.* Cambridge, UK: Cambridge University Press, 2003.
4. A. Hassibi and T. H. Lee, A programmable 0.18 μm CMOS electrochemical sensor microarray for biomolecular detection. *IEEE Sensors Journal,* vol. 6, no. 6, pp. 1380–1388, Dec. 2006.
5. M. Schienle, C. Paulus, A. Frey, F. Hofmann, B. Holzapfl, P. Schinder-Bauer, and R. Thewes, A fully electronic DNA Sensor with 128 positions and in-pixel A/D conversion. *IEEE Journal Solid-State Circuits,* vol. 39, no. 12, pp. 2438–2445, Dec. 2004.
6. F. Heer, M. Keller, G. Yu, J. Janata, M. Josowicz, and A. Hierlemann, CMOS electro-chemical DNA-detection array with on-chip ADC. *IEEE ISSCC Digest Technical Papers,* pp.168–169, Feb. 2008.
7. A. Hassibi, R. Navid, R. W. Dutton, and T. H. Lee, Comprehensive study of noise processes in electrode electrolyte interfaces. *Journal Applied Physics,* vol. 96, no. 2, pp. 1074–1082, July 2004.
8. B. Alberts, D. Bray, K. Hopkin, A. Johnson, J. Lewis, M. Raff, K. Roberts, and P. Walter, *Essential cell biology,* 2nd ed. New York: Garland Science/Taylor & Francis Group, 2003.
9. H. Lee, Y. Liu, D. Ham, and R. M. Westervelt, Integrated cell manipulation system—CMOS/microfluidic hybrid. *Lab on a Chip,* vol. 7, no. 3, pp. 331–337, March 2007.
10. G. Li, V. Joshi, R. L. White, S. X. Wang, J. T. Kemp, C. Webb, R. W. Davis, and S. Sun, Detection of single micron-sized magnetic bead and magnetic nanoparticles using spin valve sensors for biological applications. *Journal Applied Physics,* vol. 93, no. 10, pp. 7557–7559, May. 2003.
11. G. Li, S. Sun, R. Wilson, R. White, N. Pourmand, and S. X. Wang, Spin valve sensors for ultrasensitive detection of superparamagnetic nanoparticles for biological applications. *IEEE Journal Sensors and Actuators A,* vol. 126, pp. 98–106, Nov. 2006.
12. S. Han, H. Yu, B. Murmann, N. Pourmand, and S. X. Wang, A high-density magnetoresistive biosensor array with drift-compensation mechanism. *IEEE ISSCC Digest Technical Papers,* pp.168–169, Feb. 2007.
13. T. Aytur, P. R. Beatty, and B. Boser, An immunoassay platform based on CMOS Hall sensors. *Solid-State Sensor, Actuator and Microsystems Workshop,* Hilton Head Island, SC, pp. 126–129, June 2002.

14. P. Besse, G. Boero, M. Demierre, V. Pott, and R. Popovic, Detection of a single magnetic microbead using a miniaturized silicon Hall sensor. *Applied Physics Letters,* vol. 80, no. 22, pp. 4199–4201, June 2002.

15. Y. Liu, N. Sun, H. Lee, R. Weissleder, and D. Ham, CMOS mini nuclear magnetic resonance system and its application for biomolecular sensing. *IEEE ISSCC Digest Technical Papers,* pp. 140–141, Feb. 2008.

16. H. Wang, A. Hajimiri, and Y. Chen, Effective-inductance-change based magnetic particle sensing. US Patent no. 20090267596 A1, March 7, 2008.

17. H. Wang and A. Hajimiri, Ultrasensitive magnetic particle sensor system. US Provisional Patent CIT-5224-P, Sept. 15, 2008.

18. H. Wang, Y. Chen, A. Hassibi, A. Scherer, and A. Hajimiri, A frequency-shift CMOS magnetic biosensor array with single-bead sensitivity and no external magnet. *IEEE ISSCC Digest Technical Papers,* pp. 438–439, Feb. 2009.

19. D. Ham and A. Hajimiri, Virtual damping and Einstein relation in oscillators. *IEEE Journal Solid-State Circuits,* vol. 38, no. 3, pp. 407–418, March 2003.

20. K. H. J. Buschow and F. R. de Boer, *Physics of magnetism and magnetic materials.* New York: Springer, 2003.

21. D. E. Bray and R. K. Stanley, *Nondestructive evaluation.* New York: Taylor & Francis Group, 1997.

22. A. Hajimiri and T. H. Lee, *The design of low noise oscillators.* New York: Springer, 1999.

23. D. A. Howe, D. W. Allan, and J. A. Barnes, Properties of signal sources and measurement methods. *Proceedings of the 35th Annual Symposium on Frequency Control,* 1981.

24. C. Liu and J. A. McNeil, Jitter in oscillators with 1/f noise sources. *IEEE ISCAS Digest Technical Papers,* pp. 773–776, May 2004.

25. P. Talbot, A. M. Konn, and C. Brosseau, Electromagnetic characterization of fine-scale particulate composite materials. *Journal of Magnetism and Magnetic Materials,* vol. 249, pp. 481–485, 2002.

26. C. Brosseau, J. B. Youssef, P. Talbot, and A. M. Konn, Electromagnetic and magnetic properties of multicomponent metal oxides heterostructures: Nanometer versus micrometer-sized particles. *Journal of Applied Physics,* vol. 93, no. 11, pp. 9243–9256, June 2003.

27. A. Hajimiri and T. H. Lee, Design issues in CMOS differential LC oscillators. *IEEE Journal Solid-State Circuits,* vol. 34, no. 5, pp. 717–724, May 1999.

28. http://www.ansoft.com

29. A. Babakhani, X. Guan, A. Komijani, A. Natarajan, and A. Hajimjri, A 77-GHz phased-array transceiver with on-chip antennas in silicon: Receiver and antennas. *IEEE Journal Solid-State Circuits,* vol. 41, no. 12, pp. 2795–2806, Dec. 2006.

30. H. Wang and A. Hajimiri, Low cost bounding technique for integrated circuit chips and PDMS devices. US Provisional Patent CIT-5320-P, Feb. 25, 2009.

31. http://www.invitrogen.com

32. B. Jang, P. Cao, A. Chevalier, A. Ellington, and A. Hassibi, A CMOS fluorescent-based biosensor microarray. *IEEE ISSCC Digest Technical Papers,* pp. 436–437, Feb. 2009.

33. H. Wang, A. Hajimiri, and S. Kosai, Noise suppression techniques in high-precision long-term frequency/timing measurements. US Provisional Patent CIT-5318-P, Feb. 20, 2009.

34. H. Wang, S. Kosai, C. Sideris, and A. Hajimiri, An ultrasensitive CMOS magnetic biosensor array with correlated double counting (CDC) noise suppression. *IEEE MTT-S International Microwave Symposium (IMS),* May 2010.

35. H. Wang and A. Hajimiri, Effective-inductance-change magnetic sensor with spatially uniform transducer gain. US Provisional Patent CIT-5108-P, March 7, 2008.

36. H. Wang, C. Sideris, and A. Hajimiri, A frequency-shift based CMOS magnetic biosensor with spatially uniform sensor transducer gain. *IEEE CICC Digest Technical Papers,* pp. 1–4, Sept. 2010.

Index